Birkhäuser

Lebesgue Points and Summability of Higher Dimensional Fourier Series

Ferenc Weisz

Lebesgue Points and Summability of Higher Dimensional Fourier Series

Birkhäuser

Ferenc Weisz
Department of Numerical Analysis
Eötvös Loránd University
Budapest, Hungary

ISBN 978-3-030-74635-3 ISBN 978-3-030-74636-0 (eBook)
https://doi.org/10.1007/978-3-030-74636-0

Mathematics Subject Classification: 42B08, 42A24, 42A50, 42B25, 42B30, 42A38, 46B70

This book is published under the imprint Birkhäuser, www.birkhauser-science.com by the registered company Springer Nature Switzerland AG
The registered company address is: Gewerbestrasse 11, 6330 Cham, Switzerland

To Márti
for her patience
and love

Preface

The main purpose of this book is to investigate the summability of higher dimensional Fourier series and to generalize the concept of Lebesgue points.

It is a basic question in Fourier analysis whether the partial sums

$$s_n f(x) := \sum_{|k| \leq n} \widehat{f}(k) e^{\imath kx} \quad (n \in \mathbb{N})$$

converge to the integrable function $f \in L_1(\mathbb{T})$, where $\mathbb{T}$ denotes the torus and the Fourier coefficients are defined by

$$\widehat{f}(k) = \frac{1}{2\pi} \int_{\mathbb{T}} f(x) e^{-\imath kx} dx \quad (\imath := \sqrt{-1}).$$

One of the deepest results in harmonic analysis is Carleson's theorem ([51], [174]), i.e., for $f \in L_p(\mathbb{T})$, $1 < p < \infty$,

$$\lim_{n \to \infty} s_n f = f \quad \text{a.e.}$$

The convergence holds also in the $L_p(\mathbb{T})$-norm. In this book, we do not prove Carleson's theorem as it is investigated exhaustively in several books (e.g. Arias de Reyna [9] or Grafakos [143] or Muscalu and Schlag [242]).

This convergence does not hold for $p = 1$. However, using a summability method, we can generalize these results. In this book, we will focus on the well known Fejér and Cesàro summability. The most known result in summability theory is Lebesgue's theorem [197] about the Fejér means [107], i.e., the Fejér means of an integrable function converge almost everywhere to the function:

$$\lim_{n \to \infty} \frac{1}{n} \sum_{j=0}^{n-1} s_j f(x) = f(x) \quad \text{a.e.}$$

The set of convergence was also characterized. A point $x \in \mathbb{T}$ is called a Lebesgue point of f if

$$\lim_{h\to 0}\frac{1}{2h}\int_{-h}^{h}|f(x+t)-f(x)|dt = 0.$$

Lebesgue [197] proved that almost every point is a Lebesgue point of $f \in L_1(\mathbb{T})$ and the Fejér means converge to f in each Lebesgue point. The same holds for the Cesàro means.

In this book, these results will be generalized to d-dimensional Fourier series and to different summability means. The generalization of the Lebesgue points is not straightforward. We will investigate six types of generalizations.

We will consider different summation methods for d-dimensional trigonometric Fourier series. Basically, two types of summations will be introduced. In the first one, we take the sum in the partial sums and in the summability means over the balls of ℓ_q which is called ℓ_q-summability. In the literature, the cases $q = 1, 2, \infty$, i.e., the triangular, circular and cubic summability are investigated. In the second version of summation, we take the sum over rectangles which is called rectangular summability. In this case, three types of convergence and maximal operators are considered: the restricted (over a cone and over a cone-like set) and the unrestricted ones. Under the first one, we mean the convergence over the diagonal or more generally, over a cone or over a cone-like set. The unrestricted convergence is taken over $\mathbb{N}^d$. In each version, the three most known summability methods, the Fejér, Cesàro and Riesz means will be investigated in details. The Fejér summation is a special case of the Cesàro method. Moreover, in each type of summability, we will deal with the so-called θ-summation as well, which is a general summability method generated by a single function $\theta : \mathbb{R} \to \mathbb{R}$. This summation contains all well known summability methods, such as the Fejér, Riesz, Weierstrass, Abel, Picard, Bessel, Rogosinski, de La Vallée-Poussin summations, however, it does not contain the Cesàro summation. We consider norm convergence and almost everywhere convergence of the different summability means.

We introduce two types of Hardy spaces. For the ℓ_q- and restricted rectangular summability, we use the Hardy space $H_p^{\square}(\mathbb{T}^d)$ and for the unrestricted summability, the Hardy space $H_p(\mathbb{T}^d)$. We do not verify the results about the Hardy spaces, e.g., we give the atomic decompositions and the equivalence of the different norms without proofs, because the readers can find them in several books (see e.g. Grafakos [143], Yang et al. [361] and Weisz [346]). We prove that the maximal operators of the summability means are bounded from the corresponding Hardy space $H_p^{\square}(\mathbb{T}^d)$ or $H_p(\mathbb{T}^d)$ to $L_p(\mathbb{T}^d)$, whenever $p > p_0$ for some $p_0 < 1$. The critical index p_0 depends on the summability method and on the dimension. For $p = 1$, we obtain a weak type inequality by interpolation, which implies the almost everywhere convergence of the summability means. The one-dimensional version of the almost everywhere convergence and the weak type inequality are proved usually with the help of a Calderon-Zygmund type decomposition lemma. However, in two- or higher dimensions, this lemma can not be used for all cases investigated in this monograph. Our method, that can also be applied well in higher dimensions, can be regarded as a new method to prove the almost everywhere convergence and weak type inequalities.

In Chap. 1, we give the basic results about the one-dimensional Fourier series. We prove the Lebesgue theorem mentioned above and some convergence results for the partial sums of the one-dimensional Fourier series. We prove norm–and almost everywhere convergence and boundedness results for the ℓ_q-summability in Chap. 2, and for the restricted and unrestricted rectangular summability in Chap. 3. Yet in the same chapter, we give a sufficient and necessary condition for the norm convergence of the rectangular θ-means. In particular, if the function θ is in the Feichtinger's algebra $S_0(\mathbb{R}^d)$ used in the theory of Gabor analysis, then norm convergence of the θ-means holds.

In Chap. 4, we introduce six types of Lebesgue points for higher dimensional functions. We need different Lebesgue points and different Hardy-Littlewood maximal operators for the different summability methods. We study the boundedness of all Hardy-Littlewood maximal operators on the $L_p(\mathbb{T}^d)$ spaces. We will show that the summability means converge to the integrable function in each Lebesgue point and almost every point is a Lebesgue point. For the ℓ_2-θ-summability, we give a sufficient and necessary condition for the convergence in Lebesgue points.

This book was aimed to be written so that it is as nearly self-contained as possible. However, it is assumed that the reader has some basic knowledge on analysis, functional analysis and on Hardy spaces. Besides the classical results, recent results of the last 20–30 years are studied. For simplicity, we will prove all results for the two-dimensional case. If needed, after the theorems we will give a guide how we can prove them for higher dimensions and where we can find the proofs. I am sure, in this way the book is more understandable, easier to read and it can reach a wider readership. So I hope, the book will be useful not only for researchers but also for graduate, postgraduate and Ph.D. students.

Budapest, Hungary Ferenc Weisz

Contents

List of Figures

Chapter 1
One-Dimensional Fourier Series

In this chapter, we present some theorems for one-dimensional Fourier series and for the Hardy-Littlewood maximal function. In Sect. 1.1, we introduce the $L_p(\mathbb{T})$ spaces and prove some basic inequalities. In Sect. 1.2, we prove that the partial sums of the Fourier series are uniformly bounded on the $L_p(\mathbb{T})$ spaces when $1 < p < \infty$. As a consequence, we obtain the norm convergence of the partial sums. We do not give the proof of the almost everywhere convergence because it can be found at several places, e.g., in Carleson [51], Grafakos [143], Arias de Reyna [9], Muscalu and Schlag [242], Lacey [192], or Demeter [80].

In the next section, the Hardy-Littlewood maximal function is considered and we prove that it is bounded on the $L_p(\mathbb{T})$ spaces $(1 < p \leq \infty)$ and is of weak type $(1, 1)$. Lebesgue's differentiation theorem is also proved. We introduce the Lebesgue points and show that almost every point is a Lebesgue point.

It was proved by Fejér [107] that the Fejér means of the one-dimensional Fourier series of a continuous function converge uniformly to the function. A similar problem for integrable functions was investigated by Lebesgue [197]. He proved that for every integrable function f,

$$\frac{1}{n}\sum_{k=0}^{n-1} s_k f(x) \to f(x) \quad \text{as} \quad n \to \infty$$

at each Lebesgue point of f, thus almost everywhere, where $s_k f$ denotes the kth partial sum of the Fourier series of f. Later, Riesz [260], Butzer and Nessel [47], Stein and Weiss [293], and Torchinsky [310] proved the same convergence result for the Riesz, Weierstrass, Picard, Bessel, and de La Vallée-Poussin summations. In Sects. 1.4 and 1.5, we will generalize these results to Cesàro summability.

F. Weisz, *Lebesgue Points and Summability of Higher Dimensional Fourier Series*,
https://doi.org/10.1007/978-3-030-74636-0_1

1.1 The L_p Spaces

Let us denote the set of complex numbers, the set of real numbers, the set of rational numbers, the set of integers, the set of non-negative integers, and the set of positive integers by $\mathbb{C}$, $\mathbb{R}$, $\mathbb{Q}$, $\mathbb{Z}$, $\mathbb{N}$, and $\mathbb{P}$, respectively. The subsets of $\mathbb{R}$ and $\mathbb{Q}$ containing only positive numbers are denoted by $\mathbb{R}_+$ and $\mathbb{Q}_+$, respectively. $\mathbb{T}$ denotes the torus, which can be identified naturally with the interval $[-\pi, \pi)$.

In this book, the constants C are absolute constants and the constants C_p are depending only on p and may denote different constants in different contexts.

Definition 1.1.1 The space $L_p(\mathbb{X})$ is consisting of all Lebesgue measurable functions $f : \mathbb{X} \to \mathbb{C}$, for which

$$\|f\|_p := \left(\int_{\mathbb{X}} |f|^p \, d\lambda \right)^{1/p}, \qquad \text{if } 0 < p < \infty$$

and

$$\|f\|_\infty := \sup_{\mathbb{X}} |f|, \qquad \text{if } p = \infty,$$

where $\mathbb{X} \subset \mathbb{R}$ is an arbitrary Lebesgue measurable set and λ denotes the Lebesgue measure.

Two functions in $L_p(\mathbb{X})$ will be considered equal if they are equal λ-almost everywhere. It is known that $L_p(\mathbb{X})$ is a Banach space if $1 \le p \le \infty$ and a complete quasi-normed space if $0 < p < 1$. We also use the notation $|I|$ for the Lebesgue measure of the set I. Most often we will use the notation $\mathbb{X} = \mathbb{R}$ or $\mathbb{X} = \mathbb{T}$. The functions from the $L_p(\mathbb{T})$ space can be extended to $\mathbb{R}$ such that they are periodic with respect to 2π. In case of $\mathbb{X} = \mathbb{Z}$, the corresponding space will be denoted by $\ell_p(\mathbb{Z})$ and it is consisting of all complex sequences $c = (c_k, k \in \mathbb{Z})$, for which

$$\|c\|_{\ell_p} := \left(\sum_{k \in \mathbb{Z}} |c_k|^p \right)^{1/p}, \qquad \text{if } 0 < p < \infty$$

and

$$\|c\|_{\ell_\infty} := \sup_{k \in \mathbb{Z}} |c_k|, \qquad \text{if } p = \infty.$$

The space of continuous functions with the supremum norm is denoted by $C(\mathbb{X})$ and $C_c(\mathbb{R})$ denotes the space of continuous functions having compact support. We will use the notation $C_0(\mathbb{R})$ for the space of continuous functions vanishing at infinity, i.e.,

$$C_0(\mathbb{R}) := \left\{ f : \mathbb{R} \to \mathbb{C} : f \in C(\mathbb{R}), \lim_{|x| \to \infty} f(x) = 0 \right\}.$$

We also introduce the notion of weak $L_p(\mathbb{T})$ spaces.

Definition 1.1.2 A measurable function f is in the weak $L_p(\mathbb{T})$ space, or, in other words, in the $L_{p,\infty}(\mathbb{T})$ $(0 < p < \infty)$ space if

$$\|f\|_{p,\infty} := \sup_{\rho>0} \rho\,\lambda(|f| > \rho)^{1/p} < \infty.$$

In case of $p = \infty$, let $L_{p,\infty}(\mathbb{T}) := L_\infty(\mathbb{T})$.

The weak $L_p(\mathbb{T})$ spaces are quasi-norm spaces because

$$\|f\|_{p,\infty} = 0 \iff f = 0 \qquad \text{a.e.}$$

$$\|cf\|_{p,\infty} = |c|\|f\|_{p,\infty} \qquad (c \in \mathbb{C}),$$

$$\|f + g\|_{p,\infty} \le c_p(\|f\|_{p,\infty} + \|g\|_{p,\infty}),$$

where $c_p = \max(2, 2^{1/p})$.

We show that the weak $L_p(\mathbb{T})$ spaces are larger than the $L_p(\mathbb{T})$ spaces.

Proposition 1.1.3 *If* $0 < p < \infty$, *then* $L_p(\mathbb{T}) \subset L_{p,\infty}(\mathbb{T})$ *and*

$$\|f\|_{p,\infty} \le \|f\|_p.$$

Proof It is easy to see that

$$\int_{\mathbb{T}} |f(x)|^p\,dx \ge \int_{\{x:|f(x)|>\rho\}} |f(x)|^p\,dx \ge \rho^p \lambda\left(|f| > \rho\right),$$

which proves the proposition. ■

If $h(x) := |x|^{-1/p}$, then obviously $h \notin L_p(\mathbb{R})$, but $h \in L_{p,\infty}(\mathbb{R})$ because

$$\rho^p \lambda\left(\left\{x : |x|^{-1/p} > \rho\right\}\right) = 2\rho^p \rho^{-p} = 2.$$

Thus, the inclusion $L_p(\mathbb{R}) \subset L_{p,\infty}(\mathbb{R})$ is proper if $0 < p < \infty$. Recall that the weak space $L_{p,\infty}(\mathbb{R})$ is also complete for each p.

1.2 Convergence of Fourier Series

We introduce the trigonometric Fourier series and show that the partial sums of a function $f \in L_p(\mathbb{T})$ $(1 < p < \infty)$ converge almost everywhere as well in the $L_p(\mathbb{T})$-norm to the function f.

Definition 1.2.1 For an integrable function $f \in L_1(\mathbb{T})$, its kth Fourier coefficient is defined by

$$\widehat{f}(k) = \frac{1}{2\pi} \int_{\mathbb{T}} f(x) e^{-\imath kx} \, dx \qquad (k \in \mathbb{Z}).$$

The formal trigonometric series

$$\sum_{k \in \mathbb{Z}} \widehat{f}(k) e^{\imath kx} \qquad (x \in \mathbb{T})$$

is called the Fourier series of f.

Definition 1.2.2 For $f \in L_1(\mathbb{T})$ and $n \in \mathbb{N}$, the nth partial sum $s_n f$ of the Fourier series of f and the nth Dirichlet kernel D_n are introduced by

$$s_n f(x) := \sum_{k=-n}^{n} \widehat{f}(k) e^{\imath kx}$$

and

$$D_n(t) := \sum_{k=-n}^{n} e^{\imath kt},$$

respectively.

We get immediately that

$$\begin{aligned} s_n f(x) &= \sum_{k=-n}^{n} \frac{1}{2\pi} \int_{\mathbb{T}} f(t) e^{\imath k(x-t)} \, dt \\ &= \frac{1}{2\pi} \int_{\mathbb{T}} f(x-t) D_n(t) \, dt \qquad (n \in \mathbb{N}) \end{aligned} \tag{1.2.1}$$

(see Fig. 1.1).

Lemma 1.2.3 *For all $n \in \mathbb{N}$ and $t \in \mathbb{T}$, $t \neq 0$,*

$$D_n(t) = \frac{\sin((n+1/2)t)}{\sin(t/2)}. \tag{1.2.2}$$

Proof Using some simple trigonometric identities, we obtain

$$\begin{aligned} D_n(t) &= 1 + 2 \sum_{k=1}^{n} \cos(kt) \\ &= \frac{1}{\sin(t/2)} \left(\sin(t/2) + 2 \sum_{k=1}^{n} \cos(kt) \sin(t/2) \right) \\ &= \frac{1}{\sin(t/2)} \left(\sin(t/2) + \sum_{k=1}^{n} (\sin((k+1/2)t) - \sin((k-1/2)t)) \right), \end{aligned}$$

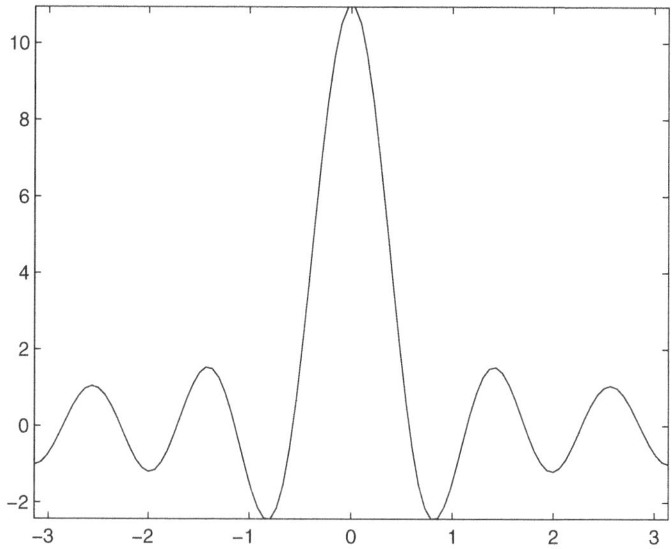

Fig. 1.1 Dirichlet kernel D_n for $n = 5$

which shows the lemma. ∎

The next lemma follows easily from this.

Lemma 1.2.4 *For all $n \in \mathbb{N}$ and $t \in \mathbb{T}$, $t \neq 0$, we have*

$$|D_n| \leq 2n + 1 \qquad \textit{and} \qquad |D_n(t| \leq C/t.$$

It is easy to see that the $L_1(\mathbb{T})$-norms of D_n are not uniformly bounded, more exactly $\|D_n\|_1 \sim \log n$.

Before proving the norm convergence of the partial sums, we need some other definitions and results. We follow the proof of Grafakos [143].

Definition 1.2.5 For some $n \in \mathbb{N}$, the function

$$\sum_{k=-n}^{n} c_k e^{ikx} \qquad (x \in \mathbb{R})$$

is said to be a trigonometric polynomial.

It is a well-known result that the trigonometric polynomials are dense in $L_p(\mathbb{T})$ for any $1 \leq p < \infty$.

Definition 1.2.6 For a trigonometric polynomial f define the conjugate function $\widetilde{f}$ by

$$\widetilde{f}(x) := -\imath \sum_{k\in\mathbb{Z}} \operatorname{sign}(k)\widehat{f}(k)e^{\imath kx}.$$

Now we show that $\widetilde{f}$ is bounded on $L_p(\mathbb{T})$ $(1 < p < \infty)$ (see Riesz [261, 262]).

Theorem 1.2.7 *If* $1 < p < \infty$, *then*

$$\left\|\widetilde{f}\right\|_p \le C_p \|f\|_p \qquad (f \in L_p(\mathbb{T})).$$

Proof First suppose that f is a real trigonometric polynomial and $\widehat{f}(0) = 0$. It is easy to see that $\widetilde{f}$ is also real valued and $f + \imath \widetilde{f}$ contains only positive frequencies. Since $\int_{\mathbb{T}} e^{\imath kx}\, dx = 0$ $(k \neq 0)$, we have

$$\int_{\mathbb{T}} \left(f(x) + \imath \widetilde{f}(x)\right)^{2k} dx = 0,$$

where k is a positive natural number. Taking the real part of the integral and using that f and $\widetilde{f}$ are real valued, we obtain

$$\begin{aligned} 0 &= \sum_{j=0}^{k} (-1)^{k-j} \binom{2k}{2j} \int_{\mathbb{T}} f(x)^{2j} \widetilde{f}(x)^{2k-2j}\, dx \\ &= (-1)^k \int_{\mathbb{T}} \widetilde{f}(x)^{2k}\, dx + \sum_{j=1}^{k} (-1)^{k-j} \binom{2k}{2j} \int_{\mathbb{T}} f(x)^{2j} \widetilde{f}(x)^{2k-2j}\, dx. \end{aligned}$$

This and Hölder's inequality imply that

$$\begin{aligned} \|\widetilde{f}\|_{2k}^{2k} &\le \sum_{j=1}^{k} \binom{2k}{2j} \int_{\mathbb{T}} f(x)^{2j} \widetilde{f}(x)^{2k-2j}\, dx \\ &\le \sum_{j=1}^{k} \binom{2k}{2j} \|f\|_{2k}^{2j} \, \|\widetilde{f}\|_{2k}^{2k-2j}. \end{aligned}$$

Let $R = \left\|\widetilde{f}\right\|_{2k} / \|f\|_{2k}$ and divide by $\|f\|_{2k}^{2k}$ to obtain

$$R^{2k} - \sum_{j=1}^{k} \binom{2k}{2j} R^{2k-2j} \le 0.$$

Then R is smaller than the largest root in absolute value of the polynomial on the left-hand side, say $R \le C_{2k}$, in other words

$$\left\|\widetilde{f}\right\|_p \le C_p \|f\|_p \qquad \text{for } p = 2k. \tag{1.2.3}$$

If $\widehat{f}(0) \neq 0$, then apply this inequality to $f - \widehat{f}(0)$. Since $|\widehat{f}(0)| \le \|f\|_p$, we get the preceding inequality with $2C_p$. Every general trigonometric polynomial can be written as the sum of two real-valued trigonometric polynomials. Therefore, (1.2.3) holds for every trigonometric polynomials and by density for all $f \in L_p(\mathbb{T})$, $p = 2k$. By interpolation (see, e.g., Berg and Löfström [33] or Weisz [346]), (1.2.3) holds for all $2 \le p < \infty$. Finally, observe that the adjoint operator of $f \mapsto \widetilde{f}$ is $f \mapsto -\widetilde{f}$, which implies by duality that (1.2.3) holds also for $1 < p \le 2$. ∎

Definition 1.2.8 For a trigonometric polynomial f, the Riesz projections P^+ and P^- are defined by

$$P^+ f(x) \sim \sum_{k=1}^{\infty} \widehat{f}(k) e^{\imath kx}$$

and

$$P^- f(x) \sim \sum_{k=-\infty}^{-1} \widehat{f}(k) e^{\imath kx}.$$

Observe that $f = P^+ f + P^- f + \widehat{f}(0)$ and $\widetilde{f} = -\imath P^+ f + \imath P^- f$.

Theorem 1.2.9 *If* $1 < p < \infty$ *and* $f \in L_p(\mathbb{T})$, *then*

$$\left\| P^+ f \right\|_p \le C_p \|f\|_p$$

and

$$\left\| P^- f \right\|_p \le C_p \|f\|_p .$$

Proof Since

$$P^+ f = \frac{1}{2}(f + \imath \widetilde{f}) - \frac{1}{2}\widehat{f}(0),$$

and $|\widehat{f}(0)| \le \|f\|_p$, the first inequality follows from Theorem 1.2.7. The second one can be proved similarly. ∎

The following theorem is a fundamental result and it can be found in most books about trigonometric Fourier series (e.g., Zygmund [367], Bary [19], Torchinsky [310], or Grafakos [143]). It is due to Riesz [260].

Theorem 1.2.10 *If* $f \in L_p(\mathbb{T})$ *for some* $1 < p < \infty$, *then*

$$\sup_{n \in \mathbb{N}} \|s_n f\|_p \le C_p \|f\|_p \tag{1.2.4}$$

and

$$\lim_{n \to \infty} s_n f = f \qquad \text{in the } L_p(\mathbb{T})\text{-norm.} \tag{1.2.5}$$

Proof Define

$$P_n^+ g(x) = \sum_{k=0}^{2n} \widehat{g}(k) e^{\imath kx}.$$

It is easy to see that

$$\sum_{k=-n}^{n} \widehat{f}(k) e^{\imath kx} = e^{-\imath nx} \sum_{k=0}^{2n} (\widehat{f(\cdot) e^{\imath n(\cdot)}})(k) e^{\imath kx}.$$

This implies that the norm of $s_n : L_p(\mathbb{T}) \to L_p(\mathbb{T})$ is equal to the norm of $P_n^+ : L_p(\mathbb{T}) \to L_p(\mathbb{T})$.

We have

$$\begin{aligned} P_n^+ f &= \sum_{k=0}^{\infty} \widehat{f}(k) e^{\imath kx} - \sum_{k=2n+1}^{\infty} \widehat{f}(k) e^{\imath kx} \\ &= \sum_{k=0}^{\infty} \widehat{f}(k) e^{\imath kx} - e^{\imath(2n+1)x} \sum_{k=0}^{\infty} \widehat{f}(k+2n+1) e^{\imath kx} \\ &= P^+ f(x) - e^{\imath(2n+1)x} P^+ (e^{-\imath(2n+1)(\cdot)} f) - \widehat{f}(0)(1 - e^{\imath(2n+1)x}) \end{aligned}$$

for all trigonometric polynomials. By density this yields that

$$\left\| P_n^+ f \right\|_p \leq \left(2 \left\| P^+ \right\| + 2\right) \|f\|_p$$

for all $f \in L_p(\mathbb{R})$ and $n \in \mathbb{N}$, which proves (1.2.4). The convergence (1.2.5) is clearly valid for all trigonometric polynomials and so the convergence follows for all $f \in L_p(\mathbb{T})$ $(1 < p < \infty)$ by density. ■

Since the L_1-norms of D_n are not uniformly bounded, Theorem 1.2.10 is not true for $p = 1$ and $p = \infty$.

One of the deepest results in harmonic analysis is Carleson's theorem that the partial sums of the Fourier series converge almost everywhere to $f \in L_p(\mathbb{T})$ $(1 < p \leq \infty)$. Since the proof can be found in many papers and books (see, e.g., Carleson [51], Hunt [174], Arias de Reyna [9], Grafakos [143], Muscalu and Schlag [242], Lacey [192], or Demeter [80]), we present the result without proof.

Definition 1.2.11 We denote by

$$s_* f := \sup_{n \in \mathbb{N}} |s_n f|$$

the maximal operator of the partial sums.

Theorem 1.2.12 *If $f \in L_p(\mathbb{T})$ for some $1 < p < \infty$, then*

$$\|s_* f\|_p \leq C_p \|f\|_p$$

and if $1 < p \leq \infty$, *then*

$$\lim_{n\to\infty} s_n f = f \qquad a.e.$$

The inequality of Theorem 1.2.12 does not hold if $p = 1$ or $p = \infty$, and the almost everywhere convergence does not hold if $p = 1$. Du Bois Reymond [84] and Fejér [108] proved the existence of a continuous function $f \in C(\mathbb{T})$ and a point $x_0 \in \mathbb{T}$ such that the partial sums $s_n f(x_0)$ diverge as $n \to \infty$. Kolmogorov gave an integrable function $f \in L_1(\mathbb{T})$, whose Fourier series diverges almost everywhere or even everywhere (see Kolmogorov [186, 187], Zygmund [367], or Grafakos [143]).

Since there are many function spaces contained in $L_1(\mathbb{T})$ but containing $L_p(\mathbb{T})$ $(1 < p \leq \infty)$, it is natural to ask whether there is a "largest" subspace of $L_1(\mathbb{T})$ for which almost everywhere convergence holds. The next result, due to Antonov [7], generalizes Theorem 1.2.12.

Theorem 1.2.13 *If*

$$\int_{\mathbb{T}} |f(x)| \log^+ |f(x)| \log^+ \log^+ \log^+ |f(x)| \, dx < \infty, \tag{1.2.6}$$

then

$$\lim_{n\to\infty} s_n f = f \qquad a.e.$$

Note that $\log^+ u = \max(0, \log u)$. It is easy to see that if $f \in L_p(\mathbb{T})$ $(1 < p \leq \infty)$, then f satisfies (1.2.6). If f satisfies (1.2.6), then of course $f \in L_1(\mathbb{T})$. For the converse direction, Konyagin [188] obtained the next result.

Theorem 1.2.14 *If the non-decreasing function* $\phi : \mathbb{R}_+ \to \mathbb{R}_+$ *satisfies the condition*

$$\phi(u) = o\Big(u\sqrt{\log u}/\sqrt{\log\log u}\Big) \qquad as\ u \to \infty,$$

then there exists an integrable function f *such that*

$$\int_{\mathbb{T}} \phi(|f(x)|) \, dx < \infty$$

and

$$\limsup_{n\to\infty} s_n f(x) = \infty \qquad for\ all\ x \in \mathbb{T},$$

i.e., the Fourier series of f *diverges everywhere.*

For example, if $\phi(u) = u \log^+ \log^+ u$, then there exists a function f such that its Fourier series diverges everywhere and

$$\int_{\mathbb{T}} |f(x)| \log^+ \log^+ |f(x)|\, dx < \infty.$$

1.3 Hardy-Littlewood Maximal Function and Lebesgue Points

Before continuing our investigations about the convergence of the Fourier series, we have to introduce the Hardy-Littlewood maximal function. We will prove that it is bounded on $L_p(\mathbb{T})$ for $1 < p \le \infty$ and it is of weak type $(1, 1)$. Using this result, we obtain Lebesgue's differentiation theorem and the theorem about the Lebesgue points.

Definition 1.3.1 For $f \in L_1(\mathbb{T})$, the Hardy-Littlewood maximal function is defined by

$$Mf(x) := \sup_{x \in I} \frac{1}{|I|} \int_I |f|\, d\lambda \qquad (x \in \mathbb{T}),$$

where the supremum is taken over all open intervals I containing x.

We can also define the centered maximal function,

$$M_c f(x) := \sup_{h>0} \frac{1}{|I(x,h)|} \int_{I(x,h)} |f|\, d\lambda \qquad (x \in \mathbb{T}),$$

where $I(x, h)$ $(x \in \mathbb{T}, h > 0)$ denotes the interval with center x and radius h:

$$I(x, h) := \{y \in \mathbb{T} : |x - y| < h\}.$$

Obviously, $M_c f \le Mf$. If $x \in I(y, h)$, then $I(y, h) \subset I(x, 2h)$ and so $Mf \le 2M_c f$. Let $rI(x, h) := I(x, rh)$ for $r > 0$.

Lemma 1.3.2 (Vitali covering lemma) *Let be given finitely many open intervals I_j and let $E = \bigcup_j I_j$. Then there exists a finite subcollection $I_1, \ldots, I_m$ of disjoint intervals, such that*

$$\sum_{k=1}^{m} |I_k| \ge \frac{|E|}{3}.$$

Proof Let I_1 be an interval of the collection $\{I_j\}$ with maximal radius. Next choose I_2 to have maximal radius among the subcollection of intervals disjoint with I_1. We continue this process until we can go no further. Then the intervals $I_1, \ldots, I_m$ are disjoint. Observe that $3I_k$ contains all intervals of the original collection that intersect I_k $(k = 1, \ldots, m)$. From this, it follows that $\cup_{k=1}^m 3I_k$ contains all intervals from the original collection. Thus

$$|E| \le \left| \bigcup_{k=1}^{m} 3I_k \right| \le \sum_{k=1}^{m} |3I_k| \le 3 \sum_{k=1}^{m} |I_k|,$$

which shows the lemma. ■

Theorem 1.3.3 *The maximal operator M is of weak type $(1, 1)$, i.e.,*

$$\sup_{\rho>0} \rho\lambda(Mf > \rho) \le 3\|f\|_1 \qquad (f \in L_1(\mathbb{T})). \tag{1.3.1}$$

Moreover, if $1 < p \le \infty$, then

$$\|Mf\|_p \le C_p \|f\|_p \qquad (f \in L_p(\mathbb{T})). \tag{1.3.2}$$

Proof Let $E \subset \{Mf > \rho\}$ be a compact subset. For each $x \in \{Mf > \rho\}$, there exists an open interval I_x such that $x \in I_x$ and

$$\rho < \frac{1}{|I_x|} \int_{I_x} |f| \, d\lambda. \tag{1.3.3}$$

Since $x \in I_x$, we can select a finite collection of these intervals covering E. By Lemma 1.3.2, we can choose a finite disjoint subcollection $I_1, \ldots, I_m$ of this covering with

$$|E| \le 3 \sum_{k=1}^{m} |I_k|.$$

Since each I_k satisfies (1.3.3), adding these inequalities, we obtain

$$|E| < \frac{3}{\rho} \sum_{k=1}^{m} \int_{I_k} |f| \, d\lambda \le \frac{3}{\rho} \int_{\{Mf>\rho\}} |f| \, d\lambda.$$

Taking the supremum over all compact sets $E \subset \{Mf > \rho\}$, we conclude

$$\lambda(Mf > \rho) \le \frac{3}{\rho} \int_{\{Mf>\rho\}} |f| \, d\lambda \le \frac{3}{\rho} \int_{\mathbb{T}} |f| \, d\lambda,$$

which gives exactly (1.3.1).

For $p = \infty$, obviously

$$\frac{1}{|I|} \int_I |f| \, d\lambda \le \|f\|_\infty,$$

and so

$$\|Mf\|_\infty \le \|f\|_\infty \qquad (f \in L_\infty(\mathbb{R})).$$

Now the theorem follows easily for $1 < p < \infty$ by interpolation (see, e.g., Bergh and Löfström [33]). ∎

The boundedness on $L_\infty(\mathbb{T})$ and the weak type (1, 1) boundedness of M imply a finer version of (1.3.1).

Theorem 1.3.4 *We have*

$$\rho\lambda(Mf > 2\rho) \le 3\int_\rho^\infty \lambda(|f| > t)\,dt \qquad (\rho > 0).$$

Proof Let us decompose f into the sum of $f_0 \in L_1(\mathbb{T})$ and $f_1 \in L_\infty(\mathbb{T})$ as follows. For an arbitrary $\rho > 0$, set

$$f_{1,\rho}(t) := \begin{cases} f(t), & \text{if } |f(t)| \le \rho; \\ \rho/\operatorname{sign} f(t), & \text{otherwise.} \end{cases}$$

and

$$f_{0,\rho}(t) = f(t) - f_{1,\rho}(t).$$

Then $\|f_{1,\rho}\|_\infty \le \rho$. Since

$$Mf \le Mf_{0,\rho} + Mf_{1,\rho} \qquad \text{and} \qquad \left\|Mf_{1,\rho}\right\|_\infty \le C_\infty \left\|f_{1,\rho}\right\|_\infty \le \rho,$$

we have

$$\{Mf > 2\rho\} \subset \{Mf_{0,\rho} > \rho\} \cup \{Mf_{1,\rho} > \rho\} = \{Mf_{0,\rho} > \rho\}.$$

Hence

$$\begin{aligned} \rho\lambda(Mf > 2\rho) &\le \rho\lambda(Mf_{0,\rho} > \rho) \\ &\le 3\left\|f_{0,\rho}\right\|_1 \\ &= 3\int_{\{|f|>\rho\}} (|f| - \rho)\,d\lambda \\ &= 3\int_{\mathbb{T}}\int_0^\infty 1_{\{|f|>t>\rho\}}\,dt\,d\lambda \\ &= 3\int_\rho^\infty \lambda(|f| > t)\,dt \end{aligned}$$

as desired. ∎

It is known that inequality (1.3.2) does not hold for $p = 1$. However, we can prove that

$$\|Mf\|_1 \le C + C\left\||f|\left(\log^+|f|\right)\right\|_1,$$

where $\log^+ u := \max(0, \log u)$. We generalize this inequality as follows.

Theorem 1.3.5 *For every $k \in \mathbb{P}$ and $f \in L_1(\log L)^k(\mathbb{T})$,*

$$\left\| Mf \left(\log^+ Mf\right)^{k-1} \right\|_1 \le C + C \left\| |f| \left(\log^+ |f|\right)^k \right\|_1 .$$

Proof First, we handle the case $k > 1$. Observe that

$$\begin{aligned} \left\| |f| \left(\log^+ |f|\right)^{k-1} \right\|_1 &= \int_0^\infty \lambda(|f| > \rho) \frac{d(\rho(\log^+ \rho)^{k-1})}{d\rho}\, d\rho \\ &= \int_1^\infty \lambda(|f| > \rho) \frac{d(\rho(\log^+ \rho)^{k-1})}{d\rho}\, d\rho. \end{aligned}$$

Theorem 1.3.4 implies

$$\begin{aligned} &\int_1^\infty \lambda(|Mf| > \rho) \frac{d(\rho(\log^+ \rho)^{k-1})}{d\rho}\, d\rho \\ &\quad \le \int_1^\infty \frac{6}{\rho} \int_{\rho/2}^\infty \lambda(|f| > t)\, dt \frac{d(\rho(\log^+ \rho)^{k-1})}{d\rho}\, d\rho \\ &\quad = 6 \int_{1/2}^\infty \lambda(|f| > t) \int_1^{2t} \frac{1}{\rho} \frac{d(\rho(\log^+ \rho)^{k-1})}{d\rho}\, d\rho\, dt. \end{aligned}$$

Since

$$\frac{d(\rho(\log^+ \rho)^{k-1})}{d\rho} = (\log^+ \rho)^{k-1} + (k-1)(\log^+ \rho)^{k-2},$$

we conclude

$$\begin{aligned} \int_1^{2t} \frac{1}{\rho} \frac{d(\rho(\log^+ \rho)^{k-1})}{d\rho}\, d\rho &= \frac{1}{k}(\log^+(2t))^k + (\log^+(2t))^{k-1} \\ &= \frac{1}{2k} \frac{d(2t(\log^+ 2t)^k)}{dt}. \end{aligned}$$

Therefore

$$\begin{aligned} &\int_1^\infty \lambda(|Mf| > \rho) \frac{d(\rho(\log^+ \rho)^{k-1})}{d\rho}\, d\rho \\ &\quad \le \frac{3}{k} \int_{1/2}^\infty \lambda(|f| > t) \frac{d(2t(\log^+ 2t)^k)}{dt}\, dt \\ &\quad = C \left\| 2|f| \left(\log^+ |2f|\right)^k \right\|_1 \\ &\quad \le C + C \left\| |f| \left(\log^+ |f|\right)^k \right\|_1, \end{aligned} \tag{1.3.4}$$

which completes the proof for $k > 1$.

Let $k = 1$ and notice that $\lambda(Mf \le 1) \le 1$. Then

$$\int_{\{Mf>1\}} Mf(t)\,dt = \int_0^\infty \lambda(Mf > \max(\rho, 1))\,d\rho$$
$$= \int_1^\infty \lambda(Mf > \rho)\,d\rho + \lambda(Mf > 1).$$

Moreover,

$$\lambda(Mf > 1) \le 3\|f\|_1 = 3\int_{\{|f|\le e\}} |f|\,d\lambda + 3\int_{\{|f|>e\}} |f|\,d\lambda$$
$$\le C + C\int_{\mathbb{T}} |f|\log^+|f|\,d\lambda.$$

Since (1.3.4) holds for $k = 1$, too, we obtain

$$\|Mf\|_1 = \int_{\{Mf\le 1\}} Mf(t)\,dt + \int_{\{Mf>1\}} Mf(t)\,dt$$
$$\le C + C\int_{\mathbb{T}} |f|\log^+|f|\,d\lambda,$$

as we stated in the theorem. ■

Now we present a density theorem due to Marcinkiewicz and Zygmund [234]. Let $L_0(\mathbb{T})$ denote the set of measurable functions and $X \subset L_0(\mathbb{T})$. Let the operators $T, T_n : X \to L_0(\mathbb{T})$ $(n \in \mathbb{N})$ be given. Moreover, we introduce the maximal operator by

$$T_* f(x) := \sup_{n\in\mathbb{N}} |T_n f(x)| \qquad (f \in X, x \in \mathbb{T}).$$

Theorem 1.3.6 *Let X be a normed space of measurable functions and $S \subset X$ be dense in X. Suppose that T and T_n $(n \in \mathbb{N})$ are linear operators and*

$$\lim_{n\to\infty} T_n f = Tf \qquad a.e.$$

for all $f \in S$. If

$$\sup_{\rho>0} \rho\,\lambda(|Tf| > \rho) \le C\|f\|_X \qquad (f \in X) \tag{1.3.5}$$

and

$$\sup_{\rho>0} \rho\,\lambda(T_* f > \rho) \le C\|f\|_X \qquad (f \in X), \tag{1.3.6}$$

then for every $f \in X$,

$$\lim_{n\to\infty} T_n f = Tf \qquad a.e.$$

Proof Fix $f \in X$ and set

$$\xi := \limsup_{n\to\infty} |T_n f - Tf|.$$

It is sufficient to show that $\xi = 0$ a.e. Choose a sequence $f_m \in S$ ($m \in \mathbb{N}$) such that

$$\lim_{m\to\infty} \|f - f_m\|_X = 0.$$

By the triangle inequality,

$$\xi \le \limsup_{n\to\infty} |T_n(f - f_m)| + \limsup_{n\to\infty} |T_n f_m - Tf_m| + |T(f_m - f)|$$

for all $m \in \mathbb{N}$. Since $f_m \in S$, we have

$$\limsup_{n\to\infty} |T_n f_m - Tf_m| = \lim_{n\to\infty} |T_n f_m - Tf_m| = 0 \qquad \text{a.e.,}$$

so

$$\xi \le T_*(f_m - f) + |T(f_m - f)| \qquad \text{a.e.}$$

Applying inequalities (1.3.5) and (1.3.6), we obtain

$$\begin{aligned}\lambda(\xi > 2\rho) &\le \lambda(T_*(f_m - f) > \rho) + \lambda(|T(f_m - f)| > \rho)\\ &\le C\rho^{-1}\|f_m - f\|_X + C\rho^{-1}\|f_m - f\|_X\end{aligned}$$

for all $\rho > 0$ and $m \in \mathbb{N}$. Since $f_m \to f$ in the X-norm as $m \to \infty$, we get that

$$\lambda(\xi > 2\rho) = 0$$

for all $\rho > 0$. This implies immediately that $\xi = 0$ almost everywhere. ∎

The next theorem can be proved in the same way.

Theorem 1.3.7 *Let X be a normed space of measurable functions and $S \subset X$ be dense in X. Suppose that T_n is a sublinear operator for every $n \in \mathbb{N}$ and*

$$\lim_{n\to\infty} T_n f = 0 \qquad \text{a.e.}$$

for all $f \in S$. If

$$\sup_{\rho>0} \rho\,\lambda(T_* f > \rho) \le C\|f\|_X \qquad (f \in X),$$

then for every $f \in X$,

$$\lim_{n\to\infty} T_n f = 0 \qquad \text{a.e.}$$

Now we can state Lebesgue's differentiation theorem mentioned before.

Corollary 1.3.8 *For all* $f \in L_1(\mathbb{T})$,

$$\lim_{r\to 0} \frac{1}{2r} \int_{x-r}^{x+r} f(t)\, dt = f(x) \qquad a.e.\ x \in \mathbb{T}.$$

Proof Let $r_n > 0$ $(n \in \mathbb{N})$ and $\lim_{n\to\infty} r_n = 0$. Define

$$Tf(x) := f(x) \qquad \text{and} \qquad T_n f(x) := \frac{1}{2r_n} \int_{x-r_n}^{x+r_n} f(t)\, dt \qquad (x \in \mathbb{T}).$$

These operators are linear and

$$\sup_{\rho>0} \rho\, \lambda(|Tf| > \rho) = \sup_{\rho>0} \rho\, \lambda(|f| > \rho) \le \sup_{\rho>0} \int_{\{|f|>\rho\}} |f|\, d\lambda \le \|f\|_1$$

implies (1.3.5). Inequality (1.3.6) follows from Theorem 1.3.3. The result obviously holds for continuous functions. If S denotes the set of continuous functions, then S is dense in $L_1(\mathbb{T})$. Now Theorem 1.3.6 implies Corollary 1.3.8. ■

Similarly, we get

Corollary 1.3.9 *For all* $f \in L_1(\mathbb{T})$,

$$\lim_{x\in I, |I|\to 0} \frac{1}{|I|} \int_I f\, d\lambda = f(x) \qquad a.e.\ x \in \mathbb{T}.$$

Corollary 1.3.8 implies that $|f(x)| \le Mf(x)$ for almost every $x \in \mathbb{T}$, and so the converse of (1.3.2) is also true:

$$\|f\|_p \le \|Mf\|_p \qquad (1 \le p \le \infty).$$

Now we introduce the concept of Lebesgue points. Corollary 1.3.8 can be written in the form

$$\lim_{h\to 0} \frac{1}{2h} \int_{-h}^{h} f(x-t)\, dt = f(x)$$

for almost every $x \in \mathbb{T}$ and $f \in L_1(\mathbb{T})$. Thus

$$\lim_{h\to 0} \frac{1}{2h} \int_{-h}^{h} (f(x-t) - f(x))\, dt = 0$$

for almost every $x \in \mathbb{T}$, which is equivalent to

$$\lim_{h\to 0} \frac{1}{2h} \left| \int_{-h}^{h} (f(x-t) - f(x))\, dt \right| = 0$$

for almost every $x \in \mathbb{T}$. Though the definition of the Lebesgue point is a stronger condition, we prove in the next theorem that almost every point is a Lebesgue point.

Definition 1.3.10 A point $x \in \mathbb{T}$ is called a Lebesgue point of $f \in L_1(\mathbb{T})$ if

$$\lim_{h \to 0} \frac{1}{2h} \int_{-h}^{h} |f(x-t) - f(x)| \, dt = 0.$$

Theorem 1.3.11 *Almost every point $x \in \mathbb{R}$ is a Lebesgue point of $f \in L_1(\mathbb{T})$.*

Proof For all rational numbers q let

$$G_q := \left\{ x \in \mathbb{R} : \lim_{h \to 0} \frac{1}{2h} \int_{-h}^{h} |f(x-t) - q| \, dt = |f(x) - q| \right\}.$$

Applying Corollary 1.3.8 to the function $|f(\cdot) - q|$, we can see that $B_q := \mathbb{R} \setminus G_q$ is of Lebesgue measure 0. Observe that f is almost everywhere finite. Set $N := \{x \in \mathbb{R} : |f(x)| = \infty\}$. Then the set

$$B := N \bigcup \left(\bigcup_{q \in \mathbb{Q}} B_q \right)$$

has Lebesgue measure 0. We show that the points of $G := \mathbb{T} \setminus B$ are Lebesgue points. Let $\epsilon > 0$ and $x \in G$ be arbitrary. Choose $q \in \mathbb{Q}$ such that

$$|f(x) - q| < \frac{\epsilon}{2}.$$

Then

$$\begin{aligned} \frac{1}{2h} & \int_{-h}^{h} |f(x-t) - f(x)| \, dt \\ & \leq \frac{1}{2h} \int_{-h}^{h} |f(x-t) - q| \, dt + \frac{1}{2h} \int_{-h}^{h} |q - f(x)| \, dt \\ & = \frac{1}{2h} \int_{-h}^{h} |f(x-t) - q| \, dt + |q - f(x)|. \end{aligned}$$

Since $x \notin B_q$, we have

$$\limsup_{h \to \infty} \frac{1}{2h} \int_{-h}^{h} |f(x-t) - f(x)| \, dt \leq 2 |f(x) - q| < \epsilon.$$

Thus, every $x \in G$ is a Lebesgue point of f. ■

Lemma 1.3.12 *If x is a Lebesgue point of $f \in L_1(\mathbb{T})$, then $f(x)$ and $Mf(x)$ are finite.*

Proof $f(x)$ is clearly finite. For $\epsilon = 1$ there exists $\delta > 0$ such that for all $|h| < \delta$,

$$\frac{1}{2h} \int_{-h}^{h} |f(x-t) - f(x)| \, dt < 1.$$

Thus

$$\frac{1}{2h} \int_{-h}^{h} |f(x-t)| \, dt \le \frac{1}{2h} \int_{-h}^{h} |f(x-t) - f(x)| \, dt + |f(x)| < 1 + |f(x)|.$$

On the other hand,

$$\frac{1}{2h} \int_{-h}^{h} |f(x-t)| \, dt \le \frac{1}{2\delta} \|f\|_1$$

for all $|h| \ge \delta$. ■

1.4 Summability of One-Dimensional Fourier Series

Though Theorems 1.2.10 and 1.2.12 are not true for $p = 1$ and $p = \infty$, with the help of some summability methods they can be generalized. Obviously, summability means have better convergence properties than the original Fourier series. Summability is intensively studied in the literature (see, e.g., the books Stein and Weiss [293], Butzer and Nessel [47], Trigub and Belinsky [319], Grafakos [143] and Weisz [332, 346], and the references therein).

One of the first investigated summability methods is the Fejér method. In 1904, Fejér [107] investigated the arithmetic means of the partial sums, the so-called Fejér means $\sigma_n f$. He proved for an integrable function $f \in L_1(\mathbb{T})$ that if the left and right limits $f(x-0)$ and $f(x+0)$ exist at a point x, then the Fejér means converge to $(f(x-0) + f(x+0))/2$, that is,

$$\lim_{n \to \infty} \sigma_n f(x) = \frac{f(x-0) + f(x+0)}{2}. \tag{1.4.1}$$

One year later Lebesgue [197] extended this theorem and obtained that the convergence holds for every $f \in L_1(\mathbb{T})$ and every Lebesgue points, i.e.,

$$\lim_{n \to \infty} \sigma_n f(x) = f(x) \tag{1.4.2}$$

at each Lebesgue point of f, thus almost everywhere. In this section, we generalize these results.

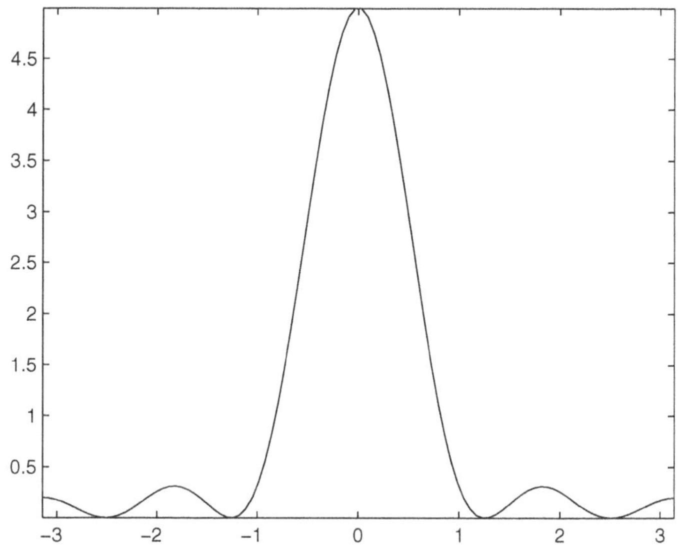

Fig. 1.2 Fejér kernel K_n for $n = 5$

Definition 1.4.1 For $f \in L_1(\mathbb{T})$ and $n \in \mathbb{N}$, the nth Fejér means $\sigma_n f$ of the Fourier series of f and the nth Fejér kernel K_n are introduced by

$$\sigma_n f(x) := \sum_{k=-n}^{n} \left(1 - \frac{|k|}{n}\right) \widehat{f}(k) e^{\imath kx}$$

and

$$K_n(t) := \sum_{k=-n}^{n} \left(1 - \frac{|k|}{n}\right) e^{\imath kt},$$

respectively.

One can see that

$$\sigma_n f(x) = \frac{1}{2\pi} \int_{\mathbb{T}} f(x-t) K_n(t)\, dt$$

(see Fig. 1.2). We will prove the next result in Lemma 1.4.12.

Lemma 1.4.2 *For $f \in L_1(\mathbb{T})$ and $n \in \mathbb{N}$, we have*

$$\sigma_n f(x) = \frac{1}{n} \sum_{j=0}^{n-1} s_j f(x)$$

and

$$K_n(t) = \frac{1}{n} \sum_{j=0}^{n-1} D_j(t).$$

Lemma 1.4.3 *For $n \geq 1$ and $t \in \mathbb{T}$, $t \neq 0$,*

$$\sum_{k=0}^{n-1} \sin(k+1/2)t = \frac{1-\cos(nt)}{2\sin(t/2)} \tag{1.4.3}$$

and

$$\sum_{k=0}^{n-1} \cos(k+1/2)t = \frac{\sin(nt)}{2\sin(t/2)}. \tag{1.4.4}$$

Proof Adding the equalities

$$2\sin(t/2)\sin(k+1/2)t = \cos(kt) - \cos(k+1)t$$

and

$$2\sin(t/2)\cos(k+1/2)t = \sin(k+1)t - \sin(kt),$$

we obtain the lemma. ■

Lemma 1.4.4 *For $n \geq 1$ and $t \in \mathbb{T}$, $t \neq 0$, we have*

$$K_n(t) = \frac{1}{n}\left(\frac{\sin(nt/2)}{\sin(t/2)}\right)^2.$$

Proof By Lemmas 1.2.3 and 1.4.2,

$$K_n(t) = \frac{1}{n}\sum_{k=0}^{n-1} D_k(t) = \frac{1}{n}\sum_{j=0}^{n-1} \frac{\sin((k+1/2)t)}{\sin(t/2)}.$$

The lemma follows from (1.4.3). ■

Corollary 1.4.5 *For $n \geq 1$ and $-\pi \leq t \leq \pi$, $t \neq 0$,*

$$|K_n(t)| \leq 2n-1 \quad \textit{and} \quad |K_n(t)| \leq \frac{C}{n|t|^2}.$$

Proof The inequalities follow from Lemmas 1.2.4 and 1.4.4. ■

Now we generalize the Fejér summability.

Definition 1.4.6 For $\alpha \neq -1, -2, \ldots$, let $A_{-1}^{\alpha} := 0$ and

$$A_n^\alpha := \binom{n+\alpha}{n} = \frac{(\alpha+1)(\alpha+2)\cdots(\alpha+n)}{n!} \qquad (n \in \mathbb{N}).$$

Obviously, $A_0^\alpha = 1$ and if $\alpha = 0$, then $A_n^0 = 1$, if $\alpha = 1$, then $A_n^1 = n + 1$ $(n \in \mathbb{N})$.

Lemma 1.4.7 *For any $n \in \mathbb{N}$, $\alpha, \beta \neq -1, -2, \ldots$, we have*

$$A_n^{\alpha+\beta+1} = \sum_{k=0}^{n} A_k^\alpha A_{n-k}^\beta .$$

Proof It is known that, for any $x \in \mathbb{C}$, $|x| < 1$,

$$(1-x)^{-\alpha-1} = \sum_{n=0}^{\infty} \binom{-\alpha-1}{n} (-x)^n .$$

From this, it follows that

$$(1-x)^{-\alpha-1} = \sum_{n=0}^{\infty} \binom{n+(-n-\alpha-1)}{n} (-x)^n = \sum_{n=0}^{\infty} A_n^\alpha x^n . \tag{1.4.5}$$

Similarly,

$$(1-x)^{-\beta-1} = \sum_{n=0}^{\infty} A_n^\beta x^n$$

and

$$(1-x)^{-\alpha-\beta-2} = \sum_{n=0}^{\infty} A_n^{\alpha+\beta+1} x^n .$$

However, the last series can be obtained also by multiplying the first two:

$$(1-x)^{-\alpha-\beta-2} = (1-x)^{-\alpha-1}(1-x)^{-\beta-1} = \sum_{n=0}^{\infty} \left(\sum_{k=0}^{n} A_k^\alpha A_{n-k}^\beta \right) x^n ,$$

which implies the desired result. ∎

Lemma 1.4.8 *For any $n \in \mathbb{N}$, $\alpha \neq -1, -2, \ldots$, we have*

$$A_n^\alpha = \sum_{k=0}^{n} A_k^{\alpha-1}, \qquad A_n^\alpha - A_{n-1}^\alpha - A_n^{\alpha-1}.$$

Proof We obtain the first equality by replacing α by $\alpha - 1$ and β by 0 in Lemma 1.4.7. The second one follows easily from the first one. ∎

Lemma 1.4.9 *For any $n, N \in \mathbb{N}$, $-N < \alpha \leq N$ and $\alpha \neq -1, -2, \ldots$, there exist $c_N, C_N > 0$ such that*

$$c_N n^\alpha < |A_n^\alpha| < C_N n^\alpha. \tag{1.4.6}$$

Proof By Taylor's formula, for $x \in (-1, 1)$ there exists $\xi \in (0, x)$ such that

$$\ln(1+x) = x - \frac{1}{2(1+\xi)^2} x^2.$$

This implies that $\ln(1+x) = x + O(x^2)$ if $-N/(N+1) < x < 1$. Then

$$\begin{aligned} \ln |A_n^\alpha| &= \sum_{k=1}^{n} \ln \left| 1 + \frac{\alpha}{k} \right| \\ &= \sum_{k=1}^{N} \ln \left| 1 + \frac{\alpha}{k} \right| + \alpha \sum_{k=N+1}^{n} \frac{1}{k} + \alpha^2 \sum_{k=N+1}^{n} O\left(\frac{1}{k^2}\right) \\ &= \sum_{k=1}^{N} \ln \left| 1 + \frac{\alpha}{k} \right| + \alpha\Big(\ln n + O(1)\Big) + \alpha^2 O(1), \end{aligned}$$

that is,

$$A_n^\alpha = n^\alpha O(1).$$

This proves the lemma. ■

Definition 1.4.10 For $f \in L_1(\mathbb{T})$, $n \in \mathbb{N}$ and $\alpha \geq 0$, the nth Cesàro means $\sigma_n^\alpha f$ of the Fourier series of f and the nth Cesàro kernel K_n^α are introduced by

$$\sigma_n^\alpha f(x) := \frac{1}{A_{n-1}^\alpha} \sum_{k=-n}^{n} A_{n-1-|k|}^\alpha \widehat{f}(k) e^{\imath kx}$$

and

$$K_n^\alpha(t) := \frac{1}{A_{n-1}^\alpha} \sum_{k=-n}^{n} A_{n-1-|k|}^\alpha e^{\imath kt},$$

respectively.

Note that the Cesàro means are also called (C, α)-means. Obviously, for $\alpha = 1$, we get back the Fejér means and for $\alpha = 0$, the partial sums. The definition of the Cesàro kernels implies

Lemma 1.4.11 *For $\alpha \geq 0$ and $n \in \mathbb{N}$, we have*

$$\frac{1}{2\pi} \int_{\mathbb{T}} K_n^\alpha(t)\, dt = 1.$$

One can see that

$$\sigma_n^\alpha f(x) = \frac{1}{2\pi} \frac{1}{A_{n-1}^\alpha} \sum_{k=-n}^{n} A_{n-1-|k|}^\alpha \int_{\mathbb{T}} f(t) e^{\imath k(x-t)}\, dt$$
$$= \frac{1}{2\pi} \int_{\mathbb{T}} f(x-t) K_n^\alpha(t)\, dt \qquad (n \in \mathbb{N}). \tag{1.4.7}$$

Lemma 1.4.12 *For $f \in L_1(\mathbb{T})$, $\alpha > 0$ and $n \in \mathbb{N}$, we have*

$$\sigma_n^\alpha f(x) = \frac{1}{A_{n-1}^\alpha} \sum_{j=0}^{n-1} A_{n-1-j}^{\alpha-1} s_j f(x)$$

and

$$K_n^\alpha(t) = \frac{1}{A_{n-1}^\alpha} \sum_{j=0}^{n-1} A_{n-1-j}^{\alpha-1} D_j(t).$$

Proof By Lemma 1.4.8,

$$K_n^\alpha(t) = \frac{1}{A_{n-1}^\alpha} \sum_{k=-n}^{n} A_{n-1-|k|}^\alpha e^{\imath kt}$$
$$= \frac{1}{A_{n-1}^\alpha} \sum_{k=-n}^{n} \sum_{j=|k|}^{n-1} A_{n-1-j}^{\alpha-1} e^{\imath kt}$$
$$= \frac{1}{A_{n-1}^\alpha} \sum_{j=0}^{n-1} A_{n-1-j}^{\alpha-1} D_j(t),$$

which shows the lemma. ∎

The following lemma shows that if the (C, α) means $(\alpha > -1)$ are convergent then the $(C, \alpha + h)$ means $(h > 0)$ are convergent, too.

Lemma 1.4.13 *For $\alpha > -1$ and $h > 0$, we have*

$$\sigma_n^{\alpha+h} f = \frac{1}{A_{n-1}^{\alpha+h}} \sum_{k=1}^{n} A_{n-k}^{h-1} A_{k-1}^\alpha \sigma_k^\alpha f.$$

Proof Indeed, by Lemmas 1.4.7 and 1.4.12,

$$\begin{aligned}\frac{1}{A_{n-1}^{\alpha+h}}\sum_{k=1}^{n} A_{n-k}^{h-1}A_{k-1}^{\alpha}\sigma_k^{\alpha} f &= \frac{1}{A_{n-1}^{\alpha+h}}\sum_{k=1}^{n} A_{n-k}^{h-1}\sum_{j=0}^{k-1} A_{k-1-j}^{\alpha-1}s_j f \\ &= \frac{1}{A_{n-1}^{\alpha+h}}\sum_{j=0}^{n-1} s_j f \sum_{k=j+1}^{n} A_{n-k}^{h-1}A_{k-1-j}^{\alpha-1} \\ &= \frac{1}{A_{n-1}^{\alpha+h}}\sum_{j=0}^{n-1} A_{n-1-j}^{\alpha+h-1} s_j f,\end{aligned}$$

which gives the result. ■

For Cesàro means, instead of inequalities (1.4.3) and (1.4.4), we will use the following lemma.

Lemma 1.4.14 *For* $0 < \alpha \le 1$, $n \ge 1$, *and* $t \in \mathbb{T}$, $t \ne 0$,

$$\left|\sum_{k=0}^{n-1} A_{n-1-k}^{\alpha-1}\sin((k+1/2)t)\right| \le \frac{C}{|\sin(t/2)|^{\alpha}} + \frac{Cn^{\alpha-1}}{|\sin(t/2)|} \tag{1.4.8}$$

and

$$\left|\sum_{k=0}^{n-1} A_{n-1-k}^{\alpha-1}\cos((k+1/2)t)\right| \le \frac{C}{|\sin(t/2)|^{\alpha}} + \frac{Cn^{\alpha-1}}{|\sin(t/2)|}. \tag{1.4.9}$$

Proof The inequalities follow from Lemma 1.4.3 for $\alpha = 1$. Let $0 < \alpha < 1$. Suppose that $-\pi \le t \le \pi$. Then

$$\sum_{k=0}^{n-1} A_{n-1-k}^{\alpha-1}\sin((k+1/2)t) = \Im\left(\sum_{k=0}^{n-1} A_{n-1-k}^{\alpha-1}e^{\imath(k+1/2)t}\right)$$

and

$$\begin{aligned}\sum_{k=0}^{n-1} A_{n-1-k}^{\alpha-1}e^{\imath(k+1/2)t} &= e^{\imath(n-1/2)t}\sum_{k=0}^{n-1} A_{n-1-k}^{\alpha-1}e^{\imath(k+1-n)t} \\ &= e^{\imath(n-1/2)t}\sum_{j=0}^{n-1} A_j^{\alpha-1}e^{-\imath jt},\end{aligned} \tag{1.4.10}$$

where $\Im$ denotes the imaginary part of the function. We know that

$$\sum_{j=0}^{\infty} A_j^{\alpha-1}x^j = (1-x)^{-\alpha} \tag{1.4.11}$$

for $x \in \mathbb{C}$, $|x| < 1$. However, this holds also for $|x| = 1$, $x \neq 1$. Indeed, $A_n^{\alpha-2} \leq 0$ and so by Lemma 1.4.8, $(A_n^{\alpha-1})_{n\in\mathbb{N}}$ is non-increasing. The left-hand side of (1.4.11) is convergent because Abel rearrangement implies that

$$\begin{aligned}\left|\sum_{j=n}^{m} A_j^{\alpha-1} x^j\right| &= \left|\sum_{j=n}^{m-1}\left(A_j^{\alpha-1} - A_{j+1}^{\alpha-1}\right)\left(\sum_{i=n}^{j} x^i\right) + A_m^{\alpha-1}\left(\sum_{i=n}^{m} x^i\right)\right| \\ &\leq A_n^{\alpha-1} \sup_{n\leq j\leq m}\left|\frac{x^n - x^{j+1}}{1-x}\right| \\ &\leq \frac{2A_n^{\alpha-1}}{|1-x|} \to 0 \end{aligned} \tag{1.4.12}$$

as $n \to \infty$. The last convergence follows from Lemma 1.4.9. Similarly, if $0 < r < 1$ is near to 1, say $r_0 < r < 1$, then

$$\sum_{j=0}^{\infty} A_j^{\alpha-1} r^j x^j = (1-rx)^{-\alpha}$$

and

$$\left|\sum_{j=n}^{m} A_j^{\alpha-1} r^j x^j\right| \leq \frac{2A_n^{\alpha-1}}{|1-rx|} \leq \frac{4A_n^{\alpha-1}}{|1-x|} \to 0$$

for $|x| = 1$, $x \neq 1$. This implies that

$$\begin{aligned}\left|(1-x)^{-\alpha} - \sum_{j=0}^{n-1} A_j^{\alpha-1} x^j\right| &\leq \left|(1-x)^{-\alpha} - (1-rx)^{-\alpha}\right| \\ &\quad + \left|(1-rx)^{-\alpha} - \sum_{j=0}^{n-1} A_j^{\alpha-1} r^j x^j\right| + \left|\sum_{j=0}^{n-1} A_j^{\alpha-1} r^j x^j - \sum_{j=0}^{n-1} A_j^{\alpha-1} x^j\right| \\ &< 2\epsilon + \left|(1-rx)^{-\alpha} - \sum_{j=0}^{n-1} A_j^{\alpha-1} r^j x^j\right| < 3\epsilon \end{aligned}$$

if r is near enough to 1 and n is large enough. Thus (1.4.11) is true for $|x| \leq 1, x \neq 1$.

Using (1.4.10), (1.4.11), (1.4.12), and (1.4.6), we get that

$$\begin{aligned}\left|\sum_{k=0}^{n-1} A_{n-1-k}^{\alpha-1} e^{\imath(k+1/2)t}\right| &= \left|(1-e^{-\imath t})^{-\alpha} - \sum_{j=n}^{\infty} A_j^{\alpha-1} e^{-\imath jt}\right| \\ &\le |(1-e^{-\imath t})^{-\alpha}| + 2A_n^{\alpha-1}|(1-e^{-\imath t})^{-1}| \\ &\le \frac{C}{|t|^\alpha} + \frac{Cn^{\alpha-1}}{|t|},\end{aligned}$$

which proves (1.4.8). Inequality (1.4.9) can be handled similarly. ■

The derivatives of the left-hand sides of (1.4.8) and (1.4.9) can be estimated as follows.

Lemma 1.4.15 *For $0 < \alpha \le 1$, $n \ge 1$ and $t \in \mathbb{T}$, $t \neq 0$,*

$$\left|\sum_{k=0}^{n-1} kA_{n-1-k}^{\alpha-1} \sin((k+1/2)t)\right| \le \frac{Cn}{|\sin(t/2)|^\alpha} + \frac{Cn^\alpha}{|\sin(t/2)|}$$

and

$$\left|\sum_{k=0}^{n-1} kA_{n-1-k}^{\alpha-1} \cos((k+1/2)t)\right| \le \frac{Cn}{|\sin(t/2)|^\alpha} + \frac{Cn^\alpha}{|\sin(t/2)|}.$$

Proof We apply Lemma 1.4.14 to obtain

$$\begin{aligned}\left|\sum_{k=0}^{n-1} kA_{n-1-k}^{\alpha-1} \sin((k+1/2)t)\right| &\le \sum_{j=1}^{n-1}\left|\sum_{k=j}^{n-1} A_{n-1-k}^{\alpha-1} \sin((k+1/2)t)\right| \\ &\le \sum_{j=1}^{n-1}\left(\frac{C}{|\sin(t/2)|^\alpha} + \frac{Cj^{\alpha-1}}{|\sin(t/2)|}\right) \\ &\le \frac{Cn}{|\sin(t/2)|^\alpha} + \frac{Cn^\alpha}{|\sin(t/2)|}.\end{aligned}$$

The second inequality can be shown in the same way. ■

The following theorem will be used several times in this book. It can also be found in Zygmund [367].

Theorem 1.4.16 *For $0 < \alpha \le 1$, $n \ge 1$ and $-\pi \le t \le \pi$, $t \neq 0$,*

$$\left|K_n^\alpha(t)\right| \le 2n-1 \quad \text{and} \quad \left|K_n^\alpha(t)\right| \le \frac{C}{n^\alpha |t|^{\alpha+1}}. \tag{1.4.13}$$

Proof For $\alpha = 1$, the theorem is exactly Corollary 1.4.5. Let $0 < \alpha < 1$. The first inequality follows from Lemma 1.2.4 and 1.4.12. By Lemmas 1.2.3 and 1.4.12,

$$\begin{aligned}K_n^\alpha(t) &= \frac{1}{A_{n-1}^\alpha}\sum_{k=0}^{n-1} A_{n-1-k}^{\alpha-1} D_k(t)\\ &= \frac{1}{A_{n-1}^\alpha}\sum_{k=0}^{n-1} A_{n-1-k}^{\alpha-1} \frac{\sin((k+1/2)t)}{\sin(t/2)}.\end{aligned}$$

Now, by Lemma 1.4.14,

$$|K_n^\alpha(t)| \le \frac{1}{A_{n-1}^\alpha |\sin(t/2)|}\left(\frac{C}{|t|^\alpha}+\frac{Cn^{\alpha-1}}{|t|}\right) \le \frac{C}{n^\alpha |t|^{\alpha+1}}+\frac{C}{n|t|^2}.$$

If $|t| \ge 1/n$, then

$$\frac{1}{n|t|^2} \le \frac{1}{n^\alpha |t|^{\alpha+1}}.$$

If $|t| < 1/n$, then the first inequality of (1.4.13) implies the second one. ∎

1.5 Convergence at Lebesgue Points of the Cesàro Means

Now we are ready to generalize Lebesgue's theorem given in (1.4.2) for Cesàro summability. But first we introduce the Herz spaces which, as we will see later, are very closely connected to the concept of Lebesgue points.

Definition 1.5.1 The Herz space $E_\infty(\mathbb{T})$ contains all functions f for which

$$\|f\|_{E_\infty} := \sum_{k=-\infty}^{0} 2^k \left\| f 1_{P_k} \right\|_\infty < \infty,$$

where $P_k := I(0, 2^k\pi) \setminus I(0, 2^{k-1}\pi)$, $(k \in \mathbb{Z})$.

Recall that $I(x,h) := \{y \in \mathbb{T} : |x-y| < h\}$. The Cesàro kernels are all in $E_\infty(\mathbb{T})$ for $0 < \alpha \le 1$.

Theorem 1.5.2 *If $0 < \alpha \le 1$, then $K_n^\alpha \in E_\infty(\mathbb{T})$ and*

$$\sup_{n\in\mathbb{N}} \left\| K_n^\alpha \right\|_{E_\infty} \le C_\alpha.$$

Proof By Theorem 1.4.16, $|K_n^\alpha(t)| \le g_n^\alpha(t)$, where

$$g_n^\alpha(t) := C \min\left(n, \frac{1}{n^\alpha |t|^{\alpha+1}}\right).$$

Since g_n^α is non-increasing and integrable, we obtain

$$\left\| K_n^\alpha \right\|_{E_\infty} = \sum_{k=-\infty}^{0} 2^k \left\| K_n^\alpha 1_{P_k} \right\|_\infty \leq C_\alpha \int_{\mathbb{T}} g_n^\alpha(t)\, dt$$
$$\leq C_\alpha \int_0^{1/n} n\, d\lambda + C_\alpha n^{-\alpha} \int_{1/n}^{\pi} |t|^{-\alpha-1}\, dt \leq C_\alpha,$$

which shows the desired result. ■

In the same way, we obtain

Corollary 1.5.3 *If* $0 < \alpha \leq 1$, *then* $K_n^\alpha \in L_1(\mathbb{T})$ *and*

$$\sup_{n\in\mathbb{N}} \left\| K_n^\alpha \right\|_1 \leq C_\alpha.$$

Theorem 1.5.4 *If* $0 < \alpha < \infty$, *then*

$$\lim_{n\to\infty} \sigma_n^\alpha f(x) = f(x)$$

for all Lebesgue points of $f \in L_1(\mathbb{T})$.

Proof First suppose that $0 < \alpha \leq 1$. Set

$$G(u) := \int_{-u}^{u} |f(x-t) - f(x)|\, dt \qquad (u > 0).$$

Since x is a Lebesgue point of f, for all $\epsilon > 0$, there exists $m \in \mathbb{Z}$ such that

$$\frac{G(u)}{2u} \leq \epsilon \quad \text{if} \quad 0 < u \leq 2^m. \tag{1.5.1}$$

It follows from Lemma 1.4.11 and (1.4.7) that

$$\sigma_n^\alpha f(x) - f(x) = \frac{1}{2\pi} \int_{\mathbb{T}} (f(x-t) - f(x)) K_n^\alpha(t)\, dt.$$

Thus

$$\begin{aligned} |\sigma_n^\alpha f(x) - f(x)| &\leq C \int_{\mathbb{T}^d} |f(x-t) - f(x)| \left|K_n^\alpha(t)\right| dt \\ &= C \int_{-2^m\pi}^{2^m\pi} |f(x-t) - f(x)| \left|K_n^\alpha(t)\right| dt \\ &\quad + C \int_{\mathbb{T}\setminus(-2^m\pi, 2^m\pi)} |f(x-t) - f(x)| \left|K_n^\alpha(t)\right| dt \\ &=: A_1(x) + A_2(x). \end{aligned}$$

We estimate $A_1(x)$ by

$$\begin{aligned}A_1(x) &= C \sum_{k=-\infty}^{m} \int_{P_k} |f(x-t) - f(x)| \left|K_n^\alpha(t)\right| \, dt \\ &\le C \sum_{k=-\infty}^{m} \sup_{P_k} \left|K_n^\alpha\right| \int_{P_k} |f(x-t) - f(x)| \, dt \\ &\le C \sum_{k=-\infty}^{m} \sup_{P_k} \left|K_n^\alpha\right| G(2^k \pi).\end{aligned}$$

Then, by (1.5.1),

$$A_1(x) \le C\epsilon \sum_{k=-\infty}^{m} 2^k \sup_{P_k} \left|K_n^\alpha\right| \le C\epsilon \left\|K_n^\alpha\right\|_{E_\infty(\mathbb{T})} \le C\epsilon.$$

On the other hand, Theorem 1.4.16 implies

$$\begin{aligned}A_2(x) &\le C \sup_{\mathbb{T}\setminus(-2^m\pi, 2^m\pi)} \left|K_n^\alpha\right| \int_{\mathbb{T}\setminus(-2^m\pi, 2^m\pi)} |f(x-t) - f(x)| \, dt \\ &\le \frac{C}{n^\alpha 2^{m(\alpha+1)}} \left(\|f\|_1 + |f(x)|\right),\end{aligned}$$

which tends to 0 as $n \to \infty$. Finally, for $1 < \alpha < \infty$, the result follows from Lemma 1.4.13. ■

We can weaken the definition of Lebesgue points and we can suppose that

$$\lim_{h \to 0+0} \frac{1}{h} \int_0^h |f(x-t) + f(x+t) - 2f(x)| \, dt = 0. \tag{1.5.2}$$

By a triangle inequality, it is clear that if x is a Lebesgue point then (1.5.2) holds. The following result can be proved similar to Theorem 1.5.4.

Theorem 1.5.5 *If $0 < \alpha < \infty$, $f \in L_1(\mathbb{T})$ and (1.5.2) holds for a point $x \in \mathbb{T}$, then*

$$\lim_{n \to \infty} \sigma_n^\alpha f(x) = f(x).$$

Proof Using (1.4.7), Lemma 1.4.11 and that the function K_n^α is even, we obtain

$$\begin{aligned}\sigma_n^\alpha f(x) - f(x) &= \frac{1}{2\pi} \int_{-\pi}^{\pi} (f(x-t) - f(x)) K_n^\alpha(t) \, dt \\ &= \frac{1}{2\pi} \int_0^{\pi} (f(x-t) + f(x+t) - 2f(x)) K_n^\alpha(t) \, dt.\end{aligned}$$

Thus

$$|\sigma_n^\alpha f(x) - f(x)| = \frac{1}{2\pi} \int_0^\pi |f(x-t) + f(x+t) - 2f(x)| \left|K_n^\alpha(t)\right| dt$$

and the proof can be finished as in Theorem 1.5.4. ∎

Now we can generalize Fejér's theorem given in (1.4.1).

Corollary 1.5.6 *Suppose that $0 < \alpha < \infty$, $f \in L_1(\mathbb{T})$ and that the left and right limits $f(x-0)$ and $f(x+0)$ exist at a point x. Then*

$$\lim_{n\to\infty} \sigma_n^\alpha f(x) = \frac{f(x-0) + f(x+0)}{2}.$$

Proof Choosing

$$f(x) := \frac{f(x-0) + f(x+0)}{2},$$

we can easily see that (1.5.2) holds. The corollary follows from Theorem 1.5.5. ∎

If f is continuous at a point x, then we get

Corollary 1.5.7 *Suppose that $0 < \alpha < \infty$, $f \in L_1(\mathbb{T})$ and f is continuous at a point x. Then*

$$\lim_{n\to\infty} \sigma_n^\alpha f(x) = f(x).$$

In the next theorem, we verify the norm convergence of the Cesàro means.

Theorem 1.5.8 *Suppose that $0 < \alpha < \infty$ and $1 \le p < \infty$. If $f \in L_p(\mathbb{T})$, then*

$$\sup_{n\in\mathbb{N}} \left\|\sigma_n^\alpha f\right\|_p \le C_\alpha \|f\|_p$$

and

$$\lim_{n\to\infty} \sigma_n^\alpha f = f \qquad \text{in the } L_p(\mathbb{T})\text{-norm.}$$

Proof Again, it is enough to show the result for $0 < \alpha \le 1$. By (1.4.7), Minkowski inequality and Corollary 1.5.3,

$$\left\|\sigma_n^\alpha f(\cdot)\right\|_p \le \frac{1}{2\pi} \int_{\mathbb{T}} \|f(\cdot - t)\|_p \left|K_n^\alpha(t)\right| dt \le C_\alpha \|f\|_p.$$

The convergence obviously holds for all trigonometric polynomials and so it holds also for all $f \in L_p(\mathbb{T})$ $(1 \le p < \infty)$ by density. ∎

We get the next corollary with the same proof.

Corollary 1.5.9 *If* $0 < \alpha < \infty$ *and* $f \in C(\mathbb{T})$, *then*

$$\sup_{n \in \mathbb{N}} \left\| \sigma_n^\alpha f \right\|_\infty \leq C_\alpha \, \|f\|_\infty$$

and

$$\lim_{n \to \infty} \sigma_n^\alpha f = f \qquad \textit{uniformly.}$$

Chapter 2
ℓ_q-Summability of Higher Dimensional Fourier Series

Here, we study the theory of multi-dimensional Fourier series. In the first section, we introduce different versions of the partial sums of the d-dimensional Fourier series and the corresponding Dirichlet kernels, i.e., the cubic, triangular, circular and rectangular partial sums and Dirichlet kernels. We show that the cubic, triangular and rectangular partial sums converge in the $L_p(\mathbb{T}^d)$-norm to the function $(1 < p < \infty)$. The multi-dimensional version of Carleson's theorem is also considered.

The summability of Fourier series can be generalized for higher dimensions basically in two ways. In this chapter, we study the ℓ_q-summability of higher dimensional Fourier series. As in the literature, we investigate the three cases $q = 1$, $q = 2$ and $q = \infty$. The other type of summability, the so-called rectangular summability will be investigated in the next chapter. For each type, we investigate the Cesàro and Riesz summation. In Sect. 2.2, we present the basic definitions of the ℓ_q-summability and prove some estimations for the ℓ_q-Cesàro and Riesz kernels. In the next section, we prove that the ℓ_q-Cesàro means and ℓ_q-Riesz means of $f \in L_p(\mathbb{T}^d)$ $(1 \le p < \infty)$ converge to f in the $L_p(\mathbb{T}^d)$-norm.

In Sect. 2.4, we prove the basic results for Fourier series of distributions. We introduce the Hardy spaces $H_p^\square(\mathbb{T}^d)$ and present the atomic decomposition of these spaces. We verify also sufficient conditions for an operator to be bounded from $H_p^\square(\mathbb{T}^d)$ to $L_p(\mathbb{T}^d)$. Applying this result, we show that the maximal operator of the ℓ_q-Cesàro and Riesz means are bounded from $H_p^\square(\mathbb{T}^d)$ to $L_p(\mathbb{T}^d)$ for any $p > p_0$, where $p_0 < 1$ is depending on the summation and on the dimension. This result implies the almost everywhere convergence of the summability means. In Sect. 2.6, we introduce a general summability method, the so-called θ-summability generated by a single function θ and prove similar results for the ℓ_q-θ-means. In the last section, as special cases, we present some summability methods, such as the de La Vallée-Poussin, Jackson-de La Vallée-Poussin, Rogosinski, Weierstrass, Picard and Bessel summations.

F. Weisz, *Lebesgue Points and Summability of Higher Dimensional Fourier Series*,
https://doi.org/10.1007/978-3-030-74636-0_2

2.1 Higher Dimensional Partial Sums

In this section, we generalize the results of Sect. 1.2, we introduce four types of partial sums of the d-dimensional trigonometric Fourier series and study their $L_p(\mathbb{T}^d)$-norm and almost everywhere convergence of a function $f \in L_p(\mathbb{T}^d)$.

We introduce the following notations. For $x = (x_1, \dots, x_d) \in \mathbb{R}^d$ and $u = (u_1, \dots, u_d) \in \mathbb{R}^d$ set

$$u \cdot x := \sum_{k=1}^{d} u_k x_k, \qquad \|x\|_p := \left(\sum_{k=1}^{d} |x_k|^p \right)^{1/p} \qquad (1 \le p < \infty)$$

and

$$\|x\|_\infty := \sup_{k=1,\dots,d} |x_k|, \qquad |x| := \|x\|_2 .$$

Definition 2.1.1 The functions

$$e^{\imath k \cdot x} = \prod_{j=1}^{d} e^{\imath k_j x_j}$$

are called d-dimensional trigonometric system, where $k = (k_1, \dots, k_d) \in \mathbb{Z}^d$, $x = (x_1, \dots, x_d) \in \mathbb{T}^d$.

Definition 2.1.2 For an integrable function $f \in L_1(\mathbb{T}^d)$, its kth d-dimensional Fourier coefficient is defined by

$$\widehat{f}(k) = \frac{1}{(2\pi)^d} \int_{\mathbb{T}^d} f(x) e^{-\imath k \cdot x} \, dx \qquad (k \in \mathbb{Z}^d).$$

The formal trigonometric series

$$\sum_{k \in \mathbb{Z}^d} \widehat{f}(k) e^{\imath k \cdot x} \qquad (x \in \mathbb{T}^d)$$

is called the d-dimensional Fourier series of f.

We will generalize the one-dimensional partial sums in Definition 1.2.2 for higher dimensional functions in two ways. In the first generalization, we take the sum over the indices $\|k\|_q \le n$ instead of $k = -n, \dots, n$, where $1 \le q \le \infty$. These sums are called ℓ_q-partial sums. In the second generalization, we take the sum in each dimension, i.e., over the indices $|k_1| \le n_1, \dots, |k_d| \le n_d$. Here, we call the sums rectangular partial sums. The most natural choices $q = 2$, $q = 1$, $q = \infty$ and the rectangular partial sums are investigated in several papers and books (for $q = 2$, see e.g. Stein and Weiss [290, 293], Davis and Chang [76], Grafakos [143, 145, 146], Lu and Yan [229], Feichtinger and Weisz [103, 104], for $q = 1$, Berens, Li and Xu

[30–32, 356], Weisz [336, 337], for $q = \infty$, Marcinkiewicz [233], Zhizhiashvili [366], Weisz [332, 342, 346], for the rectangular sums, Zygmund [367] and Weisz [332, 342, 346]).

Definition 2.1.3 For $f \in L_1(\mathbb{T}^d)$, $1 \le q \le \infty$ and $n \in \mathbb{N}$, the nth ℓ_q-partial sum $s_n^q f$ of the Fourier series of f and the nth ℓ_q-Dirichlet kernel D_n^q are given by

$$s_n^q f(x) := \sum_{k \in \mathbb{Z}^d,\, \|k\|_q \le n} \widehat{f}(k) e^{\imath k \cdot x}$$

and

$$D_n^q(u) := \sum_{k \in \mathbb{Z}^d,\, \|k\|_q \le n} e^{\imath k \cdot u},$$

respectively.

The next lemma follows easily from the definition.

Lemma 2.1.4 *For all $n \in \mathbb{N}$, $1 \le q \le \infty$ and $t \in \mathbb{T}^d$, we have*

$$|D_n^q(t)| \le Cn^d.$$

The partial sums are called triangular if $q = 1$, circular if $q = 2$ and cubic if $q = \infty$ (see Figs. 2.1, 2.2, 2.3 and 2.4).

Definition 2.1.5 For $f \in L_1(\mathbb{T}^d)$ and $n = (n_1, \ldots, n_d) \in \mathbb{N}^d$, the nth rectangular partial sum $s_n f$ of the Fourier series of f and the nth rectangular Dirichlet kernel D_n are given by

$$s_n f(x) := \sum_{|k_1| \le n_1} \cdots \sum_{|k_d| \le n_d} \widehat{f}(k) e^{\imath k \cdot x}$$

and

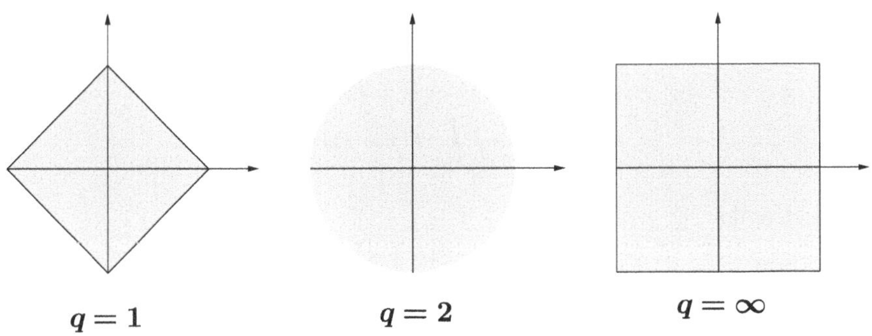

Fig. 2.1 Regions of the ℓ_q-partial sums for $d = 2$

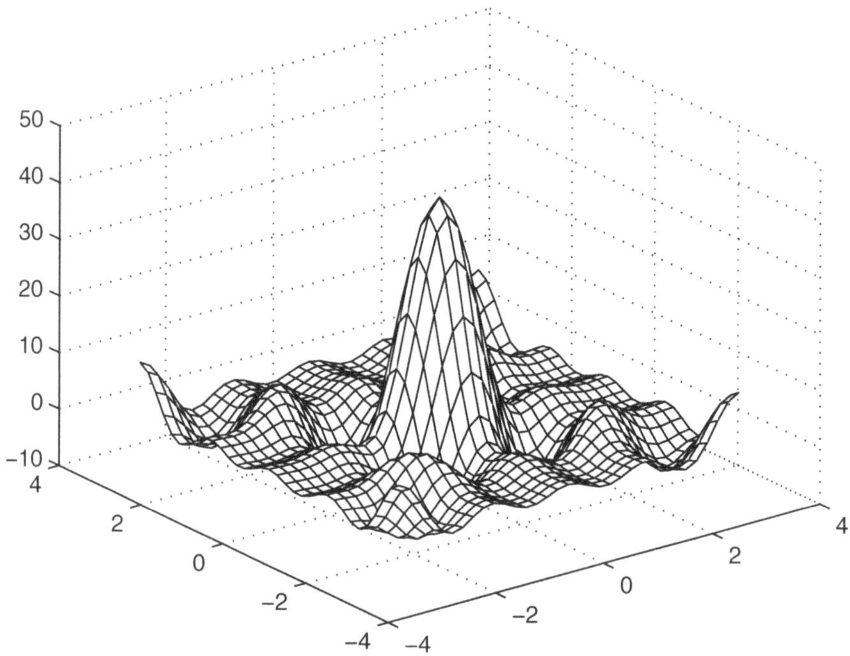

Fig. 2.2 The Dirichlet kernel D_n^q with $d = 2, q = 1, n = 4$

$$D_n(u) := \sum_{|k_1| \le n_1} \cdots \sum_{|k_d| \le n_d} e^{\imath k \cdot u},$$

respectively.

Similar to (1.2.1), we obtain

Lemma 2.1.6 *For $f \in L_1(\mathbb{T}^d)$ and $n \in \mathbb{N}$,*

$$s_n^q f(x) = \frac{1}{(2\pi)^d} \int_{\mathbb{T}^d} f(x-t) D_n^q(t)\, dt$$

and

$$s_n f(x) = \frac{1}{(2\pi)^d} \int_{\mathbb{T}^d} f(x-t) D_n(t)\, dt.$$

It is clear that

$$D_n(u) = D_{n_1}(u_1) \cdots D_{n_d}(u_d),$$

where D_{n_j} is the one-dimensional Dirichlet kernel (see Fig. 2.5).

Definition 2.1.7 For some $n = (n_1, \ldots, n_d) \in \mathbb{N}^d$, the function

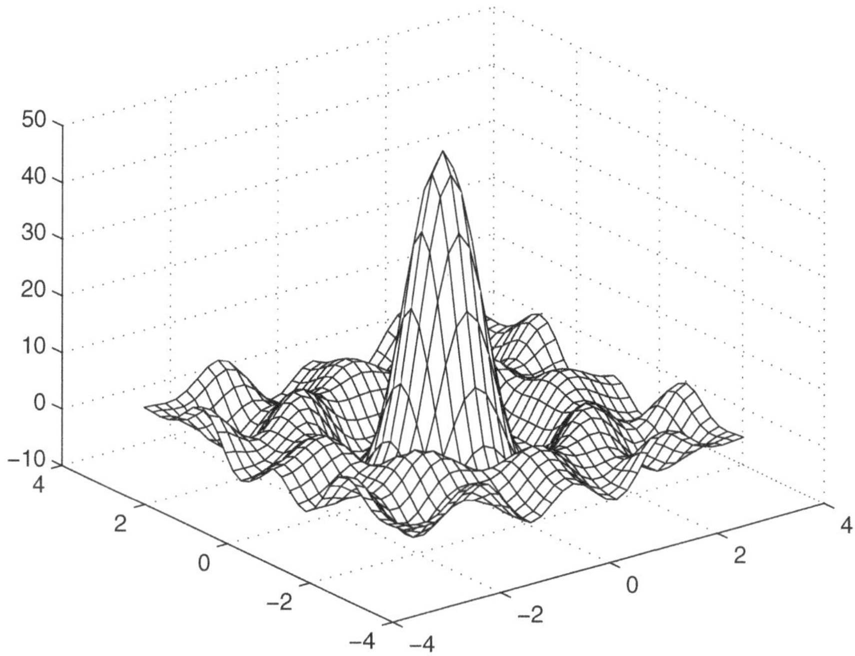

Fig. 2.3 The Dirichlet kernel D_n^q with $d = 2, q = 2, n = 4$

$$\sum_{k_1=-n_1}^{n_1} \cdots \sum_{k_d=-n_d}^{n_d} c_k e^{\imath k \cdot x} \qquad (x \in \mathbb{T}^d)$$

is said to be a trigonometric polynomial.

By iterating the one-dimensional result, we get easily the L_p-norm convergence for the rectangular partial sums.

Theorem 2.1.8 *If $f \in L_p(\mathbb{T}^d)$ for some $1 < p < \infty$, then*

$$\sup_{n \in \mathbb{N}^d} \|s_n f\|_p \le C_p \|f\|_p$$

and

$$\lim_{n \to \infty} s_n f = f \qquad \textit{in the } L_p(\mathbb{T}^d)\textit{-norm.}$$

Here, $n \to \infty$ means the Pringsheim convergence, i.e., $\min(n_1, \ldots, n_d) \to \infty$.

Proof By Theorem 1.2.10,

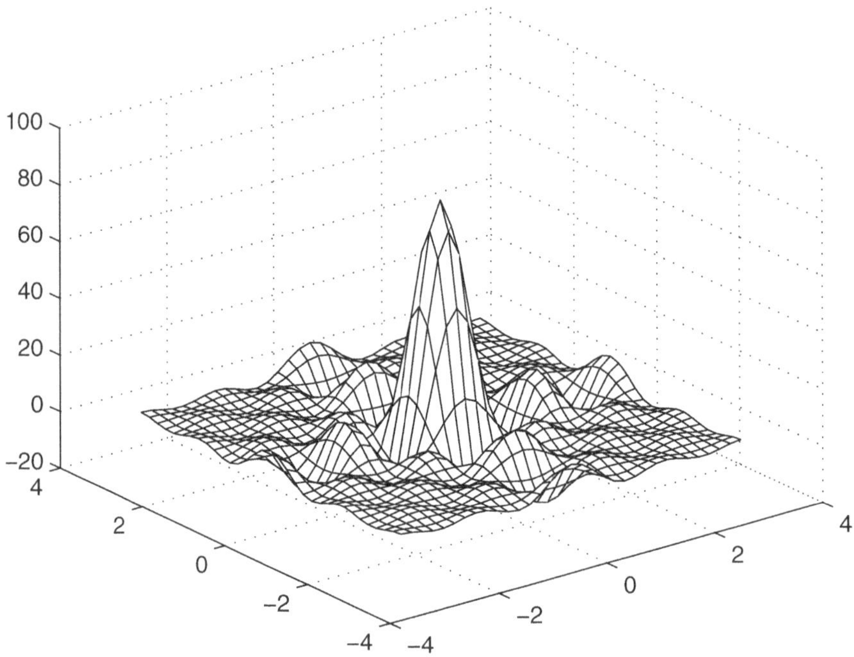

Fig. 2.4 The Dirichlet kernel D_n^q with $d = 2, q = \infty, n = 4$

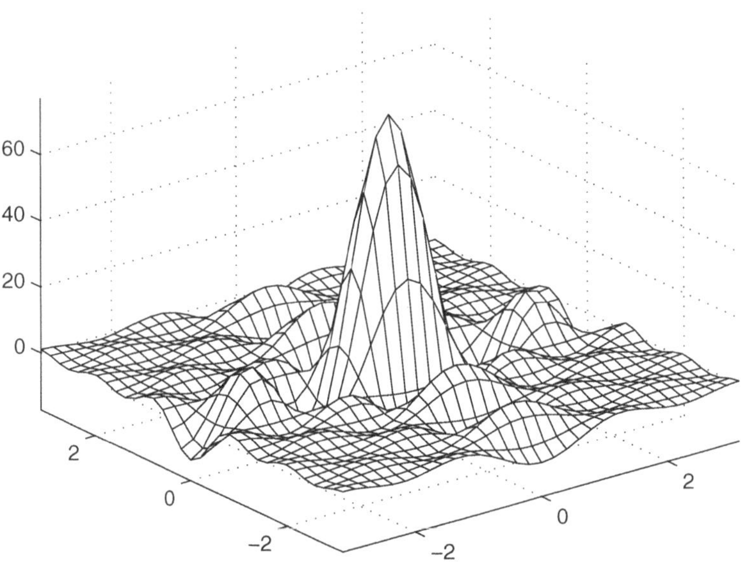

Fig. 2.5 The rectangular Dirichlet kernel with $d = 2, n_1 = 3, n_2 = 5$

$$\begin{aligned}
&\int_{\mathbb{T}} |s_n f(x)|^p \, dx_1 \\
&\quad = \int_{\mathbb{T}} \left| \int_{\mathbb{T}} \left(\int_{\mathbb{T}} f(t) D_{n_2}(x_2 + t_2) \, dt_2 \right) D_{n_1}(x_1 + t_1) \, dt_1 \right|^p \, dx_1 \\
&\quad \le C_p \int_{\mathbb{T}} \left| \int_{\mathbb{T}} f(t) D_{n_2}(x_2 + t_2) \, dt_2 \right|^p \, dt_1.
\end{aligned}$$

Again by the same theorem,

$$\begin{aligned}
\int_{\mathbb{T}} \int_{\mathbb{T}} |s_n f(x)|^p \, dx_1 \, dx_2 &\le C_p \int_{\mathbb{T}} \int_{\mathbb{T}} \left| \int_{\mathbb{T}} f(t) D_{n_2}(x_2 + t_2) \, dt_2 \right|^p \, dx_2 \, dt_1 \\
&\le C_p \int_{\mathbb{T}} \int_{\mathbb{T}} |f(t)|^p \, dt_2 \, dt_1,
\end{aligned}$$

which gives the desired inequality of Theorem 2.1.8. The convergence is a consequence of this inequality and of the density of trigonometric polynomials. ■

In the next theorem, we present the norm convergence of the triangular and cubic partial sums. We omit the proof since it can be found at several places of the literature (see e.g., Fefferman [93], Grafakos [143] or Weisz [346]).

Theorem 2.1.9 *If $q = 1, \infty$ and $f \in L_p(\mathbb{T}^d)$ for some $1 < p < \infty$, then*

$$\sup_{n \in \mathbb{N}} \left\| s_n^q f \right\|_p \le C_p \|f\|_p$$

and

$$\lim_{n \to \infty} s_n^q f = f \qquad \text{in the } L_p(\mathbb{T}^d)\text{-norm.}$$

If $q = 2$, then the same result is valid for $p = 2$.

Since the characteristic function of the unit ball is not an $L_p(\mathbb{R}^d)$ $(1 < p \ne 2 < \infty, d \ge 2)$ multiplier (see Fefferman [95] or Grafakos [143, p. 743] or Lu and Yan [229, p. 743]), we have

Theorem 2.1.10 *If $d \ge 2$, $q = 2$ and $1 < p \ne 2 < \infty$, then the preceding theorem is not true.*

The analogue of Carleson's theorem holds also for the triangular and cubic partial sums in higher dimensions (see Fefferman [93, 94] and Grafakos [143, p. 231]), but it does not hold for the circular and rectangular partial sums.

Definition 2.1.11 We denote by

$$s_*^q f := \sup_{n \in \mathbb{N}} \left| s_n^q f \right|$$

the maximal operator of the ℓ_q-partial sums.

Theorem 2.1.12 *If $q = 1, \infty$ and $f \in L_p(\mathbb{T}^d)$ for some $1 < p < \infty$, then*

$$\left\| s_*^q f \right\|_p \leq C_p \|f\|_p$$

and if $1 < p \leq \infty$, then

$$\lim_{n \to \infty} s_n^q f = f \qquad a.e.$$

Theorem 2.1.12 does not hold for circular partial sums (see Stein and Weiss [293, p. 268]).

Theorem 2.1.13 *If $q = 2$ and $p < 2d/(d+1)$, then there exists a function $f \in L_p(\mathbb{T}^d)$ whose circular partial sums $s_n^q f$ diverge almost everywhere.*

This means that for a general function in $L_p(\mathbb{T}^d)$ $(p < 2)$ almost everywhere convergence of the circular partial sums is not true if the dimension is sufficiently large. It is an open problem, whether Theorem 2.1.12 holds for $p = 2$ and for circular partial sums. A counterexample, which proves the next result, can be found in Fefferman [94].

Theorem 2.1.14 *There exists a continuous function f such that for the rectangular partial sums $s_n f$,*

$$\lim_{n \to \infty} s_n f(x) = f(x)$$

does not hold for any $x \in \mathbb{T}^d$.

The generalization of Theorem 1.2.13 for higher dimensions was proved by Antonov [8].

Theorem 2.1.15 *If $q = \infty$ and*

$$\int_{\mathbb{T}^d} |f(x)|(\log^+ |f(x)|)^d \log^+ \log^+ \log^+ |f(x)| \, dx < \infty,$$

then

$$\lim_{n \to \infty} s_n^q f = f \qquad a.e.$$

2.2 The ℓ_q-Summability Kernels

As in the one-dimensional case, Theorems 2.1.8, Theorem 2.1.9 and the inequality in Theorem 2.1.12 do not hold for $p = 1$ and $p = \infty$. Using a summability method, we can extend the theorems to $p = 1$ and $p = \infty$ again. Now we introduce the ℓ_q-summability means and kernels and show some results for the kernels. We concentrate on the two-dimensional kernels.

Definition 2.2.1 For $f \in L_1(\mathbb{T}^d)$, $1 \le q \le \infty$ and $n \in \mathbb{N}$, the nth ℓ_q-Fejér means $\sigma_n^q f$ of the Fourier series of f and the nth ℓ_q-Fejér kernel K_n^q are introduced by

$$\sigma_n^q f(x) := \sum_{k\in\mathbb{Z}^d,\ \|k\|_q\le n} \left(1 - \frac{\|k\|_q}{n}\right) \widehat{f}(k) e^{\imath k\cdot x}$$

and

$$K_n^q(t) := \sum_{k\in\mathbb{Z}^d,\ \|k\|_q\le n} \left(1 - \frac{\|k\|_q}{n}\right) e^{\imath k\cdot t},$$

respectively.

We generalize this definition as we did for the one-dimensional Fourier series and introduce the ℓ_q-Cesàro means.

Definition 2.2.2 Let $f \in L_1(\mathbb{T}^d)$, $n \in \mathbb{N}$, $\alpha \ge 0$ and $q = 1$ or $q = \infty$. The nth ℓ_q-Cesàro means $\sigma_n^{q,\alpha} f$ of the Fourier series of f and the nth ℓ_q-Cesàro kernel $K_n^{q,\alpha}$ are introduced by

$$\sigma_n^{q,\alpha} f(x) := \frac{1}{A_{n-1}^{\alpha}} \sum_{k\in\mathbb{Z}^d,\ \|k\|_q\le n} A_{n-1-\|k\|_q}^{\alpha} \widehat{f}(k) e^{\imath k\cdot x}$$

and

$$K_n^{q,\alpha}(t) := \frac{1}{A_{n-1}^{\alpha}} \sum_{k\in\mathbb{Z}^d,\ \|k\|_q\le n} A_{n-1-\|k\|_q}^{\alpha} e^{\imath k\cdot t},$$

respectively.

We also call the Cesàro means ℓ_q-(C,α)-means. For $\alpha = 1$, we get back the ℓ_q-Fejér means and for $\alpha = 0$, the ℓ_q-partial sums. We introduce also a second generalization of the Fejér summation. For the circular summability (i.e., for $q = 2$), we will investigate rather this generalization.

Definition 2.2.3 For $f \in L_1(\mathbb{T}^d)$, $1 \le q \le \infty$, $n \in \mathbb{N}$ and $0 < \alpha, \gamma < \infty$, the nth ℓ_q-Riesz means $\sigma_n^{q,\alpha,\gamma} f$ of the Fourier series of f and the nth ℓ_q-Riesz kernel $K_n^{q,\alpha,\gamma}$ are given by

$$\sigma_n^{q,\alpha,\gamma} f(x) := \sum_{k\in\mathbb{Z}^d,\ \|k\|_q\le n} \left(1 - \left(\frac{\|k\|_q}{n}\right)^{\gamma}\right)^{\alpha} \widehat{f}(k) e^{\imath k\cdot x}$$

and

$$K_n^{q,\alpha,\gamma}(t) := \sum_{k\in\mathbb{Z}^d,\ \|k\|_q\le n} \left(1 - \left(\frac{\|k\|_q}{n}\right)^{\gamma}\right)^{\alpha} e^{\imath k\cdot t},$$

respectively.

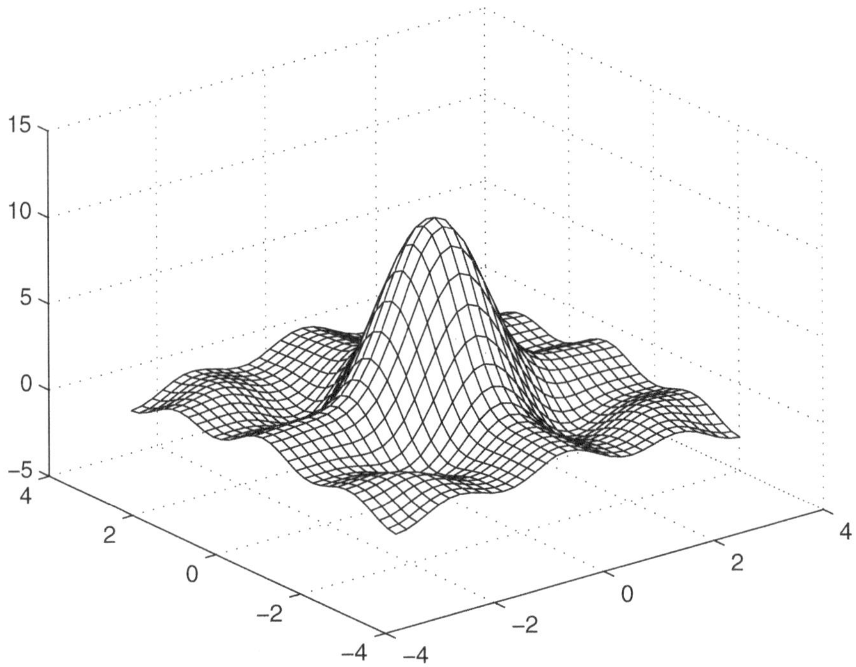

Fig. 2.6 The Fejér kernel K_n^q with $d = 2, q = 1, n = 4$

We will always suppose that $0 \le \alpha < \infty$, $1 \le \gamma < \infty$. If $\alpha = \gamma = 1$, we get back the ℓ_q-Fejér means. In the case $q = 2$, let $\gamma \in \mathbb{N}$. If $\alpha = 0$, we get the partial sums and if $q = \gamma = 2$, $\alpha > 0$, the means are called Bochner-Riesz means. The cubic summability (when $q = \infty$) is also called Marcinkiewicz summability (see Figs. 2.6, 2.7, 2.8, 2.9 and 2.10).

The following two lemmas follow the definition.

Lemma 2.2.4 *Let $0 \le \alpha, \gamma < \infty$ and $n \in \mathbb{N}$. If $q = 1$ or $q = \infty$, then*

$$\frac{1}{(2\pi)^d} \int_{\mathbb{T}^d} K_n^{q,\alpha}(t)\, dt = 1.$$

If $1 \le q \le \infty$, then

$$\frac{1}{(2\pi)^d} \int_{\mathbb{T}^d} K_n^{q,\alpha,\gamma}(t)\, dt = 1.$$

Lemma 2.2.5 *Under the same conditions as in Lemma 2.2.4,*

$$|K_n^{q,\alpha}(t)| \le Cn^d \qquad \text{and} \qquad |K_n^{q,\alpha,\gamma}(t)| \le Cn^d \qquad (t \in \mathbb{T}^d).$$

Proof We have

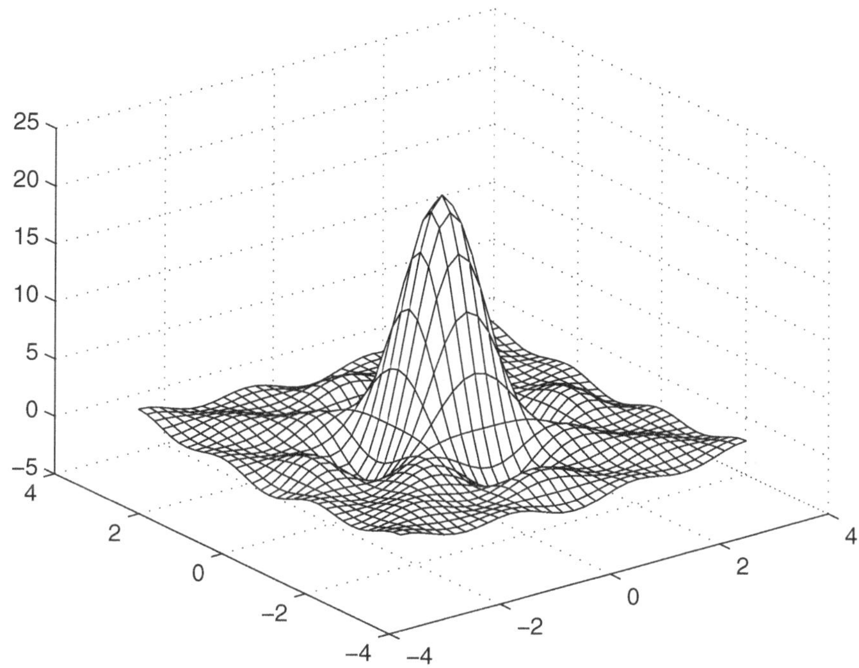

Fig. 2.7 The Fejér kernel K_n^q with $d = 2, q = \infty, n = 4$

$$|K_n^{q,\alpha}(t)| \le \frac{1}{A_{n-1}^{\alpha}} \sum_{k\in\mathbb{Z}^d,\ \|k\|_q\le n} A_{n-1-\|k\|_q}^{\alpha} \le C\frac{1}{A_{n-1}^{\alpha}} \sum_{j=0}^{n} A_{n-1-j}^{\alpha} j^{d-1} \le Cn^d.$$

The second inequality can be shown in the same way. ■

One can easily see that

Lemma 2.2.6 *Let $f \in L_1(\mathbb{T}^d)$, $n \in \mathbb{N}$ and $0 < \alpha, \gamma < \infty$. If $q = 1$ or $q = \infty$, then*

$$\sigma_n^{q,\alpha} f(x) = \frac{1}{(2\pi)^d} \int_{\mathbb{T}^d} f(x-t) K_n^{q,\alpha}(t)\, dt.$$

If $1 \le q \le \infty$, then

$$\sigma_n^{q,\alpha,\gamma} f(x) = \frac{1}{(2\pi)^d} \int_{\mathbb{T}^d} f(x-t) K_n^{q,\alpha,\gamma}(t)\, dt.$$

Lemma 2.2.7 *For $f \in L_1(\mathbb{T}^d)$, $\alpha > 0$, $q = 1, \infty$ and $n \in \mathbb{N}$, we have*

$$\sigma_n^{q,\alpha} f(x) = \frac{1}{A_{n-1}^{\alpha}} \sum_{j=0}^{n-1} A_{n-1-j}^{\alpha-1} s_j^q f(x)$$

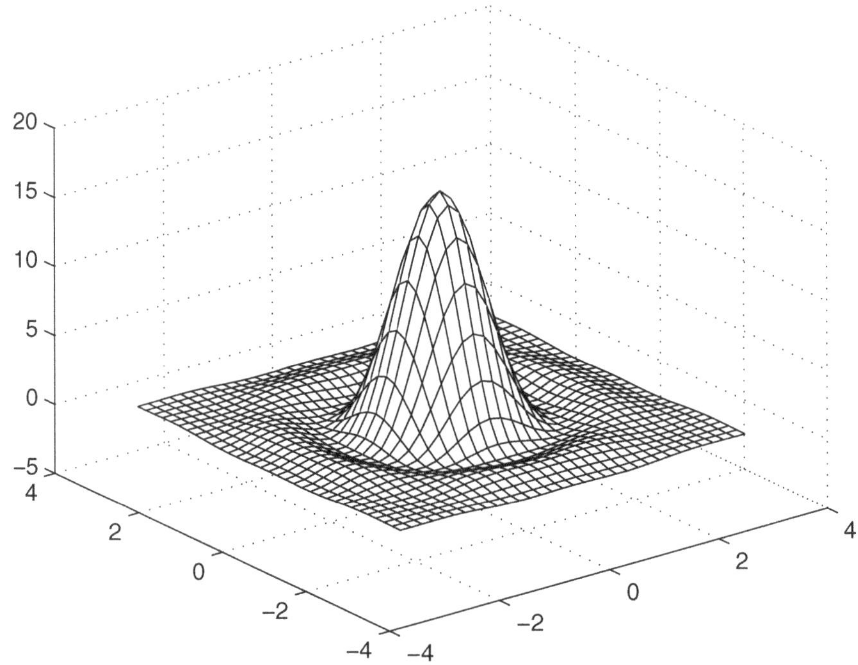

Fig. 2.8 The Fejér kernel K_n^q with $d = 2, q = 2, n = 4$

and

$$K_n^{q,\alpha}(t) = \frac{1}{A_{n-1}^{\alpha}} \sum_{j=0}^{n-1} A_{n-1-j}^{\alpha-1} D_j^q(t).$$

Proof Since $\|k\|_q$ is an integer, Lemma 1.4.8 implies that

$$\begin{aligned} K_n^{q,\alpha}(t) &= \frac{1}{A_{n-1}^{\alpha}} \sum_{k \in \mathbb{Z}^d,\, \|k\|_q \le n} A_{n-1-\|k\|_q}^{\alpha} e^{\imath k \cdot t} \\ &= \frac{1}{A_{n-1}^{\alpha}} \sum_{k \in \mathbb{Z}^d,\, \|k\|_q \le n} \sum_{j=\|k\|_q}^{n-1} A_{n-1-j}^{\alpha-1} e^{\imath k \cdot t} \\ &= \frac{1}{A_{n-1}^{\alpha}} \sum_{j=0}^{n-1} A_{n-1-j}^{\alpha-1} D_j^q(t), \end{aligned}$$

which shows the lemma. ■

Obviously, the ℓ_q-Fejér means are the arithmetic means of the ℓ_q-partial sums when $q = 1, \infty$:

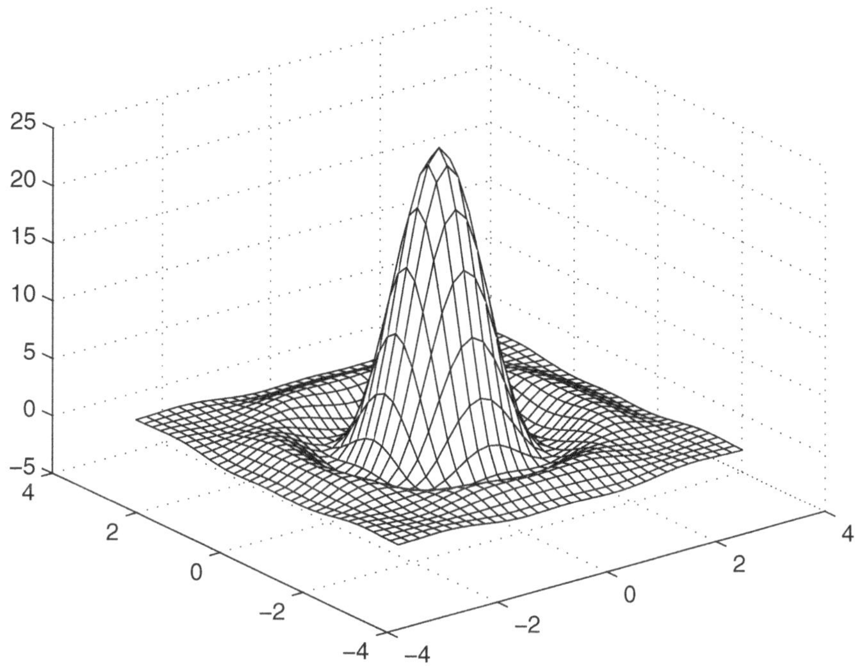

Fig. 2.9 The Bochner-Riesz kernel $K_n^{q,\alpha}$ with $d = 2, q = 2, n = 4, \alpha = 1, \gamma = 2$

$$\sigma_n^q f(x) = \frac{1}{n} \sum_{k=0}^{n-1} s_k^q f(x).$$

Similar to Lemma 1.4.13, we have

Lemma 2.2.8 *For $\alpha > -1$, $q = 1, \infty$ and $h > 0$, we have*

$$\sigma_n^{q,\alpha+h} f = \frac{1}{A_{n-1}^{\alpha+h}} \sum_{k=1}^{n} A_{n-k}^{h-1} A_{k-1}^{\alpha} \sigma_k^{q,\alpha} f.$$

The proofs of the results presented later are very different for the cases $q = 1, 2, \infty$ because the kernel functions are very different. In the next subsections, we give some estimations for the kernels. Since we will prove later the results basically for $d = 2$, we present these estimations in the two-dimensional case.

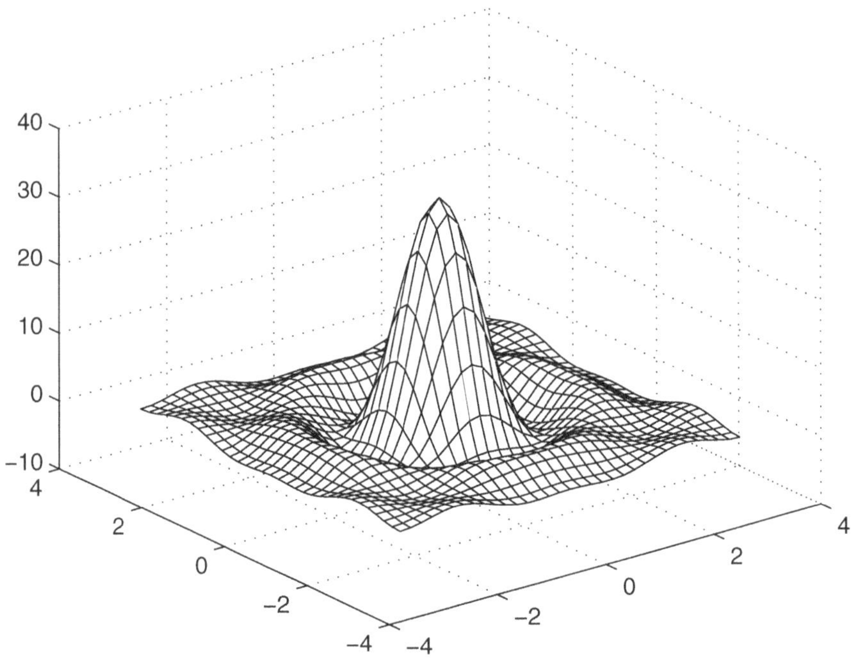

Fig. 2.10 The Bochner-Riesz kernel $K_n^{q,\alpha}$ with $d = 2, q = 2, n = 4, \alpha = 1/2, \gamma = 2$

2.2.1 Kernel Functions for q = 1

For the triangular Dirichlet kernel, we need the notion of the divided difference, which is usually used in numerical analysis.

Definition 2.2.9 The nth divided difference of a one-dimensional function f at the (pairwise distinct) knots $x_1, \dots, x_n \in \mathbb{R}$ is introduced inductively as

$$[x_1]f := f(x_1), \qquad [x_1, \dots, x_n]f := \frac{[x_1, \dots, x_{n-1}]f - [x_2, \dots, x_n]f}{x_1 - x_n}.$$

One can see that the difference is a symmetric function of the nodes. The following theorem is proved in DeVore and Lorentz [82, p. 120]), so we omit the proof.

Theorem 2.2.10 *We have*

$$[x_1, \dots, x_n]f = \sum_{k=1}^{n} \frac{f(x_k)}{\prod_{j=1, j\neq k}^{n}(x_k - x_j)}. \tag{2.2.1}$$

If f is $(n-1)$-times continuously differentiable on $[a, b]$ and $x_i \in [a, b]$, then there exists $\xi \in [a, b]$ such that

$$[x_1, \ldots, x_n]f = \frac{f^{(n-1)}(\xi)}{(n-1)!}. \tag{2.2.2}$$

To give an explicit form of the triangular Dirichlet kernel, we will need the following trigonometric identities.

Lemma 2.2.11 *For all $n \in \mathbb{N}$ and $0 \le x, y \le \pi$,*

$$\sum_{k=0}^{n} \epsilon_k \cos(ky) \sin((n-k+1/2)x)$$
$$= \sin(x/2) \frac{\cos(x/2)\cos((n+1/2)x) - \cos(y/2)\cos((n+1/2)y)}{\cos x - \cos y} \tag{2.2.3}$$

and

$$\sum_{k=0}^{n} \epsilon_k \cos(ky) \cos((n-k+1/2)x)$$
$$= \cos(x/2) \frac{\sin(y/2)\sin((n+1/2)y) - \sin(x/2)\sin((n+1/2)x)}{\cos x - \cos y}, \tag{2.2.4}$$

where $\epsilon_0 := 1/2$ and $\epsilon_k := 1$, $k \ge 1$.

Proof By trigonometric identities,

$$\begin{aligned}
\sum_{k=0}^{n} \epsilon_k \cos(ky) \sin((n-k+1/2)x)& \\
&= \sin((n+1/2)x) \sum_{k=0}^{n} \epsilon_k \cos(ky)\cos(kx) \\
&\quad - \cos((n+1/2)x) \sum_{k=0}^{n} \epsilon_k \cos(ky)\sin(kx) \\
&= \frac{1}{2}\sin((n+1/2)x) \sum_{k=0}^{n} \Big(\epsilon_k \cos(k(x-y)) + \epsilon_k \cos(k(x+y)\Big) \\
&\quad - \frac{1}{2}\cos((n+1/2)x) \sum_{k=0}^{n} \Big(\epsilon_k \sin(k(x-y)) + \epsilon_k \sin(k(x+y))\Big).
\end{aligned}$$

Similarly to (1.2.2), we can show that

$$\sum_{k=0}^{n} \epsilon_k \sin(kx) = \frac{\cos(x/2) - \cos((n+1/2)x)}{2\sin(x/2)}.$$

Using this and (1.2.2), we conclude

$$\begin{aligned}
&\sum_{k=0}^{n} \epsilon_k \cos(ky)\sin((n-k+1/2)x) \\
&\quad = \frac{1}{4}\sin((n+1/2)x)\left(\frac{\sin((n+1/2)(x-y))}{\sin((x-y)/2)} + \frac{\sin((n+1/2)(x+y))}{\sin((x+y)/2)}\right) \\
&\quad\quad - \frac{1}{4}\cos((n+1/2)x)\left(\frac{\cos((x-y)/2) - \cos((n+1/2)(x-y))}{\sin((x-y)/2)}\right. \\
&\quad\quad \left. + \frac{1}{4}\frac{\cos((x+y)/2) - \cos((n+1/2)(x+y))}{\sin((x+y)/2)}\right).
\end{aligned}$$

Since

$$\begin{aligned}
&\sin((n+1/2)x)\sin((n+1/2)(x-y)) \\
&\quad + \cos((n+1/2)x)\cos((n+1/2)(x-y)) = \cos((n+1/2)y)
\end{aligned}$$

and

$$\begin{aligned}
&\sin((n+1/2)x)\sin((n+1/2)(x+y)) \\
&\quad + \cos((n+1/2)x)\cos((n+1/2)(x+y)) = \cos((n+1/2)y),
\end{aligned}$$

we conclude that

$$\begin{aligned}
&\sum_{k=0}^{n} \epsilon_k \cos(ky)\sin((n-k+1/2)x) \\
&\quad = \frac{1}{4}\frac{\cos((n+1/2)y) - \cos((n+1/2)x)\cos((x-y)/2)}{\sin((x-y)/2)} \\
&\quad\quad + \frac{1}{4}\frac{\cos((n+1/2)y) - \cos((n+1/2)x)\cos((x+y)/2)}{\sin((x+y)/2)} \\
&\quad = \frac{1}{4}\frac{\cos((n+1/2)y)\Big(\sin((x+y)/2) + \sin((x-y)/2)\Big)}{\sin((x-y)/2)\sin((x+y)/2)} \\
&\quad\quad - \frac{1}{4}\frac{\cos((n+1/2)x)}{\sin((x-y)/2)\sin((x+y)/2)} \times \\
&\quad\quad \times \Big(\cos((x-y)/2)\sin((x+y)/2) + \cos((x+y)/2)\sin((x-y)/2)\Big).
\end{aligned}$$

Using again some trigonometric identities, we get that

$$\begin{aligned}
&\sum_{k=0}^{n} \epsilon_k \cos(ky)\sin((n-k+1/2)x) \\
&\quad = \frac{1}{2}\frac{2\cos((n+1/2)y)\sin(x/2)\cos(y/2)}{\cos y - \cos x}
\end{aligned}$$

$$
-\frac{1}{2}\frac{\cos((n+1/2)x)\sin x}{\cos y - \cos x}
$$

$$
= \sin(x/2)\frac{\cos(x/2)\cos((n+1/2)x) - \cos(y/2)\cos((n+1/2)y)}{\cos x - \cos y}.
$$

Formula (2.2.4) can be shown in the same way. ∎

Define the function G_n by

$$
G_n(\cos x) := (-1)^{[(d-1)/2]} 2\cos(x/2)(\sin x)^{d-2}\mathrm{soc}\,((n+1/2)x),
$$

where the function soc is defined by

$$
\mathrm{soc}\, x := \begin{cases} \cos x, & \text{if } d \text{ is even;} \\ \sin x, & \text{if } d \text{ is odd.} \end{cases}
$$

The following representation of the triangular Dirichlet kernel was proved by Herriot [165] and Berens and Xu [30, 356].

Lemma 2.2.12 *For $x \in \mathbb{T}^d$,*

$$
\begin{aligned}
D_n^1(x) &= [\cos x_1, \ldots, \cos x_d] G_n \\
&= (-1)^{[(d-1)/2]} 2 \sum_{k=1}^{d} \frac{\cos(x_k/2)(\sin x_k)^{d-2}\mathrm{soc}\,((n+1/2)x_k)}{\prod_{j=1, j\neq k}^{d}(\cos x_k - \cos x_j)}.
\end{aligned} \tag{2.2.5}
$$

Proof We will prove this lemma for all dimensions because the main idea of the proof is induction with respect to the dimension. First, we note that the second equality follows from the definition of G_n and from the property of the divided difference described in (2.2.1). In this proof, let us denote the Dirichlet kernel by $D_{d,n}^1(x) := D_n^1(x)$. We have seen in (1.2.2) that in the one-dimensional case

$$
\begin{aligned}
D_{1,n}^1(x) = D_n^1(x) &= \frac{\sin((n+1/2)x)}{\sin(x/2)} \\
&= 2\cos(x/2)(\sin x)^{-1}\sin((n+1/2)x),
\end{aligned}
$$

thus (2.2.5) holds for $d = 1$. Suppose the lemma is true for integers up to d and let d be even. It is easy to see that

$$
\begin{aligned}
D_{d+1,n}^1(x) &= 2^{d+1} \sum_{j\in\mathbb{N}^d,\ \|j\|_1 \le n} \epsilon_{j_1}\cos(j_1 x_1)\cdots\epsilon_{j_{d+1}}\cos(j_{d+1}x_{d+1}) \\
&= 2\sum_{l=0}^{n} \epsilon_l \cos(l x_{d+1}) D_{d,n-l}(x_1, \ldots, x_d)
\end{aligned}
$$

$$= (-1)^{[(d-1)/2]} 4 \sum_{k=1}^{d} \frac{\cos(x_k/2)(\sin x_k)^{d-2}}{\prod_{j=1, j\neq k}^{d}(\cos x_k - \cos x_j)}$$

$$\sum_{l=0}^{n} \epsilon_l \cos(l x_{d+1}) \cos((n-l+1/2)x_k),$$

where $\epsilon_0 := 1/2$ and $\epsilon_l := 1$, $l \geq 1$. Using (2.2.4), we obtain

$$\begin{aligned} D^1_{d+1,n}(x) &= -(-1)^{[(d-1)/2]} 4 \sum_{k=1}^{d} \frac{\cos(x_k/2)(\sin x_k)^{d-2}}{\prod_{j=1, j\neq k}^{d+1}(\cos x_k - \cos x_j)} \\ &\quad \cos(x_k/2)\sin(x_k/2)\sin((n+1/2)x_k) \\ &\quad + (-1)^{[(d-1)/2]} 4 \sum_{k=1}^{d} \frac{\cos(x_k/2)(\sin x_k)^{d-2}}{\prod_{j=1, j\neq k}^{d+1}(\cos x_k - \cos x_j)} \\ &\quad \cos(x_k/2)\sin(x_{d+1}/2)\sin((n+1/2)x_{d+1}) \\ &= -(-1)^{[(d-1)/2]} 2 \left(\sum_{k=1}^{d} \frac{\cos(x_k/2)(\sin x_k)^{d-1}\sin((n+1/2)x_k)}{\prod_{j=1, j\neq k}^{d+1}(\cos x_k - \cos x_j)} \right. \\ &\quad - \sin(x_{d+1}/2)\sin((n+1/2)x_{d+1}) \times \\ &\quad \left. \times \sum_{k=1}^{d} \frac{(1+\cos x_k)(\sin x_k)^{d-2}}{\prod_{j=1, j\neq k}^{d+1}(\cos x_k - \cos x_j)} \right). \end{aligned} \tag{2.2.6}$$

Since d is even, the function $h(t) := (1+t)(1-t^2)^{(d-2)/2}$ is a polynomial of degree $d-1$. Then, by (2.2.2),

$$\begin{aligned} 0 &= [\cos x_1, \ldots, \cos x_{d+1}]h \\ &= \sum_{k=1}^{d} \frac{(1+\cos x_k)(\sin x_k)^{d-2}}{\prod_{j=1, j\neq k}^{d+1}(\cos x_k - \cos x_j)} + \frac{(1+\cos x_{d+1})(\sin x_{d+1})^{d-2}}{\prod_{j=1, j\neq d+1}^{d+1}(\cos x_{d+1} - \cos x_j)}. \end{aligned}$$

This and (2.2.6) imply

$$\begin{aligned} D^1_{d+1,n}(x) &= -(-1)^{[(d-1)/2]} 2 \left(\sum_{k=1}^{d} \frac{\cos(x_k/2)(\sin x_k)^{d-1}\sin((n+1/2)x_k)}{\prod_{j=1, j\neq k}^{d+1}(\cos x_k - \cos x_j)} \right. \\ &\quad \left. + \sin(x_{d+1}/2)\sin((n+1/2)x_{d+1}) \frac{(1+\cos x_{d+1})(\sin x_{d+1})^{d-2}}{\prod_{j=1, j\neq d+1}^{d+1}(\cos x_{d+1} - \cos x_j)} \right) \\ &= (-1)^{[d/2]} 2 \sum_{k=1}^{d+1} \frac{\cos(x_k/2)(\sin x_k)^{d-1}\sin((n+1/2)x_k)}{\prod_{j=1, j\neq k}^{d+1}(\cos x_k - \cos x_j)}, \end{aligned}$$

which proves the result if d is even. If d is odd, the lemma can be proved similarly. ∎

As a special case, for $d = 2$, we get the next corollary.

Corollary 2.2.13 *For $x \in \mathbb{T}^2$, we have*

$$\begin{aligned} D_n^1(x_1, x_2) &= [\cos x_1, \cos x_2]G_n \\ &= 2\frac{\cos(x_1/2)\cos((n+1/2)x_1) - \cos(x_2/2)\cos((n+1/2)x_2)}{\cos x_1 - \cos x_2}. \end{aligned}$$

In what follows, we may suppose that $x \in \mathbb{T}^2$ and $\pi > x_1 > x_2 > 0$. We denote the characteristic function of a set H by 1_H, i.e.,

$$1_H(x) := \begin{cases} 1, & \text{if } x \in H; \\ 0, & \text{if } x \notin H. \end{cases}$$

Lemma 2.2.14 *If $0 < \alpha \leq 1$ and $\pi > x_1 > x_2 > 0$, then*

$$\begin{aligned} \left|K_n^{1,\alpha}(x_1, x_2)\right| \leq{}& C(x_1 - x_2)^{-1}(x_1 + x_2)^{-1}1_{\{x_2 \leq \pi/2\}} \\ &+ C(x_1 - x_2)^{-1}(2\pi - x_1 - x_2)^{-1}1_{\{x_2 > \pi/2\}}, \end{aligned} \tag{2.2.7}$$

$$\begin{aligned} &1_{\{x_2 \leq \pi/2\}}\left|K_n^{1,\alpha}(x_1, x_2)\right| \\ &\quad\leq Cn^{-\alpha}(x_1 - x_2)^{-1}(x_1 + x_2)^{-1}x_2^{-\alpha}1_{\{x_2 \leq \pi/2\}} \\ &\quad\quad + Cn^{-1}(x_1 - x_2)^{-1}(x_1 + x_2)^{-1}x_2^{-1}1_{\{x_2 \leq \pi/2\}}, \end{aligned} \tag{2.2.8}$$

$$\begin{aligned} &1_{\{x_2 > \pi/2\}}\left|K_n^{1,\alpha}(x_1, x_2)\right| \\ &\quad\leq Cn^{-\alpha}(x_1 - x_2)^{-1}(2\pi - x_1 - x_2)^{-1}x_2^{-\alpha}1_{\{x_2 > \pi/2\}} \\ &\quad\quad + Cn^{-1}(x_1 - x_2)^{-1}(2\pi - x_1 - x_2)^{-1}x_2^{-1}1_{\{x_2 > \pi/2\}}, \end{aligned} \tag{2.2.9}$$

$$\begin{aligned} 1_{\{x_2 \leq \pi/2\}}\left|K_n^{1,\alpha}(x_1, x_2)\right| \leq{}& Cn^{1-\alpha}(x_1 + x_2)^{-1}x_2^{-\alpha}1_{\{x_2 \leq \pi/2\}} \\ &+ C(x_1 + x_2)^{-1}x_2^{-1}1_{\{x_2 \leq \pi/2\}} \end{aligned} \tag{2.2.10}$$

and

$$\begin{aligned} 1_{\{x_2 > \pi/2\}}\left|K_n^{1,\alpha}(x_1, x_2)\right| \leq{}& Cn^{1-\alpha}(2\pi - x_1 - x_2)^{-1}x_2^{-\alpha}1_{\{x_2 > \pi/2\}} \\ &+ C(2\pi - x_1 - x_2)^{-1}x_2^{-1}1_{\{x_2 > \pi/2\}}. \end{aligned} \tag{2.2.11}$$

Proof By the trigonometric identity,

$$\cos a - \cos b = -2\sin((a-b)/2)\sin((a+b)/2),$$

Corollary 2.2.13 can be rewritten as

$$D_k^1(x_1,x_2) = -\frac{\cos(x_1/2)\cos((k+1/2)x_1) - \cos(x_2/2)\cos((k+1/2)x_2)}{\sin((x_1-x_2)/2)\sin((x_1+x_2)/2)}. \tag{2.2.12}$$

We will use that

$$\sin(x_1 \pm x_2)/2 \sim x_1 \pm x_2 \quad \text{if} \quad x_2 \le \pi/2$$

and

$$\sin(x_1 - x_2)/2 \sim x_1 - x_2, \quad \sin(x_1 + x_2)/2 \sim 2\pi - x_1 - x_2 \quad \text{if} \quad x_2 > \pi/2.$$

By Lemma 2.2.7 and (2.2.12), we can see that

$$\begin{aligned} K_n^{1,\alpha}(x_1,x_2) &= \frac{1}{A_{n-1}^\alpha}\sum_{k=0}^{n-1} A_{n-1-k}^{\alpha-1} \\ &\quad \frac{\cos(x_2/2)\cos((k+1/2)x_2) - \cos(x_1/2)\cos((k+1/2)x_1)}{\sin((x_1-x_2)/2)\sin((x_1+x_2)/2)} \\ &\le 2(x_1-x_2)^{-1}(x_1+x_2)^{-1}1_{\{x_2\le\pi/2\}} \\ &\quad + 2(x_1-x_2)^{-1}(2\pi-x_1-x_2)^{-1}1_{\{x_2>\pi/2\}}, \end{aligned} \tag{2.2.13}$$

which is exactly (2.2.7).

Suppose that $x_2 \le \pi/2$. By (2.2.13) and Lemma 1.4.14,

$$\begin{aligned} &\left|K_n^{1,\alpha}(x_1,x_2)\right| \\ &\le (x_1-x_2)^{-1}(x_1+x_2)^{-1}\frac{1}{A_{n-1}^\alpha} \\ &\quad \left(\left|\sum_{k=0}^{n-1} A_{n-1-k}^{\alpha-1}\cos((k+1/2)x_2)\right| + \left|\sum_{k=0}^{n-1} A_{n-1-k}^{\alpha-1}\cos((k+1/2)x_1)\right|\right) \\ &\le Cn^{-\alpha}(x_1-x_2)^{-1}(x_1+x_2)^{-1}x_2^{-\alpha} + Cn^{-1}(x_1-x_2)^{-1}(x_1+x_2)^{-1}x_2^{-1}, \end{aligned}$$

which shows (2.2.8).

Lagrange's mean value theorem and (2.2.12) imply that there exists $x_1 > \xi > x_2$, such that

$$D_k^1(x_1,x_2) = -\frac{H_k'(\xi)(x_1-x_2)}{\sin((x_1-x_2)/2)\sin((x_1+x_2)/2)},$$

where

$$H_k(t) = \cos(t/2)\cos((k+1/2)t).$$

Then (2.2.10) follows from

$$\begin{aligned}\left|K_n^{1,\alpha}(x_1,x_2)\right| &\le \frac{1}{A_{n-1}^{\alpha}}\left|\sum_{k=0}^{n-1} A_{n-1-k}^{\alpha-1}\frac{H_k'(\xi)(x_1-x_2)}{\sin((x_1-x_2)/2)\sin((x_1+x_2)/2)}\right| \\ &\le Cn(x_1-x_2)(x_1-x_2)^{-1}(x_1+x_2)^{-1}(n^{-\alpha}x_2^{-\alpha}+n^{-1}x_2^{-1}).\end{aligned}$$

The inequalities (2.2.9) and (2.2.11) for $x_2 > \pi/2$ can be proved in the same way. ■

The next estimations of the kernel function come easily from Lemma 2.2.14.

Lemma 2.2.15 *If* $0 < \alpha \le 1$, $0 \le \beta \le 1$ *and* $\pi > x_1 > x_2 > 0$, *then*

$$\begin{aligned}\left|K_n^{1,\alpha}(x_1,x_2)\right| &\le C(x_1-x_2)^{-3/2}x_2^{-1/2}1_{\{x_2\le\pi/2\}} \\ &\quad + C(x_1-x_2)^{-3/2}(\pi-x_1)^{-1/2}1_{\{x_2>\pi/2\}},\end{aligned} \tag{2.2.14}$$

$$\begin{aligned}&1_{\{x_2\le\pi/2\}}\left|K_n^{1,\alpha}(x_1,x_2)\right| \\ &\quad\le Cn^{-\alpha}(x_1-x_2)^{-1-\beta}x_2^{\beta-\alpha-1}1_{\{x_2\le\pi/2\}} \\ &\quad\quad + Cn^{-1}(x_1-x_2)^{-1-\beta}x_2^{\beta-2}1_{\{x_2\le\pi/2\}},\end{aligned} \tag{2.2.15}$$

$$\begin{aligned}&1_{\{x_2>\pi/2\}}\left|K_n^{1,\alpha}(x_1,x_2)\right| \\ &\quad\le Cn^{-\alpha}(x_1-x_2)^{-1-\beta}(\pi-x_1)^{\beta-\alpha-1}1_{\{x_2>\pi/2\}} \\ &\quad\quad + Cn^{-1}(x_1-x_2)^{-1-\beta}(\pi-x_1)^{\beta-2}1_{\{x_2>\pi/2\}},\end{aligned} \tag{2.2.16}$$

$$1_{\{x_2\le\pi/2\}}\left|K_n^{1,\alpha}(x_1,x_2)\right| \le Cn^{1-\alpha}x_2^{-\alpha-1}1_{\{x_2\le\pi/2\}} + Cx_2^{-2}1_{\{x_2\le\pi/2\}} \tag{2.2.17}$$

and

$$\begin{aligned}1_{\{x_2>\pi/2\}}\left|K_n^{1,\alpha}(x_1,x_2)\right| &\le Cn^{1-\alpha}(\pi-x_1)^{-\alpha-1}1_{\{x_2>\pi/2\}} \\ &\quad + C(\pi-x_1)^{-2}1_{\{x_2>\pi/2\}}.\end{aligned} \tag{2.2.18}$$

Proof The basic facts

$$x_1+x_2 > x_1-x_2, \qquad x_1+x_2 > x_2$$

and

$$2\pi-x_1-x_2 > x_1-x_2, \qquad 2\pi-x_1-x_2 > \pi-x_1$$

together with (2.2.7) imply

$$\left|K_n^{1,\alpha}(x_1,x_2)\right| \le 2(x_1-x_2)^{-3/2}x_2^{-1/2}1_{\{x_2\le\pi/2\}} + 2(x_1-x_2)^{-3/2}(\pi-x_1)^{-1/2}1_{\{x_2>\pi/2\}},$$

which shows (2.2.14). Since $0\le\beta\le 1$, (2.2.8) implies

$$\left|K_n^{1,\alpha}(x_1,x_2)\right| \le Cn^{-\alpha}(x_1-x_2)^{-1-\beta}x_2^{\beta-\alpha-1} + Cn^{-1}(x_1-x_2)^{-1-\beta}x_2^{\beta-2}$$

if $x_2\le\pi/2$. The other inequalities can be shown similarly. ■

Lemma 2.2.16 *If* $0<\alpha\le 1$ *and* $\pi>x_1>x_2>0$, *then*

$$1_{\{x_2\le\pi/2\}}\left|K_n^{1,\alpha}(x_1,x_2)\right| \le C(x_1-x_2)^{\alpha-1}x_2^{-\alpha-1}1_{\{x_2\le\pi/2\}} + Cx_2^{-2}1_{\{x_2\le\pi/2\}} \tag{2.2.19}$$

and

$$1_{\{x_2>\pi/2\}}\left|K_n^{1,\alpha}(x_1,x_2)\right| \le C(x_1-x_2)^{\alpha-1}(\pi-x_1)^{-\alpha-1}1_{\{x_2>\pi/2\}} + C(\pi-x_1)^{-2}1_{\{x_2>\pi/2\}}. \tag{2.2.20}$$

Proof If $\beta=0$ and $n\ge(x_1-x_2)^{-1}$, then (2.2.15) implies (2.2.19). On the other hand, (2.2.19) follows from (2.2.17) if $n<(x_1-x_2)^{-1}$. ■

In the next lemma, we estimate the partial derivatives of the kernel function.

Lemma 2.2.17 *If* $0<\alpha\le 1$, $0\le\beta\le 1$ *and* $\pi>x_1>x_2>0$, *then for* $j=1,2$,

$$1_{\{x_2\le\pi/2\}}\left|\partial_j K_n^{1,\alpha}(x_1,x_2)\right| \le Cn^{1-\alpha}(x_1-x_2)^{-1-\beta}x_2^{\beta-\alpha-1}1_{\{x_2\le\pi/2\}} + C(x_1-x_2)^{-1-\beta}x_2^{\beta-2}1_{\{x_2\le\pi/2\}} \tag{2.2.21}$$

and

$$1_{\{x_2>\pi/2\}}\left|\partial_j K_n^{1,\alpha}(x_1,x_2)\right| \le Cn^{1-\alpha}(x_1-x_2)^{-1-\beta}(\pi-x_1)^{\beta-\alpha-1}1_{\{x_2>\pi/2\}} + C(x_1-x_2)^{-1-\beta}(\pi-x_1)^{\beta-2}1_{\{x_2>\pi/2\}}. \tag{2.2.22}$$

Proof Let $x_2\le\pi/2$. By Lagrange's mean value theorem and (2.2.12),

$$
\begin{aligned}
&\partial_1 D_k^1(x_1, x_2) \\
&= \frac{1}{2}\Big(\sin(x_1/2)\cos((k+1/2)x_1) + \cos(x_1/2)(2k+1)\sin((k+1/2)x_1)\Big) \\
&\quad \sin((x_1-x_2)/2)^{-1}\sin((x_1+x_2)/2)^{-1} \\
&\quad + \frac{1}{2}(x_1-x_2)\Big(\sin((x_1-x_2)/2)^{-2}\sin((x_1+x_2)/2)^{-1}\cos((x_1-x_2)/2) \\
&\quad + \sin((x_1-x_2)/2)^{-1}\sin((x_1+x_2)/2)^{-2}\cos((x_1+x_2)/2)\Big) H_k'(\xi),
\end{aligned}
$$

where $y < \xi < x$ is a suitable number. Using the methods above,

$$
\begin{aligned}
\left|\partial_1 K_n^{1,\alpha}(x_1, x_2)\right| &= \frac{1}{A_{n-1}^{\alpha}} \left| \sum_{k=0}^{n-1} A_{n-1-k}^{\alpha-1} \partial_1 D_k^1(x_1, x_2) \right| \\
&\le C(x_1-x_2)^{-1}(x_1+x_2)^{-1}(n^{1-\alpha}x_2^{-\alpha} + x_2^{-1}) \\
&\quad + C(x_1+x_2)^{-2}(n^{1-\alpha}x_2^{-\alpha} + x_2^{-1}) \\
&\le C(x_1-x_2)^{-1-\beta}(n^{1-\alpha}x_2^{\beta-\alpha-1} + x_2^{\beta-2}),
\end{aligned}
$$

which proves (2.2.21). The case $x_2 > \pi/2$, i.e., (2.2.22), can be shown similarly. ■

2.2.2 Kernel Functions for $q = \infty$

Lemma 2.2.18 *For $x \in \mathbb{T}^d$,*

$$
D_n^{\infty}(x) = \prod_{i=1}^{d} D_n^{\infty}(x_i) = \prod_{i=1}^{d} \frac{\sin((n+1/2)x_i)}{\sin(x_i/2)}.
$$

Proof The proof follows from the definition of the cubic Dirichlet kernels and from Lemma 1.2.3. ■

To estimate the cubic Cesàro kernels, we may suppose again that $x \in \mathbb{T}^2$ and $\pi > x_1 > x_2 > 0$.

Lemma 2.2.19 *If $0 < \alpha \le 1$, $x \in \mathbb{T}^2$ and $\pi > x_1 > x_2 > 0$, then*

$$
\left|K_n^{\infty,\alpha}(x_1, x_2)\right| \le C x_1^{-1} x_2^{-1}, \tag{2.2.23}
$$

$$
\begin{aligned}
\left|K_n^{\infty,\alpha}(x_1, x_2)\right| &\le C n^{-\alpha} x_1^{-1} x_2^{-1} (x_1-x_2)^{-\alpha} \\
&\quad + C n^{-1} x_1^{-1} x_2^{-1} (x_1-x_2)^{-1}
\end{aligned} \tag{2.2.24}
$$

and

$$\left|K_n^{\infty,\alpha}(x_1,x_2)\right| \le Cn^{1-\alpha}x_1^{-1}(x_1-x_2)^{-\alpha} + Cx_1^{-1}(x_1-x_2)^{-1}. \tag{2.2.25}$$

Proof The first inequality, (2.2.23) follows easily from Lemma 1.4.8 and from

$$\begin{aligned} K_n^{\infty,\alpha}(x_1,x_2) &= \frac{1}{A_{n-1}^{\alpha}} \sum_{k=0}^{n-1} A_{n-1-k}^{\alpha-1} D_k^{\infty}(x_1,x_2) \\ &= \frac{1}{A_{n-1}^{\alpha}} \sum_{k=0}^{n-1} A_{n-1-k}^{\alpha-1} \frac{\sin((k+1/2)x_1)}{\sin(x_1/2)} \frac{\sin((k+1/2)x_2)}{\sin(x_2/2)}. \end{aligned}$$

The trigonometric identity

$$\sin a \sin b = \frac{1}{2}(\cos(a-b) - \cos(a+b)) \tag{2.2.26}$$

yields

$$\begin{aligned} &\left|K_n^{\infty,\alpha}(x_1,x_2)\right| \\ &= \frac{1}{2A_{n-1}^{\alpha}} \left| \sum_{k=0}^{n-1} A_{n-1-k}^{\alpha-1} \frac{\cos((k+1/2)(x_1-x_2)) - \cos((k+1/2)(x_1+x_2))}{\sin(x_1/2)\sin(x_2/2)} \right|. \end{aligned}$$

Observe that $\sin(x_i/2) \sim x_i$,

$$\sin(x_1 \pm x_2)/2 \sim x_1 \pm x_2 \quad \text{if} \quad x_2 \le \pi/2$$

and

$$\sin(x_1 - x_2)/2 \sim x_1 - x_2, \quad \sin(x_1 + x_2)/2 \sim 2\pi - x_1 - x_2 \quad \text{if} \quad x_2 > \pi/2.$$

Using the facts $x_1 + x_2 > x_1 - x_2$, $2\pi - x_1 - x_2 > x_1 - x_2$ and and Lemma 1.4.14, we conclude that

$$\begin{aligned} &\left|K_n^{\infty,\alpha}(x_1,x_2)\right| \\ &= \frac{C}{2A_{n-1}^{\alpha}} \frac{1}{|\sin(x_1/2)\sin(x_2/2)|} \Big(\frac{1}{|\sin(x_1-x_2)/2|^{\alpha}} + \frac{n^{\alpha-1}}{|\sin(x_1-x_2)/2|} \\ &\quad + \frac{1}{|\sin(x_1+x_2)/2|^{\alpha}} + \frac{n^{\alpha-1}}{|\sin(x_1+x_2)/2|} \Big) \\ &\le Cn^{-\alpha}x_1^{-1}x_2^{-1}(x_1-x_2)^{-\alpha} + Cn^{-1}x_1^{-1}x_2^{-1}(x_1-x_2)^{-1}, \end{aligned} \tag{2.2.27}$$

which is (2.2.24). Using Lagrange's theorem in (2.2.27) and Lemma 1.4.15, there exists $x_1 - x_2 < \xi < x_1 + x_2$ such that

$$\begin{aligned}\left|K_n^{\infty,\alpha}(x_1,x_2)\right| &= \frac{1}{A_{n-1}^{\alpha}}\left|\sum_{k=0}^{n-1}\frac{A_{n-1-k}^{\alpha-1}(k+1/2)x_2\sin((k+1/2)\xi)}{\sin(x_1/2)\sin(x_2/2)}\right| \\ &\le Cn^{1-\alpha}x_1^{-1}(x_1-x_2)^{-\alpha}+Cx_1^{-1}(x_1-x_2)^{-1}.\end{aligned}$$

This finishes the proof of the lemma. ■

Lemma 2.2.20 *If* $0<\alpha\le 1$ *and* $\pi>x_1>x_2>0$, *then*

$$\left|K_n^{\infty,\alpha}(x_1,x_2)\right| \le Cx_2^{\alpha-1}(x_1-x_2)^{-\alpha-1}+C(x_1-x_2)^{-2}. \tag{2.2.28}$$

Proof The inequality follows from (2.2.24) if $n\ge x_2^{-1}$ and from (2.2.25) if $n<x_2^{-1}$. ■

The partial derivatives of the cubic Cesàro kernels can be estimated as follows.

Lemma 2.2.21 *If* $0<\alpha\le 1$, $j=1,2$ *and* $\pi>x_1>x_2>0$, *then*

$$\left|\partial_j K_n^{\infty,\alpha}(x_1,x_2)\right| \le Cn^{1-\alpha}x_1^{-1}x_2^{-1}(x_1-x_2)^{-\alpha}.$$

Proof By Lagrange's mean value theorem and (2.2.26),

$$\begin{aligned}&\partial_1 D_k^{\infty}(x_1,x_2)\\ &=\frac{1}{2}(k+1/2)\Big(\sin((k+1/2)(x_1+x_2))-\sin((k+1/2)(x_1-x_2))\Big)\\ &\quad\sin(x_1/2)^{-1}\sin(x_2/2)^{-1}\\ &\quad+\frac{1}{4}\Big(\cos((k+1/2)(x_1-x_2))-\cos((k+1/2)(x_1+x_2))\Big)\\ &\quad\cos(x_1/2)\sin(x_1/2)^{-2}\sin(x_2/2)^{-1}\\ &=\frac{1}{2}(k+1/2)\Big(\sin((k+1/2)(x_1+x_2))-\sin((k+1/2)(x_1-x_2))\Big)\\ &\quad\sin(x_1/2)^{-1}\sin(x_2/2)^{-1}\\ &\quad+\frac{1}{4}(k+1/2)x_2\sin((k+1/2)\xi)\cos(x_1/2)\sin(x_1/2)^{-2}\sin(x_2/2)^{-1},\end{aligned}$$

where $x_1-x_2<\xi<x_1+x_2$ is a suitable number. Similarly as above,

$$\begin{aligned}\left|\partial_1 K_n^{\infty,\alpha}(x_1,x_2)\right| &= \frac{1}{A_{n-1}^{\alpha}}\left|\sum_{k=0}^{n-1}A_{n-1-k}^{\alpha-1}\partial_1 D_k^{\infty}(x_1,x_2)\right|\\ &\le Cn^{1-\alpha}x_1^{-1}x_2^{-1}(x_1-x_2)^{-\alpha}+Cx_1^{-1}x_2^{-1}(x_1-x_2)^{-1}\\ &\quad+Cn^{1-\alpha}x_1^{-2}(x_1-x_2)^{-\alpha}+Cx_1^{-2}(x_1-x_2)^{-1}\\ &\le Cn^{1-\alpha}x_1^{-1}x_2^{-1}(x_1-x_2)^{-\alpha}+Cx_1^{-1}x_2^{-1}(x_1-x_2)^{-1},\end{aligned}$$

which proves the lemma. ■

2.2.3 Kernel Functions for $q = 2$

As we mentioned before, for $q = 2$, we will consider the Riesz summability. To this, we have to introduce some special functions. For the sake of completeness, we prove some elementary properties for these functions. First, we introduce the gamma function by

$$\Gamma(x) := \int_0^\infty t^{x-1} e^{-t}\, dt \qquad (x > 0).$$

Integration by parts yields

$$\Gamma(x) = \Big[\frac{t^x e^{-t}}{x}\Big]_0^\infty + \frac{1}{x}\int_0^\infty t^x e^{-t}\, dt = \frac{1}{x}\Gamma(x+1) \qquad (x > 0).$$

Since $\Gamma(1) = 1$, we have

$$\Gamma(x+1) = x\Gamma(x) \quad (x > 0) \qquad \text{and} \qquad \Gamma(n) = (n-1)!. \tag{2.2.29}$$

After a substitution, we can see that

$$\Gamma\Big(\frac{1}{2}\Big) = \int_0^\infty t^{-1/2} e^{-t}\, dt = 2\int_0^\infty e^{-u^2}\, du = \sqrt{\pi}.$$

The beta function is defined by

$$B(x, y) := \int_0^1 s^{x-1}(1-s)^{y-1}\, ds = \int_0^1 s^{y-1}(1-s)^{x-1}\, ds,$$

where $x, y > 0$. The relationship between the beta and gamma function reads as follows:

$$\Gamma(x+y)B(x, y) = \Gamma(x)\Gamma(y). \tag{2.2.30}$$

Indeed, substituting $s = u/(1+u)$, we obtain

$$\begin{aligned}\Gamma(x+y)B(x, y) &= \Gamma(x+y)\int_0^1 s^{y-1}(1-s)^{x-1}\, ds \\ &= \Gamma(x+y)\int_0^\infty u^{y-1}\Big(\frac{1}{1+u}\Big)^{x+y} du \\ &= \int_0^\infty\int_0^\infty u^{y-1}\Big(\frac{1}{1+u}\Big)^{x+y} v^{x+y-1}e^{-v}\, dv du.\end{aligned}$$

The substitution $v = t(1+u)$ in the inner integral yields

$$\begin{aligned}\Gamma(x+y)B(x,y) &= \int_0^\infty \int_0^\infty u^{y-1} t^{x+y-1} e^{-t(1+u)}\,dt\,du \\ &= \int_0^\infty t^x e^{-t} \int_0^\infty (ut)^{y-1} e^{-tu}\,du\,dt \\ &= \int_0^\infty t^{x-1} e^{-t}\Gamma(y)\,dt \\ &= \Gamma(x)\Gamma(y),\end{aligned}$$

which shows (2.2.30).

Definition 2.2.22 For $k > -1/2$, the Bessel functions are defined by

$$J_k(t) := \frac{(t/2)^k}{\Gamma(k+1/2)\Gamma(1/2)} \int_{-1}^1 e^{\imath ts}(1-s^2)^{k-1/2}\,ds \qquad (t \in \mathbb{R}).$$

Using the Euler formulas, we can see that the Bessel functions are real-valued.

Lemma 2.2.23 *We have*

$$J_k'(t) = kt^{-1}J_k(t) - J_{k+1}(t) \qquad (t \neq 0).$$

Proof By integrating by parts and by (2.2.29), we conclude

$$\begin{aligned}\frac{d}{dt}(t^{-k}J_k(t)) &= \frac{\imath 2^{-k}}{\Gamma(k+1/2)\Gamma(1/2)} \int_{-1}^1 e^{\imath ts} s(1-s^2)^{k-1/2}\,ds \\ &= \frac{\imath 2^{-k}}{(2k+1)\Gamma(k+1/2)\Gamma(1/2)} \Bigg(-\Big[e^{\imath ts}(1-s^2)^{k+1/2}\Big]_{-1}^1 \\ &\quad + \int_{-1}^1 \imath t e^{\imath ts}(1-s^2)^{k+1/2}\,ds \Bigg) \\ &= \frac{-2^{-k-1}t}{(k+1/2)\Gamma(k+1/2)\Gamma(1/2)} \int_{-1}^1 e^{\imath ts}(1-s^2)^{k+1/2}\,ds \\ &= -t^{-k}J_{k+1}(t),\end{aligned}$$

which proves the desired result. ■

Lemma 2.2.24 *For $k > -1/2$ and $t > 0$,*

$$J_k(t) \le C_k t^k \quad \text{and} \quad J_k(t) \le C_k t^{-1/2},$$

where C_k is independent of t.

Proof Since $1 - s^2 \geq 1 - |s|$ for $|s| \leq 1$, the first estimate follows from the definition of J_k. The second one follows from the first one if $0 < t \leq 1$. So we may assume that $t > 1$. Let us integrate the complex valued function

$$e^{\imath tz}(1 - z^2)^{k-1/2} \qquad (z \in \mathbb{C})$$

over the boundary of the rectangle whose lower side is $[-1, 1]$ and whose height is $R > 0$. By Cauchy's theorem,

$$\begin{aligned} 0 = \imath \int_R^0 e^{\imath t(-1+\imath s)}(s^2 + 2\imath s)^{k-1/2}\, ds + \int_{-1}^1 e^{\imath ts}(1 - s^2)^{k-1/2}\, ds \\ + \imath \int_0^R e^{\imath t(1+\imath s)}(s^2 - 2\imath s)^{k-1/2}\, ds + \epsilon(R), \end{aligned}$$

where $\epsilon(R) \to 0$ as $R \to \infty$. Hence, taking the limit as $R \to \infty$,

$$\begin{aligned} \int_{-1}^1 e^{\imath ts}(1 - s^2)^{k-1/2}\, ds &= \imath e^{-\imath t} \int_0^\infty e^{-ts}(s^2 + 2\imath s)^{k-1/2}\, ds \\ &\quad - \imath e^{\imath t} \int_0^\infty e^{-ts}(s^2 - 2\imath s)^{k-1/2}\, ds \\ &=: I_1 + I_2. \end{aligned}$$

Observe that

$$(s^2 + 2\imath s)^{k-1/2} = (2\imath s)^{k-1/2} + \phi(s),$$

where

$$|\phi(s)| \leq C s^{k+1/2} \qquad \text{if } 0 < s \leq 1 \text{ or } s > 1 \text{ and } k \leq 3/2$$

and

$$|\phi(s)| \leq C s^{2k-1} \qquad \text{if } s > 1 \text{ and } k > 3/2.$$

Indeed, it follows from Lagrange's mean value theorem that

$$|\phi(s)| = \left|(2\imath s)^{k-1/2}\right| \left| \left(\frac{s}{2\imath} + 1\right)^{k-1/2} - 1 \right| \leq C_k s^{k+1/2} \left| \frac{\xi}{2\imath} + 1 \right|^{k-3/2},$$

where $0 < \xi < s$. Hence

$$\left|s^2 + 2\imath s\right|^{k-1/2} \leq C_k s^{k-1/2} + |\phi(s)|$$

and

$$\begin{aligned}|I_1| &\le \int_0^\infty e^{-ts}\left(C_k s^{k-1/2} + |\phi(s)|\right) ds\\ &= C_k \int_0^\infty e^{-ts} s^{k-1/2} + \int_0^1 e^{-ts}|\phi(s)|\, ds + \int_1^\infty e^{-ts}|\phi(s)|\, ds\\ &= I_{1,1} + I_{1,2} + I_{1,3}.\end{aligned}$$

By the substitution $ts = u$ and by the definition of the gamma function,

$$\begin{aligned}I_{1,1} &= C_k t^{-1} \int_0^\infty e^{-u}(u/t)^{k-1/2}\, du\\ &= C_k t^{-k-1/2} \int_0^\infty e^{-u} u^{k-1/2}\, du = C_k \Gamma(k+1/2) t^{-k-1/2}.\end{aligned}$$

The same substitution implies

$$\begin{aligned}I_{1,2} &\le \int_0^1 e^{-ts} s^{k+1/2}\, ds \le t^{-k-3/2} \int_0^\infty e^{-u} u^{k+1/2}\, ds\\ &= \Gamma(k+3/2) t^{-k-3/2} \le C_k t^{-k-1/2}.\end{aligned} \tag{2.2.31}$$

If $k \le 3/2$, then

$$I_{1,3} \le \Gamma(k+3/2) t^{-k-3/2} \le C_k t^{-k-1/2}$$

as in (2.2.31). Similarly, for $k > 3/2$,

$$\begin{aligned}I_{1,3} &\le \int_1^\infty e^{-ts} s^{2k-1}\, ds \le t^{-2k} \int_0^\infty e^{-u} u^{2k-1}\, ds\\ &= \Gamma(2k) t^{-2k} \le C_k t^{-k-1/2}.\end{aligned}$$

The integral I_2 can be estimated in the same way. ■

Lemma 2.2.25 *If $k > -1/2$, $l > -1$ and $t > 0$, then*

$$J_{k+l+1}(t) = \frac{t^{l+1}}{2^l \Gamma(l+1)} \int_0^1 J_k(ts) s^{k+1} (1-s^2)^l\, ds.$$

Proof Taking into account (2.2.30), we get that

$$\begin{aligned}J_k(t) &= \frac{2(t/2)^k}{\Gamma(k+1/2)\Gamma(1/2)} \int_0^1 \cos(ts)(1-s^2)^{k-1/2}\, ds\\ &= \sum_{j=0}^\infty (-1)^j \frac{2(t/2)^k t^{2j}}{(2j)!\Gamma(k+1/2)\Gamma(1/2)} \int_0^1 s^{2j}(1-s^2)^{k-1/2}\, ds\end{aligned}$$

$$
\begin{aligned}
&= \sum_{j=0}^{\infty} (-1)^j \frac{(t/2)^k t^{2j}}{(2j)!\Gamma(k+1/2)\Gamma(1/2)} \int_0^1 u^{j-1/2}(1-u)^{k-1/2}\,du \\
&= \sum_{j=0}^{\infty} (-1)^j \frac{(t/2)^k t^{2j}}{(2j)!\Gamma(k+1/2)\Gamma(1/2)} B(j+1/2, k+1/2) \\
&= \frac{(t/2)^k}{\Gamma(1/2)} \sum_{j=0}^{\infty} (-1)^j \frac{\Gamma(j+1/2)}{\Gamma(j+k+1)} \frac{t^{2j}}{(2j)!}.
\end{aligned} \tag{2.2.32}
$$

Thus

$$
\begin{aligned}
&\int_0^1 J_k(ts)s^{k+1}(1-s^2)^l\,ds \\
&= \int_0^1 \left(\frac{(ts/2)^k}{\Gamma(1/2)} \sum_{j=0}^{\infty} (-1)^j \frac{\Gamma(j+1/2)}{\Gamma(j+k+1)} \frac{(ts)^{2j}}{(2j)!} \right) s^{k+1}(1-s^2)^l\,ds \\
&= \frac{(t/2)^k}{\Gamma(1/2)} \sum_{j=0}^{\infty} (-1)^j \frac{\Gamma(j+1/2)}{\Gamma(j+k+1)} \frac{t^{2j}}{(2j)!} \int_0^1 s^{2k+2j+1}(1-s^2)^l\,ds.
\end{aligned}
$$

Substituting $s^2 = u$ and using (2.2.30) and (2.2.32), we conclude

$$
\begin{aligned}
&\int_0^1 J_k(ts)s^{k+1}(1-s^2)^l\,ds \\
&= \frac{(t/2)^k}{\Gamma(1/2)} \sum_{j=0}^{\infty} (-1)^j \frac{\Gamma(j+1/2)}{2\Gamma(j+k+1)} \frac{t^{2j}}{(2j)!} \int_0^1 u^{k+j}(1-u)^l\,du \\
&= \frac{(t/2)^k}{\Gamma(1/2)} \sum_{j=0}^{\infty} (-1)^j \frac{\Gamma(j+1/2)}{2\Gamma(j+k+1)} \frac{t^{2j}}{(2j)!} B(k+j+1, l+1) \\
&= \frac{2^l\Gamma(l+1)}{t^{l+1}} \frac{(t/2)^{k+l+1}}{\Gamma(1/2)} \sum_{j=0}^{\infty} (-1)^j \frac{\Gamma(j+1/2)}{\Gamma(k+l+j+2)} \frac{t^{2j}}{(2j)!} \\
&= \frac{2^l\Gamma(l+1)}{t^{l+1}} J_{k+l+1}(t),
\end{aligned}
$$

which proves the lemma. ∎

Now we can turn back to the circular Riesz means.

Definition 2.2.26 For $f \in L_1(\mathbb{R}^d)$, the Fourier transform is defined by

$$
\widehat{f}(x) := \frac{1}{(2\pi)^d} \int_{\mathbb{R}^d} f(t) e^{-\imath x \cdot t}\,dt \qquad (x \in \mathbb{R}^d).
$$

Define

$$\theta(s) := \begin{cases} (1-|s|^2)^\alpha & \text{if } |s| \le 1; \\ 0 & \text{if } |s| > 1 \end{cases} \qquad (s \in \mathbb{R})$$

and

$$\theta_0(x) := \theta(\|x\|_2) \qquad (x \in \mathbb{R}^d).$$

θ_0 is called a radial function. No we use another method than for $q = 1$ or $q = \infty$. We will express the Riesz means in terms of the Fourier transform of θ_0. As we will see in the next lemma, $\widehat{\theta_0}$ can be computed with the help of the Bessel functions.

Theorem 2.2.27 *If $\alpha > 0$ and $x \in \mathbb{R}^d$, then*

$$\widehat{\theta_0}(x) = \frac{1}{(2\pi)^{d/2}} 2^\alpha \Gamma(\alpha+1) \|x\|_2^{-d/2-\alpha} J_{d/2+\alpha}(\|x\|_2).$$

Proof The function $\theta_0 \in L_1(\mathbb{R}^d)$ because

$$\int_{\mathbb{R}^d} |\theta_0(x)|\, dx \le C \int_0^\infty |\theta(r)| r^{d-1}\, dr < \infty.$$

Using the notation $r = \|x\|_2$, $x = rx'$, $s = \|u\|_2$ and $u = su'$, we get that

$$\begin{aligned} \widehat{\theta_0}(x) &= \frac{1}{(2\pi)^d} \int_{\mathbb{R}^d} \theta_0(u) e^{-\imath x \cdot u}\, du \\ &= \frac{1}{(2\pi)^d} \int_0^\infty \theta(s) \left(\int_{\Sigma_{d-1}} e^{-\imath r s x' \cdot u'}\, du' \right) s^{d-1}\, ds, \end{aligned} \tag{2.2.33}$$

where Σ_{d-1} denotes the sphere. In the inner integral, we integrate first over the parallel

$$P_\delta := \{u' \in \Sigma_{d-1} : x' \cdot u' = \cos\delta\}$$

orthogonal to x' obtaining a function of $0 \le \delta \le \pi$, which we then integrate over $[0, \pi]$. If ω_{d-2} denotes the surface area of Σ_{d-2}, then the measure of P_δ is

$$\omega_{d-2}(\sin\delta)^{d-2} = \frac{2\pi^{(d-1)/2}}{\Gamma((d-1)/2)} (\sin\delta)^{d-2}.$$

Hence

$$\begin{aligned} \int_{\Sigma_{d-1}} e^{-\imath r s x' \cdot u'}\, du' &= \int_0^\pi e^{-\imath r s \cos\delta} \omega_{d-2}(\sin\delta)^{d-2}\, d\delta \\ &= \omega_{d-2} \int_{-1}^1 e^{\imath r s \xi} (1-\xi^2)^{(d-3)/2}\, d\xi \end{aligned}$$

$$
\begin{aligned}
&= \frac{2\pi^{(d-1)/2}}{\Gamma((d-1)/2)} \frac{\Gamma(d/2-1/2)\Gamma(1/2)}{(rs/2)^{d/2-1}} J_{d/2-1}(rs) \\
&= (2\pi)^{d/2}(rs)^{-d/2+1} J_{d/2-1}(rs).
\end{aligned}
$$

Taking into account this and (2.2.33), we conclude

$$
\begin{aligned}
\widehat{\theta_0}(x) &= \frac{1}{(2\pi)^{d/2}} r^{-d/2+1} \int_0^\infty \theta(s) J_{d/2-1}(rs) s^{d/2}\, ds \\
&= \frac{1}{(2\pi)^{d/2}} \|x\|_2^{-d/2+1} \int_0^1 J_{d/2-1}(\|x\|_2 s) s^{d/2} (1-s^2)^\alpha\, ds.
\end{aligned}
$$

Applying Lemma 2.2.25 with $k = d/2-1$, $l = \alpha$, we see that

$$
\widehat{\theta_0}(x) = \frac{1}{(2\pi)^{d/2}} \|x\|_2^{-d/2+1} J_{d/2+\alpha}(\|x\|_2) \|x\|_2^{-\alpha-1} 2^\alpha \Gamma(\alpha+1),
$$

which shows the theorem. ■

Theorem 2.2.27, Lemma 2.2.23 and 2.2.24 imply that $\widehat{\theta_0}(x)$ as well as all of its derivatives can be estimated by $\|x\|_2^{-d/2-\alpha-1/2}$.

Corollary 2.2.28 *For all $i_1, \ldots, i_d \geq 0$ and $\alpha > 0$,*

$$
|\partial_1^{i_1} \cdots \partial_d^{i_d} \widehat{\theta_0}(x)| \leq C \|x\|_2^{-d/2-\alpha-1/2} \qquad (x \neq 0).
$$

The same result holds for

$$
\theta(s) := \begin{cases} (1-|s|^\gamma)^\alpha & \text{if } |s| \leq 1; \\ 0 & \text{if } |s| > 1 \end{cases} \qquad (s \in \mathbb{R})
$$

and

$$
\theta_0(x) := \theta(\|x\|_2) \qquad (x \in \mathbb{R}^d),
$$

whenever $\gamma \in \mathbb{P}$ (see Lu [224, p. 132]). From now on, we assume that $\gamma \in \mathbb{P}$. The next result is an easy consequence of Corollary 2.2.28.

Corollary 2.2.29 *$\widehat{\theta_0} \in L_1(\mathbb{R}^d)$ if*

$$
\frac{d-1}{2} < \alpha < \infty.
$$

Now we are ready to express the Riesz means using the Fourier transform of θ_0.

Theorem 2.2.30 *If $n \in \mathbb{N}$, $f \in L_1(\mathbb{T}^d)$, $(d-1)/2 < \alpha < \infty$ and $\gamma \in \mathbb{P}$, then*

$$
\sigma_n^{2,\alpha,\gamma} f(x) = n^d \int_{\mathbb{R}^d} f(x-t) \widehat{\theta_0}(nt)\, dt
$$

for almost every $x \in \mathbb{T}^d$.

Proof If $f(t) = e^{\imath k \cdot t}$ $(k \in \mathbb{Z}^d, t \in \mathbb{T}^d)$, then

$$\begin{aligned}\sigma_n^{2,\alpha,\gamma} f(x) &= \theta_0\left(\frac{-k}{n}\right) e^{\imath k \cdot x} \\ &= e^{\imath k \cdot x} \int_{\mathbb{R}^d} e^{-\imath k \cdot t/n} \widehat{\theta_0}(t)\, dt \\ &= n^d \int_{\mathbb{R}^d} e^{\imath k \cdot (x-t)} \widehat{\theta_0}(nt)\, dt.\end{aligned}$$

The theorem holds also for trigonometric polynomials. Let f be an arbitrary element from $L_1(\mathbb{T}^d)$ and (f_k) be a sequence of trigonometric polynomials such that $f_k \to f$ in the $L_1(\mathbb{T}^d)$-norm. It follows from Lemma 2.2.6 and from the fact that $K_n^{2,\alpha,\gamma} \in L_1(\mathbb{T}^d)$ that

$$\lim_{n\to\infty} \sigma_n^{2,\alpha,\gamma} f_k = \sigma_n^{2,\alpha,\gamma} f$$

in the $L_1(\mathbb{T}^d)$ norm.

On the other hand, since $\widehat{\theta_0} \in L_1(\mathbb{R}^d)$, we have

$$\lim_{n\to\infty} \int_{\mathbb{R}^d} f_k(x-t)\widehat{\theta_0}(nt)\, dt = \int_{\mathbb{R}^d} f(x-t)\widehat{\theta_0}(nt)\, dt$$

in the $L_1(\mathbb{T}^d)$-norm. ∎

Lemma 2.2.31 *If* $n \in \mathbb{N}$, $(d-1)/2 < \alpha < \infty$ *and* $\gamma \in \mathbb{P}$, *then*

$$K_n^{2,\alpha,\gamma}(t) = (2\pi)^d n^d \sum_{k\in\mathbb{Z}^d} \widehat{\theta_0}(n(t+2k\pi)). \tag{2.2.34}$$

Proof Since f is periodic, Theorem 2.2.30 implies that

$$\begin{aligned}\sigma_n^{2,\alpha,\gamma} f(x) &= n^d \sum_{k\in\mathbb{Z}^d} \int_{2k\pi+\mathbb{T}^d} f(x-t)\widehat{\theta_0}(nt)\, dt \\ &= n^d \sum_{k\in\mathbb{Z}^d} \int_{\mathbb{T}^d} f(x-t)\widehat{\theta_0}(n(t+2k\pi))\, dt.\end{aligned}$$

The result follows from Lemma 2.2.6. ∎

2.3 Norm Convergence of the ℓ_q-Summability Means

In this section, we will prove that the Cesàro and Riesz means, $\sigma_n^{q,\alpha} f$ and $\sigma_n^{q,\alpha,\gamma} f$ are uniformly bounded on the $L_p(\mathbb{T}^d)$ spaces and they converge to the original function f in norm when $1 \leq p < \infty$, $q = 1, 2$ or $q = \infty$. Having the results of Sect. 2.2, we are ready to prove that the $L_1(\mathbb{T}^d)$-norms of the kernel functions are uniformly bounded. We start with the triangular and cubic Cesàro summability.

Theorem 2.3.1 *If* $0 < \alpha \leq 1$ *and* $q = 1$ *or* $q = \infty$, *then*

$$\sup_{n\in\mathbb{N}} \int_{\mathbb{T}^d} \left|K_n^{q,\alpha}(x)\right| dx \leq C.$$

***Proof of Theorem 2.3.1* for** $q = 1$. It is enough to integrate the kernel function over the set

$$\{(x_1, x_2) : 0 < x_2 < x_1 < \pi\}.$$

Let us decompose this set into the union $\cup_{i=1}^{10} A_i$, where

$$\begin{aligned}
A_1 &:= \{(x_1, x_2) : 0 < x_1 \leq 2/n, 0 < x_2 < x_1 < \pi, x_2 \leq \pi/2\},\\
A_2 &:= \{(x_1, x_2) : 2/n < x_1 < \pi, 0 < x_2 \leq 1/n, x_2 \leq \pi/2\},\\
A_3 &:= \{(x_1, x_2) : 2/n < x_1 < \pi, 1/n < x_2 \leq x_1/2, x_2 \leq \pi/2\},\\
A_4 &:= \{(x_1, x_2) : 2/n < x_1 < \pi, x_1/2 < x_2 \leq x_1 - 1/n, x_2 \leq \pi/2\},\\
A_5 &:= \{(x_1, x_2) : 2/n < x_1 < \pi, x_1 - 1/n < x_2 < x_1, x_2 \leq \pi/2\}\\
A_6 &:= \{(x_1, x_2) : x_2 > \pi/2, \pi - 2/n \leq x_2 < \pi, 0 < x_2 < x_1 < \pi\},\\
A_7 &:= \{(x_1, x_2) : \pi/2 < x_2 < \pi - 2/n, \pi - 1/n < x_1 < \pi\},\\
A_8 &:= \{(x_1, x_2) : \pi/2 < x_2 < \pi - 2/n, (\pi + x_2)/2 < x_1 \leq \pi - 1/n\},\\
A_9 &:= \{(x_1, x_2) : \pi/2 < x_2 < \pi - 2/n, x_2 + 1/n < x_1 \leq (\pi + x_2)/2\},\\
A_{10} &:= \{(x_1, x_2) : \pi/2 < x_2 < \pi - 2/n, x_2 < x_1 \leq x_2 + 1/n\}.
\end{aligned}$$

The sets A_i can be seen on Fig. 2.11.

By Lemma 2.2.5, we can see that

$$\int_{A_1} \left|K_n^{1,\alpha}(x_1, x_2)\right| dx + \int_{A_6} \left|K_n^{1,\alpha}(x_1, x_2)\right| dx \leq C.$$

Inequality (2.2.14) implies

$$\int_{A_2} \left|K_n^{1,\alpha}(x_1, x_2)\right| dx \leq C \int_{2/n}^{\pi} \int_0^{1/n} (x_1 - 1/n)^{-3/2} x_2^{-1/2}\, dx_2\, dx_1 \leq C$$

and

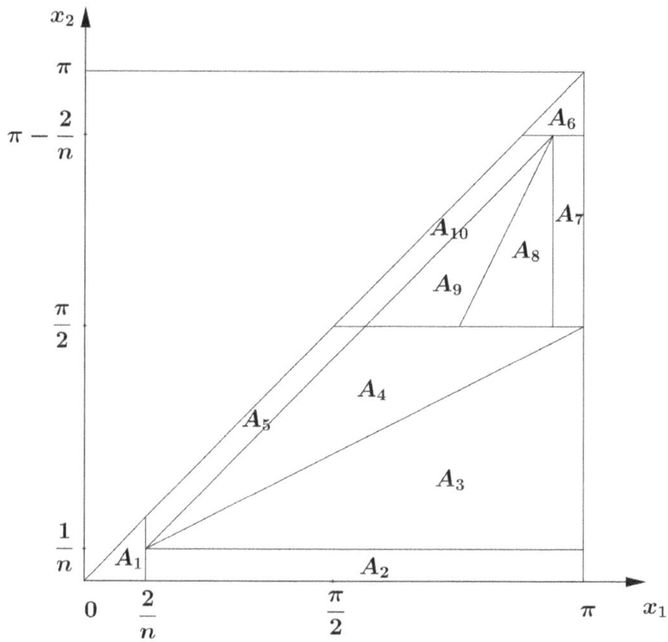

Fig. 2.11 The sets A_i

$$\int_{A_7} \left|K_n^{1,\alpha}(x_1, x_2)\right| dx$$
$$\leq C \int_{\pi/2}^{\pi-2/n} \int_{\pi-1/n}^{\pi} (\pi - 1/n - x_2)^{-3/2} (\pi - x_1)^{-1/2} \, dx_1 \, dx_2 \leq C.$$

Observe that $x_1 - x_2 \geq x_1/2$ on the set A_3. Choosing β such that $0 < \beta < \alpha$, we get from (2.2.15) that

$$\int_{A_3} \left|K_n^{1,\alpha}(x_1, x_2)\right| dx \leq Cn^{-\alpha} \int_{2/n}^{\pi} \int_{1/n}^{x_1/2} x_1^{-1-\beta} x_2^{\beta-\alpha-1} \, dx_2 \, dx_1$$
$$+ Cn^{-1} \int_{2/n}^{\pi} \int_{1/n}^{x_1/2} x_1^{-1-\beta} x_2^{\beta-2} \, dx_2 \, dx_1 \leq C.$$

Similarly, $x_1 - x_2 > (\pi - x_2)/2$ on the set A_8 and so, by (2.2.16),

$$\begin{aligned}\int_{A_8} \left|K_n^{1,\alpha}(x_1, x_2)\right| dx &\le Cn^{-\alpha} \int_{\pi/2}^{\pi-2/n} \int_{(\pi+x_2)/2}^{\pi-1/n} (\pi - x_2)^{-1-\beta}(\pi - x_1)^{\beta-\alpha-1}\, dx_1\, dx_2 \\ &\quad + Cn^{-1} \int_{\pi/2}^{\pi-2/n} \int_{(\pi+x_2)/2}^{\pi-1/n} (\pi - x_2)^{-1-\beta}(\pi - x_1)^{\beta-2}\, dx_1\, dx_2 \le C.\end{aligned}$$

We have $x_2 > x_1/2$ on A_4, hence (2.2.15) implies

$$\begin{aligned}\int_{A_4} \left|K_n^{1,\alpha}(x_1, x_2)\right| dx &\le Cn^{-\alpha} \int_{2/n}^{\pi} \int_{x_1/2}^{x_1-1/n} (x_1 - x_2)^{-1-\beta} x_1^{\beta-\alpha-1}\, dx_2\, dx_1 \\ &\quad + Cn^{-1} \int_{2/n}^{\pi} \int_{x_1/2}^{x_1-1/n} (x_1 - x_2)^{-1-\beta} x_1^{\beta-2}\, dx_2\, dx_1 \le C.\end{aligned}$$

Similarly, $\pi - x_1 \ge (\pi - x_2)/2$ on the set A_9. Thus

$$\begin{aligned}\int_{A_9} \left|K_n^{1,\alpha}(x_1, x_2)\right| dx &\le Cn^{-\alpha} \int_{\pi/2}^{\pi-2/n} \int_{x_2+1/n}^{(\pi+x_2)/2} (x_1 - x_2)^{-1-\beta}(\pi - x_2)^{\beta-\alpha-1}\, dx_1\, dx_2 \\ &\quad + Cn^{-1} \int_{\pi/2}^{\pi-2/n} \int_{x_2+1/n}^{(\pi+x_2)/2} (x_1 - x_2)^{-1-\beta}(\pi - x_2)^{\beta-2}\, dx_1\, dx_2 \le C.\end{aligned}$$

Finally, by (2.2.19),

$$\begin{aligned}\int_{A_5} \left|K_n^{1,\alpha}(x_1, x_2)\right| dx &\le C \int_{1/n}^{\pi} \int_{x_2}^{x_2+1/n} (x_1 - x_2)^{\alpha-1} x_2^{-\alpha-1}\, dx_1 dx_2 \\ &\quad + C \int_{1/n}^{\pi} \int_{x_2}^{x_2+1/n} x_2^{-2}\, dx_1 dx_2 \le C\end{aligned}$$

and

$$\begin{aligned}\int_{A_{10}} \left|K_n^{1,\alpha}(x_1, x_2)\right| dx &\le C \int_{\pi/2}^{\pi-1/n} \int_{x_1-1/n}^{x_1} (x_1 - x_2)^{\alpha-1}(\pi - x_1)^{-\alpha-1}\, dx_2 dx_1 \\ &\quad + C \int_{\pi/2}^{\pi-1/n} \int_{x_1-1/n}^{x_1} (\pi - x_1)^{-2}\, dx_2 dx_1 \le C\end{aligned}$$

which completes the proof of the theorem. ∎

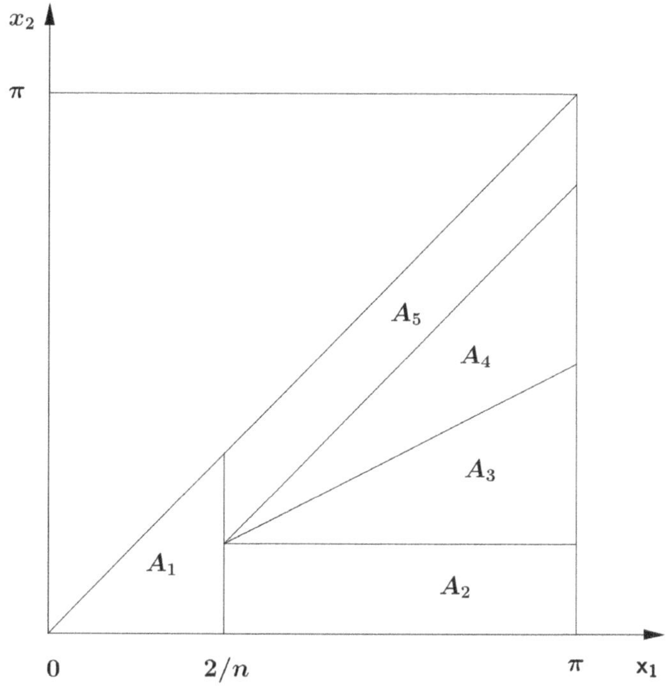

Fig. 2.12 The sets A_i

Proof of Theorem 2.3.1 **for** $q = \infty$. We integrate again over the set

$$\{(x_1, x_2) : 0 < x_2 < x_1 < \pi\}$$

and decompose this set into the union $\cup_{i=1}^{5} A_i$, where

$$\begin{aligned}
A_1 &:= \{(x_1, x_2) : 0 < x_1 \le 2/n, 0 < x_2 < x_1 < \pi\},\\
A_2 &:= \{(x_1, x_2) : 2/n < x_1 < \pi, 0 < x_2 \le 1/n\},\\
A_3 &:= \{(x_1, x_2) : 2/n < x_1 < \pi, 1/n < x_2 \le x_1/2\},\\
A_4 &:= \{(x_1, x_2) : 2/n < x_1 < \pi, x_1/2 < x_2 \le x_1 - 1/n\},\\
A_5 &:= \{(x_1, x_2) : 2/n < x_1 < \pi, x_1 - 1/n < x_2 < x_1\}
\end{aligned}$$

(see Fig. 2.12).

First of all,

$$\int_{A_1} \left| K_n^{\infty,\alpha}(x_1, x_2) \right| dx \le C.$$

By (2.2.25),

$$\left|K_n^{\infty,\alpha}(x_1,x_2)\right| \le Cn^{1-\alpha}(x_1-x_2)^{-1-\alpha} + C(x_1-x_2)^{-2}$$

and so

$$\begin{aligned}\int_{A_2} \left|K_n^{\infty,\alpha}(x_1,x_2)\right| dx_1\,dx_2 &\le Cn^{1-\alpha}\int_{2/n}^{\pi}\int_0^{1/n}(x_1-1/n)^{-1-\alpha}\,dx_2\,dx_1 \\ &+ C\int_{2/n}^{\pi}\int_0^{1/n}(x_1-1/n)^{-2}\,dx_2\,dx_1 \le C.\end{aligned}$$

Since $x_1-x_2 \ge x_1/2$ on A_3, we get from (2.2.24) that

$$\begin{aligned}\left|K_n^{\infty,\alpha}(x_1,x_2)\right| &\le Cn^{-\alpha}x_1^{-1-\alpha}x_2^{-1} + Cn^{-1}x_1^{-2}x_2^{-1} \\ &\le Cn^{-\alpha}x_1^{-1-\alpha+\beta}x_2^{-1-\beta} + Cn^{-1}x_1^{-2+\beta}x_2^{-1-\beta}\end{aligned} \tag{2.3.1}$$

for any $0<\beta<\alpha$. Thus

$$\begin{aligned}\int_{A_3} \left|K_n^{\infty,\alpha}(x_1,x_2)\right| dx_1\,dx_2 &\le Cn^{-\alpha}\int_{2/n}^{\pi}\int_{1/n}^{x_1/2}x_1^{-1-\alpha+\beta}x_2^{-1-\beta}\,dx_2\,dx_1 \\ &+ Cn^{-1}\int_{2/n}^{\pi}\int_{1/n}^{x_1/2}x_1^{-2+\beta}x_2^{-1-\beta}\,dx_2\,dx_1 \le C.\end{aligned}$$

Since $x_2 > x_1/2$ and $x_2 > x_1-x_2$ on A_4, we get from (2.2.24) that

$$\begin{aligned}\left|K_n^{\infty,\alpha}(x_1,x_2)\right| &\le Cn^{-\alpha}x_1^{-1}x_2^{-1}(x_1-x_2)^{-\alpha} + Cn^{-1}x_1^{-1}x_2^{-1}(x_1-x_2)^{-1} \\ &\le Cn^{-\alpha}x_1^{-1-\beta}(x_1-x_2)^{-1-\alpha+\beta} \\ &\quad + Cn^{-1}x_1^{-1-\beta}(x_1-x_2)^{-2+\beta}\end{aligned} \tag{2.3.2}$$

for any $0<\beta<\alpha$. Then

$$\begin{aligned}&\int_{A_4} \left|K_n^{\infty,\alpha}(x_1,x_2)\right| dx_1\,dx_2 \\ &\le Cn^{-\alpha}\int_{2/n}^{\pi}\int_{x_1/2}^{x_1-1/n}x_1^{-1-\beta}(x_1-x_2)^{-1-\alpha+\beta}\,dx_2\,dx_1 \\ &\quad + Cn^{-1}\int_{2/n}^{\pi}\int_{x_1/2}^{x_1-1/n}x_1^{-1-\beta}(x_1-x_2)^{-2+\beta}\,dx_2\,dx_1 \le C.\end{aligned}$$

Finally, $x_2 > x_1/2$ also on A_5 and so (2.2.23) implies

$$\int_{A_5} \left|K_n^{\infty,\alpha}(x_1,x_2)\right| dx_1\,dx_2 \le C\int_{2/n}^{\pi}\int_{x_1-1/n}^{x_1}x_1^{-2}\,dx_2\,dx_1 \le C,$$

which finishes the proof. ■

Now we continue with the circular Riesz summability.

Theorem 2.3.2 *If $q = 2$, $\alpha > (d-1)/2$ and $\gamma \in \mathbb{P}$, then*

$$\sup_{n\in\mathbb{N}} \int_{\mathbb{T}^d} \left|K_n^{2,\alpha,\gamma}(x)\right| dx \le C.$$

Proof Taking into account Lemma 2.2.31, we can see that

$$\int_{\mathbb{T}^d} |K_n^{2,\alpha,\gamma}(x)|\, dx \le (2\pi)^d n^d \sum_{k\in\mathbb{Z}^d} \int_{\mathbb{T}^d} |\widehat{\theta_0}(n(x+2k\pi))|\, dx = (2\pi)^d \left\|\widehat{\theta_0}\right\|_1.$$

Now the theorem follows easily from Corollary 2.2.28. ■

These imply easily

Theorem 2.3.3 *If $1 \le p < \infty$, $0 < \alpha < \infty$ and $q = 1$ or $q = \infty$, then*

$$\sup_{n\in\mathbb{N}} \left\|\sigma_n^{q,\alpha} f\right\|_p \le C\|f\|_p$$

and

$$\lim_{n\to\infty} \sigma_n^{q,\alpha} f = f \qquad \text{in the } L_p(\mathbb{T}^d)\text{-norm for all } f \in L_p(\mathbb{T}^d).$$

Proof For $0 < \alpha \le 1$, we use Minkowski's inequality and Theorem 2.3.1 to obtain

$$\begin{aligned}\left\|\sigma_n^{q,\alpha} f\right\|_p &\le \frac{1}{(2\pi)^d} \int_{\mathbb{T}^d} \|f(\cdot - t)\|_p K_n^{q,\alpha}(t)\, dt \\ &= \frac{1}{(2\pi)^d} \int_{\mathbb{T}^d} \|f\|_p K_n^{q,\alpha}(t)\, dt \\ &\le C\|f\|_p.\end{aligned}$$

For $1 < \alpha < \infty$, we can use Lemma 2.2.8. The convergence follows easily from this because the trigonometric polynomials are dense in $L_p(\mathbb{T}^d)$. ■

The next theorem can be proved in the same way.

Theorem 2.3.4 *If $1 \le p < \infty$, $q = 2$, $(d-1)/2 < \alpha < \infty$ and $\gamma \in \mathbb{P}$, then*

$$\sup_{n\in\mathbb{N}} \left\|\sigma_n^{q,\alpha,\gamma} f\right\|_p \le C\|f\|_p$$

and

$$\lim_{n\to\infty} \sigma_n^{q,\alpha,\gamma} f = f \qquad \text{in the } L_p(\mathbb{T}^d)\text{-norm for all } f \in L_p(\mathbb{T}^d).$$

Theorems 2.3.3 and 2.3.4 were proved in Berens, Li and Xu [30], Oswald [253] and Weisz [337, 338, 341] for $q = 1, \infty$ and in Bochner [36] and Stein and Weiss [293] for $q = 2$.

The situation is more complicated and not completely solved if $q = 2$ and $\alpha \leq (d-1)/2$. It is clear by the Banach-Steinhaus theorem that $\lim_{n\to\infty} \sigma_n^{q,\alpha,\gamma} f = f$ in the $L_p(\mathbb{T}^d)$-norm for all $f \in L_p(\mathbb{T}^d)$ if and only if the operators $\sigma_n^{q,\alpha,\gamma}$ are uniformly bounded from $L_p(\mathbb{T}^d)$ to $L_p(\mathbb{T}^d)$. We note that each operator $\sigma_n^{q,\alpha,\gamma}$ is bounded on $L_p(\mathbb{T}^d)$ because $K_n^{q,\alpha,\gamma} \in L_1(\mathbb{T}^d)$. For more about the norm convergence of the Bochner-Riesz means (i.e., $q = 2, \gamma = 2$) see Grafakos [143].

2.4 $H_p^\square(\mathbb{T}^d)$ Hardy Spaces

To prove almost everywhere convergence of the Cesàro and Riesz means, we will need the concept of Hardy spaces and their atomic decomposition. Before studying Hardy spaces, we have to introduce the concept of distributions.

Let $C^\infty(\mathbb{T}^d)$ denote the set of all infinitely differentiable functions on $\mathbb{T}^d$. Then $f \in C^\infty(\mathbb{T}^d)$ implies

$$\sup_{x\in\mathbb{T}^d} \left|\partial^k f(x)\right| < \infty \qquad \text{for all } k = (k_1, \ldots, k_d) \in \mathbb{N}^d,$$

where $\partial^k = \partial_1^{k_1} \cdots \partial_d^{k_d}$.

Definition 2.4.1 Let $n \in \mathbb{N}$, $f_n, f \in C^\infty(\mathbb{T}^d)$. We say that

$$\lim_{n\to\infty} f_n = f \qquad \text{in } C^\infty(\mathbb{T}^d)$$

if

$$\lim_{n\to\infty} \left\|\partial^k f_n - \partial^k f\right\|_\infty = 0 \qquad \text{for all } k \in \mathbb{N}^d.$$

Definition 2.4.2 A map $u : C^\infty(\mathbb{T}^d) \to \mathbb{C}$ is called distribution if it is linear and continuous, more exactly,

$$u(\alpha_1 f_1 + \alpha_2 f_2) = \alpha_1 u(f_1) + \alpha_2 u(f_2)$$

for all $f_1, f_2 \in C^\infty(\mathbb{T}^d)$ and $\alpha_1, \alpha_2 \in \mathbb{C}$ and

$$\lim_{n\to\infty} u(f_n) = u(f) \quad \text{if} \quad \lim_{n\to\infty} f_n = f \quad \text{in} \quad C^\infty(\mathbb{T}^d).$$

The set of distributions are denoted by $D(\mathbb{T}^d)$.

If $g \in L_p(\mathbb{T}^d)$ $(1 \leq p \leq \infty)$, then

$$u_g(f) := \frac{1}{(2\pi)^d} \int_{\mathbb{T}^d} fg \, d\lambda \qquad (f \in C^\infty(\mathbb{T}^d))$$

is a distribution. Indeed, if $\lim_{n\to\infty} f_n = f$ in $C^\infty(\mathbb{T}^d)$, then $\lim_{n\to\infty} f_n = f$ in $L_{p'}(\mathbb{T}^d)$ as well. Applying Hölder's inequality,

$$\begin{aligned} \left|u_g(f_n) - u_g(f)\right| &\le \int_{\mathbb{T}^d} |f_n(x) - f(x)| \, |g(x)| \, dx \\ &\le \|f_n - f\|_{p'} \|g\|_p \to 0, \end{aligned}$$

as $n \to \infty$. So every function from $L_p(\mathbb{T}^d)$ $(1 \le p \le \infty)$ can be identified with a distribution $u \in D(\mathbb{T}^d)$ in the previous way.

Proposition 2.4.3 *A linear functional u on $C^\infty(\mathbb{T}^d)$ is a distribution if and only if there exist $C > 0$ and $m \in \mathbb{N}$ such that*

$$|u(f)| \le C \sup_{|k| \le m} \left\|\partial^k f\right\|_\infty$$

for all $f \in C^\infty(\mathbb{T}^d)$.

Proof It is evident that the inequality ensures the continuity of u, thus u is a distribution. Conversely, suppose that u is a distribution and the inequality is not true. Then there exists $f_n \in C^\infty(\mathbb{T}^d)$ such that

$$|u(f_n)| > n \sup_{|k| \le n} \left\|\partial^k f_n\right\|_\infty .$$

Since the right-hand side is not 0, we may define

$$g_n := \frac{f_n}{n \sup_{|k| \le n} \left\|\partial^k f_n\right\|_\infty}.$$

Then $g_n \in C^\infty(\mathbb{T}^d)$ and

$$\sup_{|k| \le n} \left\|\partial^k g_n\right\|_\infty = \frac{1}{n},$$

which means that $g_n \to 0$ in $C^\infty(\mathbb{T}^d)$. On the other hand,

$$u(g_n) = \frac{u(f_n)}{n \sup_{|k| < n} \left\|\partial^k f_n\right\|_\infty} > 1.$$

This contradicts to the continuity of u, i.e., to $u(g_n) \to 0$ as $n \to \infty$. ■

Definition 2.4.4 The least integer m for which Proposition 2.4.3 holds is called the order of u.

Definition 2.4.5 The distributions u_n tend to the distribution u in the sense of distributions or in $D(\mathbb{T}^d)$ if

$$\lim_{n\to\infty} u_n(f) \to u(f) \quad \text{for all} \quad f \in C^\infty(\mathbb{T}^d).$$

The next definition extends the Fourier coefficients to distributions.

Definition 2.4.6 Let

$$e_n(x) := e^{\imath n\cdot x} \qquad (n \in \mathbb{Z}^d, x \in \mathbb{T}^d).$$

For a distribution $u \in D(\mathbb{T}^d)$, the nth Fourier coefficient is defined by

$$\widehat{u}(n) := u(e_{-n}) \qquad (n \in \mathbb{Z}^d).$$

The Fourier series, the partial sums and the summability means of u are defined in the same way as in Definitions 2.1.2, 2.1.3, 2.1.5, 2.2.2 and 2.2.3.

Theorem 2.4.7 *If $u \in C^\infty(\mathbb{T}^d)$ is of order m, then*

$$\widehat{u}(n) = O(|n|^m) \qquad as\ |n| \to \infty. \tag{2.4.1}$$

Moreover, for $1 \le q \le \infty$ and $N \in \mathbb{N}$,

$$s_N^q u = \sum_{n\in\mathbb{Z}^d,\ \|n\|_q\le N} \widehat{u}(n)e_n \to u \qquad in\ D(\mathbb{T}^d)\ as\ N \to \infty.$$

Conversely, if $c_n = O(|n|^m)$, then

$$s_N^q := \sum_{n\in\mathbb{Z}^d,\ \|n\|_q\le N} c_n e_n$$

converge to u in $D(\mathbb{T}^d)$ as $N \to \infty$ and $\widehat{u}(n) = c_n$. The same holds for the rectangular partial sums s_N.

Proof Equality (2.4.1) follows immediately from the inequality of Proposition 2.4.3 if we take therein $f = e_{-k}$. For $f \in C^\infty(\mathbb{T}^d)$,

$$s_N^q u(f) = \sum_{n\in\mathbb{Z}^d,\ \|n\|_q\le N} \widehat{u}(n)\widehat{f}(-n) = u\left(\sum_{n\in\mathbb{Z}^d,\ \|n\|_q\le N} \widehat{f}(-n)e_{-n}\right).$$

It is easy to see that $\widehat{f}(n) = O(|n|^{-k})$ for any $k \in \mathbb{N}$. Hence

$$\lim_{N\to\infty} \sum_{n\in\mathbb{Z}^d,\ \|n\|_q\le N} \widehat{f}(n)e_n = f$$

in $C^\infty(\mathbb{T}^d)$ and so

$$\lim_{N\to\infty} s_N^q u(f) = u(f).$$

Conversely, if $c_n = O(|n|^m)$, then

$$s_N^q(f) = \sum_{n\in\mathbb{Z}^d,\ \|n\|_q\le N} c_n \widehat{f}(-n) \tag{2.4.2}$$

for all $f \in C^\infty(\mathbb{T}^d)$. Since the series on the right-hand side is absolutely convergent, let

$$u(f) := \lim_{N\to\infty} s_N^q(f) = \sum_{n\in\mathbb{Z}^d} c_n \widehat{f}(-n).$$

Then u is linear and we can show easily that u is continuous as well. Writing $f = e_{-n}$ in (2.4.2), we can see that $\widehat{u}(n) = c_n$ $(n \in \mathbb{Z}^d)$. ■

Definition 2.4.8 The convolution of two functions $f, g \in L_1(\mathbb{T}^d)$ is defined by

$$(f * g)(x) := \frac{1}{(2\pi)^d}\int_{\mathbb{T}^d} f(x-t)g(t)\,dt \qquad (x \in \mathbb{T}^d).$$

It is easy to see that

$$(f * g)(x) = \frac{1}{(2\pi)^d}\int_{\mathbb{T}^d} f(t)g(x-t)\,dt \qquad (x \in \mathbb{T}^d).$$

Using Minkowski's inequality, we obtain Young's inequality. More exactly, for $f \in L_r(\mathbb{T}^d)$, $g \in L_1(\mathbb{T}^d)$ and $1 \le r \le \infty$, we have

$$\|f * g\|_r \le \|f\|_r \|g\|_1.$$

Lemma 2.4.9 *If $f, g \in L_1(\mathbb{T}^d)$, then $\widehat{f * g}(n) = \widehat{f}(n)\widehat{g}(n)$*

Proof We have,

$$\begin{aligned}
\widehat{f * g}(n) &= \frac{1}{(2\pi)^{2d}}\int_{\mathbb{T}^d}\left(\int_{\mathbb{T}^d} f(x-t)g(t)\,dt\right) e^{-\imath n\cdot x}\,dx \\
&= \frac{1}{(2\pi)^{2d}}\int_{\mathbb{T}^d}\left(\int_{\mathbb{T}^d} f(x-t)\,e^{-\imath n\cdot(x-t)}\,dx\right) g(t)\,e^{-\imath n\cdot t}\,dt \\
&= \frac{1}{(2\pi)^{2d}}\int_{\mathbb{T}^d}\left(\int_{\mathbb{T}^d} f(u)\,e^{-\imath n\cdot u}\,du\right) g(t)\,e^{-\imath n\cdot t}\,dt \\
&= \widehat{f}(n)\cdot\widehat{g}(n),
\end{aligned}$$

which finishes the proof of the lemma. ■

Now we are able to define the convolution of a distribution and function.

Definition 2.4.10 The convolution of $f \in D(\mathbb{T}^d)$ and $g \in L_1(\mathbb{T}^d)$ is defined by

$$f * g := \sum_{n \in \mathbb{N}^d} \widehat{f}(n)\widehat{g}(n)e_n \qquad \text{in } D(\mathbb{T}^d).$$

Since $\widehat{g}$ is bounded, the series is convergent by Theorem 2.4.7. Similarly, we can also define the convolution $f \in D(\mathbb{T}^d)$ and $\psi \in L_1(\mathbb{R}^d)$.

Definition 2.4.11 For $f \in D(\mathbb{T}^d)$ and $\psi \in L_1(\mathbb{R}^d)$ let

$$f * \psi := \sum_{n \in \mathbb{N}^d} \widehat{f}(n)\widehat{\psi}(n)e_n \qquad \text{in } D(\mathbb{T}^d),$$

where $\widehat{\psi}$ denotes the Fourier transform of $\psi \in L_1(\mathbb{R}^d)$.

Similar to Lemma 2.4.9,

$$f * \psi(x) = \frac{1}{(2\pi)^d} \int_{\mathbb{R}^d} f(x-u)\psi(u)\,du$$

if $f \in D(\mathbb{T}^d)$ and $\psi \in L_1(\mathbb{R}^d)$. For $t \in (0, \infty)$ and $x \in \mathbb{R}^d$, let

$$\psi_t(\xi) := t^{-d}\psi(\xi/t).$$

It is easy to see that for $f \in D(\mathbb{T}^d)$ and $\psi \in L_1(\mathbb{R}^d)$, we have

$$f * \psi_t = \sum_{n \in \mathbb{N}^d} \widehat{f}(n)\widehat{\psi}(tn)e_n \qquad \text{in } D(\mathbb{T}^d). \tag{2.4.3}$$

To define the Hardy spaces, we need the concept of Schwartz functions.

Definition 2.4.12 The function $f \in C^\infty(\mathbb{R}^d)$ is called a Schwartz function if for all $\alpha, \beta \in \mathbb{N}^d$,

$$\sup_{x \in \mathbb{R}^d} \left| x^\alpha \partial^\beta f(x) \right| = C_{\alpha,\beta} < \infty,$$

where $x^\alpha = x_1^{\alpha_1} \cdots x_d^{\alpha_d}$, $\alpha = (\alpha_1, \ldots, \alpha_d)$ and $\beta = (\beta_1, \ldots, \beta_d)$. The set of Schwartz functions are denoted by $S(\mathbb{R}^d)$.

Then $f \in D(\mathbb{T}^d)$ and $\psi \in S(\mathbb{R}^d)$ implies that (2.4.3) converges absolutely in each point as well and so $f * \psi_t \in L_\infty(\mathbb{T}^d)$.

Fix $\psi \in S(\mathbb{R}^d)$ such that $\int_{\mathbb{R}^d} \psi(x)dx \neq 0$. We define the radial maximal function and the non-tangential maximal function of $f \in D(\mathbb{T}^d)$ associated to ψ by

$$\psi^*_{\square,+}(f)(x) := \sup_{t \in (0,\infty)} |f * \psi_t(x)|$$

and

$$\psi^*_{\square,\nabla}(f)(x) := \sup_{t\in(0,\infty),|y-x|<t} |f * \psi_t(y)|,$$

respectively. For $N \in \mathbb{N}$, let

$$\mathcal{F}_N(\mathbb{R}^d) := \left\{ \psi \in S(\mathbb{R}^d) : \sup_{x\in\mathbb{R}^d, \|\beta\|_1 \le N} (1+|x|)^{N+d} |\partial^\beta \psi(x)| \le 1 \right\},$$

where $\|\beta\|_1 = \beta_1 + \cdots + \beta_d$. For any $N \in \mathbb{N}$, the radial grand maximal function and the non-tangential grand maximal function of $f \in D(\mathbb{T}^d)$ are defined by

$$f^*_{\square,+}(x) := \sup_{\psi\in\mathcal{F}_N(\mathbb{R}^d)} \sup_{t\in(0,\infty)} |f * \psi_t(y)|$$

and

$$f^*_{\square,\nabla}(x) := \sup_{\psi\in\mathcal{F}_N(\mathbb{R}^d)} \sup_{t\in(0,\infty),|y-x|<t} |f * \psi_t(y)|,$$

respectively. We fix a positive integer $N > \lfloor d(1/p-1) \rfloor$, where $\lfloor x \rfloor$ denotes the integer part of $x \in \mathbb{R}$.

Definition 2.4.13 For $0 < p < \infty$ the Hardy spaces $H_p^{\square}(\mathbb{T}^d)$ and weak Hardy spaces $H_{p,\infty}^{\square}(\mathbb{T}^d)$ consist of all distributions $f \in D(\mathbb{T}^d)$ for which

$$\|f\|_{H_p^{\square}} := \left\| \psi^*_{\square,+}(f) \right\|_p < \infty$$

and

$$\|f\|_{H_{p,\infty}^{\square}} := \left\| \psi^*_{\square,+}(f) \right\|_{p,\infty} < \infty.$$

We will see in the next theorem that the Hardy spaces are independent of ψ and N, more exactly, different functions ψ and different integers N give the same space with equivalent norms.

The d-dimensional periodic Poisson kernel is introduced by

$$P_t(x) := \sum_{k\in\mathbb{Z}^d} e^{-t\|k\|_2} e^{\imath k\cdot x} \qquad (x \in \mathbb{T}^d, t > 0).$$

Notice that $P_t \in L_1(\mathbb{T}^d)$. In the one-dimensional case, we get back the usual Poisson kernel

$$P_t(x) = \sum_{k=-\infty}^{\infty} r^{|k|} e^{\imath kx} = \frac{1-r^2}{1+r^2-2r\cos x} \qquad (x \in \mathbb{T}),$$

where $r := e^{-t}$. For $f \in D(\mathbb{T}^d)$, let

$$P^*_{\square,+}(f)(x) := \sup_{t\in(0,\infty)} |f * P_t(x)|$$

and

$$P^*_{\square,\nabla}(f)(x) := \sup_{t\in(0,\infty),|y-x|<t} |f * P_t(y)|.$$

Theorem 2.4.14 *Let* $0 < p < \infty$. *Fix* $\psi \in S(\mathbb{R}^d)$ *such that* $\int_{\mathbb{R}^d} \psi(x)dx \neq 0$ *and fix a positive integer* $N > \lfloor d(1/p - 1)\rfloor$. *Then* $f \in H_p^{\square}(\mathbb{T}^d)$ *if and only if* $\psi^*_{\square,\nabla}(f) \in L_p(\mathbb{T}^d)$ *or* $f^*_{\square,+} \in L_p(\mathbb{T}^d)$ *or* $f^*_{\square,\nabla} \in L_p(\mathbb{T}^d)$ *or* $P^*_{\square,+}(f) \in L_p(\mathbb{T}^d)$ *or* $P^*_{\square,\nabla}(f) \in L_p(\mathbb{T}^d)$. *We have the following equivalences of norms:*

$$\|f\|_{H_p^{\square}} \sim \|\psi^*_{\square,\nabla}(f)\|_p \sim \|f^*_{\square,+}\|_p \sim \|f^*_{\square,\nabla}\|_p \sim \|P^*_{\square,+}(f)\|_p \sim \|P^*_{\square,\nabla}(f)\|_p.$$

The same holds for the weak Hardy spaces:

$$\|f\|_{H^{\square}_{p,\infty}} \sim \|\psi^*_{\square,\nabla}(f)\|_{p,\infty} \sim \|f^*_{\square,+}\|_{p,\infty}$$
$$\sim \|f^*_{\square,\nabla}\|_{p,\infty} \sim \|P^*_{\square,+}(f)\|_{p,\infty} \sim \|P^*_{\square,\nabla}(f)\|_{p,\infty}.$$

Note that $\sim$ denotes the equivalence of norms and spaces, more exactly we write that $A \sim B$ if there exist positive constants c_1 and c_2 such that $c_1 A \le B \le c_2 A$.

Theorem 2.4.15 *If* $1 < p < \infty$, *then* $H_p^{\square}(\mathbb{T}^d) \sim L_p(\mathbb{T}^d)$ *and*

$$\|f\|_p \le \|f\|_{H_p^{\square}} \le C_p \|f\|_p.$$

For $p = 1$, $H_1^{\square}(\mathbb{T}^d) \subset L_1(\mathbb{T}^d) \subset H^{\square}_{1,\infty}(\mathbb{T}^d)$ *and*

$$\|f\|_1 \le \|f\|_{H_1^{\square}} \qquad (f \in H_1^{\square}(\mathbb{T}^d)),$$

$$\|f\|_{H^{\square}_{1,\infty}} \le C\|f\|_1 \qquad (f \in L_1(\mathbb{T}^d)).$$

We omit the proofs of these theorems because they are very similar to the proofs of the corresponding theorems for $H_p(\mathbb{R}^d)$, which can be found in several books and papers (e.g., in Stein [290], Grafakos [143], Lu [224], Stein [289], Stein and Weiss [293], Uchiyama [320], Fefferman and Stein [96], Weisz [346]).

We define the reflection and translation operators by

$$\breve{h}(x) := h(-x), \qquad T_x h(t) := h(t - x).$$

Theorem 2.4.16 *If* $K \in L_1(\mathbb{T}^d)$, $0 < p < \infty$ *and*

$$\lim_{k\to\infty} f_k = f \quad \textit{in the } H_p^{\square}(\mathbb{T}^d)\textit{-norm},$$

then

$$\lim_{k\to\infty} f_k * K = f * K \quad \textit{in } D(\mathbb{T}^d).$$

Proof Observe that for $f \in H_p(\mathbb{T}^d)$ and $h \in C^\infty(\mathbb{T}^d)$,

$$\begin{aligned} f * h(x) &= \sum_{n\in\mathbb{N}^d} \widehat{f}(n)\widehat{h}(n)e_n(x) \\ &= \sum_{n\in\mathbb{N}^d} \widehat{f}(n)\widehat{T_{-x}h}(n) \\ &= \sum_{n\in\mathbb{N}^d} \widehat{f}(n)e_n(T_x\breve{h}) \\ &= f(T_x\breve{h}), \end{aligned}$$

Thus

$$\left|f(\breve{h})\right| = |f * h(0)| \le \left(\sup_{x\in\mathbb{T}^d, \|\beta\|_1\le N} (1+|x|)^{N+d}|\partial^\beta h(x)|\right) f_\triangledown^*(y),$$

where $|y| < 1$ and $N > \lfloor d(1/p-1)\rfloor$. Then

$$\begin{aligned} \left|f(\breve{h})\right| &\le C\left(\sup_{x\in\mathbb{T}^d, \|\beta\|_1\le N} |\partial^\beta h(x)|\right) \inf_{|y|<1} f_\triangledown^*(y) \\ &\le C\left(\sup_{x\in\mathbb{T}^d, \|\beta\|_1\le N} |\partial^\beta h(x)|\right)\left(\int_{\mathbb{T}^d} f_\triangledown^*(y)^p\,dy\right)^{1/p} \\ &\le C\left(\sup_{\|\beta\|_1\le N} |\partial^\beta \breve{h}|\right)\|f\|_{H_p}, \end{aligned}$$

which implies that the order of f is at most N and that

$$\lim_{k\to\infty} f_k = f \quad \text{in } D(\mathbb{T}^d).$$

By Theorem 2.4.7 and by the definition of the convolution,

$$\begin{aligned} (f_k - f) * K(h) &= \sum_{n\in\mathbb{N}^d} \left(\widehat{f_k} - \widehat{f}\right)(n)\widehat{K}(n)e_n(h) \\ &= \sum_{n\in\mathbb{N}^d} \left(\widehat{f_k} - \widehat{f}\right)(n)\widehat{K}(n)\widehat{h}(-n), \end{aligned}$$

where $h \in C^\infty(\mathbb{T}^d)$ is arbitrary. Observe that the orders of f_k and f are at most N, $\widehat{K}$ is bounded and $\left|\widehat{h}(n)\right| \le C|n|^{-l}$ for any $l \in \mathbb{N}$. Then for all $\epsilon > 0$ there exists

$m \in \mathbb{N}^d$ such that

$$\left| \left(\sum_{n\in\mathbb{N}^d} - \sum_{|n|\le m} \right) \left(\widehat{f_k} - \widehat{f}\right)(n)\widehat{K}(n)\widehat{h}(-n) \right| \le \epsilon.$$

On the other hand, since

$$\lim_{k\to\infty} (f_k - f)(e_{-n}) = 0,$$

we conclude that

$$\left| \sum_{|n|\le m} \left(\widehat{f_k} - \widehat{f}\right)(n)\widehat{K}(n)\widehat{h}(-n) \right| \le \sum_{|n|\le m} \left| \left(\widehat{f_k} - \widehat{f}\right)(e_{-n}) \right| \to 0$$

as $k \to \infty$, which finishes the proof. ∎

The atomic decomposition provides a useful characterization of Hardy spaces. First, we introduce the concept of an atom.

Definition 2.4.17 A bounded function a is an $H_p^\square$-atom if there exists a cube $I \subset \mathbb{T}^d$ such that

(i) $\operatorname{supp} a \subset I$,
(ii) $\|a\|_\infty \le |I|^{-1/p}$,
(iii) $\int_I a(x)x^k\,dx = 0$ for all multi-indices $k = (k_1, \ldots, k_d)$ with $|k| \le \lfloor d(1/p - 1)\rfloor$.

In the definition, the cubes can be replaced by balls and (ii) by

(ii') $\|a\|_q \le |I|^{1/q-1/p}$ $\qquad (0 < p < q \le \infty, q > 1)$.

We could suppose that the integral in (iii) is zero for all multi-indices k for which $|k| \le N$, where $N \ge \lfloor d(1/p - 1)\rfloor$. The best possible choice of such numbers N is $\lfloor d(1/p - 1)\rfloor$. Hardy spaces have atomic decompositions. In other words, every function from the Hardy space can be decomposed into the sum of atoms (see e.g. Latter [195], Lu [224], Coifman and Weiss [62], Wilson [353, 354], Stein [290], Grafakos [143] and Weisz [346]).

Theorem 2.4.18 *A distribution $f \in D(\mathbb{T}^d)$ is in $H_p^\square(\mathbb{T}^d)$ $(0 < p \le 1)$ if and only if there exist a sequence $(a_k, k \in \mathbb{N})$ of $H_p^\square$-atoms and a sequence $(\mu_k, k \in \mathbb{N})$ of real numbers such that*

$$\sum_{k=0}^{\infty} |\mu_k|^p < \infty \quad \text{and} \quad \sum_{k=0}^{\infty} \mu_k a_k = f \quad \text{in } D(\mathbb{T}^d). \tag{2.4.4}$$

Moreover,

$$\|f\|_{H_p^\square} \sim \inf \left(\sum_{k=0}^{\infty} |\mu_k|^p \right)^{1/p},$$

where the infimum is taken over all decompositions of f of the form (2.4.4).

The following result gives a sufficient condition for an operator to be bounded from $H_p^\square(\mathbb{T}^d)$ to $L_p(\mathbb{T}^d)$. If $I \subset \mathbb{T}$ is an interval, then rI denotes the interval with the same center as I and with length $r|I|$ ($r \in \mathbb{N}$). For a rectangle

$$R = I_1 \times \cdots \times I_d \subset \mathbb{T}^d, \qquad \text{let} \qquad rR = rI_1 \times \cdots \times rI_d.$$

Instead of $2^r R$ we write R^r ($r \in \mathbb{N}$). For operators $V_n : L_1(\mathbb{T}^d) \to L_1(\mathbb{T}^d)$, we define the maximal operator

$$V_* f := \sup_{n \in \mathbb{N}^d} |V_n f|.$$

Theorem 2.4.19 *For each $n \in \mathbb{N}^d$, let $K_n \in L_1(\mathbb{T}^d)$ and $V_n f := f * K_n$. Suppose that*

$$\int_{\mathbb{T}^d \setminus rI} |V_* a|^{p_0} \, d\lambda \le C_{p_0}$$

for all $H_{p_0}^\square$-atoms a and for some fixed $r \in \mathbb{N}$ and $0 < p_0 \le 1$, where the cube I is the support of the atom. If V_* is bounded from $L_{p_1}(\mathbb{T}^d)$ to $L_{p_1}(\mathbb{T}^d)$ for some $1 < p_1 \le \infty$, then

$$\|V_* f\|_p \le C_p \|f\|_{H_p^\square} \qquad (f \in H_p^\square(\mathbb{T}^d)) \tag{2.4.5}$$

for all $p_0 \le p \le p_1$.

Proof Observe that, under the conditions of Theorem 2.4.19, the L_{p_0}-norms of $V_* a$ are uniformly bounded for all $H_{p_0}^\square$-atoms a. Indeed,

$$\begin{aligned}
\int_{\mathbb{T}^d} |V_* a|^{p_0} \, d\lambda &= \int_{rI} |V_* a|^{p_0} \, d\lambda + \int_{\mathbb{T}^d \setminus rI} |V_* a|^{p_0} \, d\lambda \\
&\le \left(\int_{rI} |V_* a|^{p_1} \, d\lambda \right)^{p_0/p_1} |rI|^{1-p_0/p_1} + C_{p_0} \\
&\le C_{p_0} \left(\int_{rI} |a|^{p_1} \, d\lambda \right)^{p_0/p_1} |I|^{1-p_0/p_1} + C_{p_0} \\
&\le C_{p_0} \left(|I|^{-p_1/p_0} |I^r| \right)^{p_0/p_1} |I|^{1-p_0/p_1} + C_{p_0} \\
&= C_{p_0}.
\end{aligned}$$

There is an atomic decomposition such that

$$f = \sum_{k=0}^{\infty} \mu_k a_k \quad \text{in the } H_{p_0}^{\square}\text{-norm} \quad \text{and} \quad \left(\sum_{k=0}^{\infty} |\mu_k|^{p_0} \right)^{1/p_0} \le C_{p_0} \|f\|_{H_{p_0}^{\square}},$$

where the convergence holds also in the $H_1^{\square}(\mathbb{T}^d)$-norm and in the $L_1(\mathbb{T}^d)$-norm if $f \in H_1^{\square}(\mathbb{T}^d)$. Since $V_n : L_1(\mathbb{T}^d) \to L_1(\mathbb{T}^d)$ is bounded, we have

$$V_n f = \sum_{k=0}^{\infty} \mu_k V_n a_k$$

and

$$|V_* f| \le \sum_{k=0}^{\infty} |\mu_k| |V_* a_k|$$

for $f \in H_1^{\square}(\mathbb{T}^d)$. Thus

$$\|V_* f\|_{p_0}^{p_0} \le \sum_{k=0}^{\infty} |\mu_k|^{p_0} \|V_* a_k\|_{p_0}^{p_0} \le C_{p_0} \|f\|_{H_{p_0}^{\square}}^{p_0} \qquad (f \in H_1^{\square}(\mathbb{T}^d)). \tag{2.4.6}$$

Obviously, the same inequality holds for the operators V_n. This and interpolation proves the theorem if $p_0 = 1$. Assume that $p_0 < 1$. Since $H_1^{\square}(\mathbb{T}^d)$ is dense in $L_1(\mathbb{T}^d)$ as well as in $H_{p_0}^{\square}(\mathbb{T}^d)$, we can extend uniquely the operators V_n and V_* such that (2.4.6) holds for all $f \in H_{p_0}^{\square}(\mathbb{T}^d)$. Let us denote these extended operators by V_n' and V_*'. Then $V_n f = V_n' f$ and $V_* f = V_*' f$ for all $f \in H_1^{\square}(\mathbb{T}^d)$. We get by interpolation from (2.4.6) that the operator

$$V_*' \quad \text{is bounded from} \quad H_{p,\infty}^{\square}(\mathbb{T}^d) \quad \text{to} \quad L_{p,\infty}(\mathbb{T}^d) \tag{2.4.7}$$

when $p_0 < p < p_1$. For the basic definitions and theorems on interpolation theory, see Bergh and Löfström [33], Bennett and Sharpley [28] or Weisz [346]. Since $p_0 < 1$, the boundedness in (2.4.7) holds especially for $p = 1$, and so Theorem 2.4.15 implies that V_*' is of weak type (1, 1):

$$\sup_{\rho>0} \rho \lambda(|V_*' f| > \rho) = \|V_*' f\|_{1,\infty} \le C \|f\|_{H_{1,\infty}^{\square}} \le C \|f\|_1 \tag{2.4.8}$$

for all $f \in L_1(\mathbb{R}^d)$. Obviously, the same holds for V_n'. Since V_n is bounded on $L_1(\mathbb{T}^d)$, if $f_k \in H_1^{\square}(\mathbb{T}^d)$ such that $\lim_{k\to\infty} f_k = f$ in the L_1-norm, then

$$\lim_{k\to\infty} V_n f_k = V_n f \qquad \text{in the } L_1(\mathbb{T}^d)\text{-norm.}$$

Inequality (2.4.8) implies that

$$\lim_{k\to\infty} V_n f_k = V_n' f \qquad \text{in the } L_{1,\infty}(\mathbb{T}^d)\text{-norm},$$

hence $V_n f = V_n' f$ for all $f \in L_1(\mathbb{T}^d)$. Similarly, for a fixed $N \in \mathbb{N}$, the operator

$$V_{N,*}f := \sup_{|n|\le N} |V_n f|$$

satisfies (2.4.8) for all $f \in H_1^\square(\mathbb{T}^d)$ and its extension $V_{N,*}'$ for all $f \in L_1(\mathbb{T}^d)$. Then

$$\begin{aligned}
\sup_{\rho>0} \rho\,\lambda(|V_{N,*}'f - V_{N,*}f| > \rho) &\le \sup_{\rho>0} \rho\,\lambda(|V_{N,*}'f - V_{N,*}'f_k| > \rho/2) \\
&\quad + \sup_{\rho>0} \rho\,\lambda(|V_{N,*}f_k - V_{N,*}f| > \rho/2) \\
&\le \sup_{\rho>0} \rho\,\lambda(|V_{N,*}'(f - f_k)| > \rho/2) \\
&\quad + \sum_{n=0}^{N} \sup_{\rho>0} \rho\,\lambda(|V_n(f_k - f)| > \rho/2N) \\
&\le C\,\|f - f_k\| \to 0
\end{aligned}$$

as $k \to \infty$. This shows the equality

$$V_{N,*}'f = V_{N,*}f \qquad \text{for all } f \in L_1(\mathbb{T}^d).$$

Moreover, for a fixed ρ,

$$\begin{aligned}
&\lambda(|V_*'f - V_{N,*}f| > \rho) \\
&\quad \le \lambda(|V_*'f - V_*'f_k| > \rho/3) + \lambda(|V_*f_k - V_{N,*}f_k| > \rho/3) \\
&\qquad + \lambda(|V_{N,*}f_k - V_{N,*}f| > \rho/3) \\
&\quad \le \lambda(V_*'(f - f_k) > \rho/3) + \lambda(V_*f_k - V_{N,*}f_k > \rho/3) \\
&\qquad + \lambda(V_{N,*}(f_k - f) > \rho/3) \\
&\quad \le \frac{C}{\rho}\|f - f_k\|_1 + \lambda(V_*f_k - V_{N,*}f_k > \rho/3) \\
&\quad < \epsilon
\end{aligned}$$

if k and N are large enough. Hence $\lim_{N\to\infty} V_{N,*}f = V_*'f$ in measure for all $f \in L_1(\mathbb{T}^d)$. On the other hand, $\lim_{N\to\infty} V_{N,*}f = V_*f$ a.e., which implies that

$$V_*f = V_*'f \quad \text{for all} \quad f \in L_1(\mathbb{T}^d).$$

Consequently, (2.4.8) holds also for V_* and (2.4.6) for all $f \in H_{p_0}^\square(\mathbb{T}^d) \cap L_1(\mathbb{T}^d)$.

Assume that $p < 1$, $f_k \in H_p^\square(\mathbb{T}^d) \cap L_1(\mathbb{T}^d)$ $(k \in \mathbb{N})$ and that $\lim_{k\to\infty} f_k = f$ in the $H_p^\square(\mathbb{T}^d)$-norm. By Theorem 2.4.16,

$$\lim_{k\to\infty} V_n f_k = V_n f \qquad \text{in } D(\mathbb{T}^d)$$

for all $n \in \mathbb{N}^d$. Since by (2.4.5), $V_n f_k$ is convergent in the $L_p(\mathbb{T}^d)$-norm as $k \to \infty$, we can identify the distribution $V_n f$ with the $L_p(\mathbb{T}^d)$-limit $\lim_{k\to\infty} V_n f_k$. Hence the same holds for $V_{N,*} f$:

$$V_{N,*} f = \lim_{k\to\infty} V_{N,*} f_k \qquad \text{in the } L_p(\mathbb{T}^d)\text{-norm.}$$

Moreover,

$$\begin{aligned}
&\|V_*' f - V_{N,*} f\|_p \\
&\quad \le \|V_*' f - V_*' f_k\|_p + \|V_* f_k - V_{N,*} f_k\|_p + \|V_{N,*} f_k - V_{N,*} f\|_p \\
&\quad \le C_p \|f - f_k\|_{H_p^\square} + \|V_* f_k - V_{N,*} f_k\|_p + \|V_{N,*} f_k - V_{N,*} f\|_p \\
&\quad < \epsilon
\end{aligned}$$

if k and N are large enough. Thus

$$\lim_{N\to\infty} V_{N,*} f = V_*' f \qquad \text{in the } L_p(\mathbb{T}^d)\text{-norm}$$

and, on the other hand,

$$\lim_{N\to\infty} V_{N,*} f = V_* f \qquad \text{a.e.,}$$

which implies that $V_* f = V_*' f$ for all $f \in H_p^\square(\mathbb{T}^d)$. Consequently, (2.4.5) holds for all $f \in H_p^\square(\mathbb{T}^d)$. ∎

Unfortunately, for a general linear operator V, the uniform boundedness of the L_{p_0}-norms of Va is not enough for the boundedness $V : H_{p_0}^\square(\mathbb{T}^d) \to L_{p_0}(\mathbb{T}^d)$ (see [41, 42, 235, 236, 259]). The next weak version of Theorem 2.4.19 can be proved similarly (see also the proof in Weisz [346]).

Theorem 2.4.20 *For each $n \in \mathbb{N}^d$, let $K_n \in L_1(\mathbb{T}^d)$ and $V_n f := f * K_n$. Suppose that*

$$\sup_{\rho>0} \rho^p \lambda\Big(\{|V_* a| > \rho\} \cap \{\mathbb{T}^d \setminus rI\}\Big) \le C_p$$

for all $H_p^\square$-atoms a and for some fixed $r \in \mathbb{N}$ and $0 < p < 1$. If V_* is bounded from $L_{p_1}(\mathbb{T}^d)$ to $L_{p_1}(\mathbb{T}^d)$ $(1 < p_1 \le \infty)$, then

$$\|V_* f\|_{p,\infty} \le C_p \|f\|_{H_p^\square} \qquad (f \in H_p^\square(\mathbb{T}^d)).$$

The weak type (1, 1) inequality follows from inequality (2.4.8).

Corollary 2.4.21 *For each $n \in \mathbb{N}^d$, let $K_n \in L_1(\mathbb{T}^d)$ and $V_n f := f * K_n$. Suppose that*

$$\int_{\mathbb{T}^d \setminus rI} |V_* a|^{p_0} \, d\lambda \leq C_{p_0}$$

for all $H_{p_0}^\square$-atoms a and for some fixed $r \in \mathbb{N}$ and $0 < p_0 < 1$, where the cube I is the support of the atom. If V_* is bounded from $L_{p_1}(\mathbb{T}^d)$ to $L_{p_1}(\mathbb{T}^d)$ for some $1 < p_1 \leq \infty$, then for all $f \in L_1(\mathbb{T}^d)$,

$$\sup_{\rho>0} \rho \, \lambda(|V_* f| > \rho) \leq C \|f\|_1.$$

Proof By Theorem 2.4.19 and interpolation,

$$V_* \quad \text{is bounded from} \quad H_{p,\infty}^\square(\mathbb{T}^d) \quad \text{to} \quad L_{p,\infty}(\mathbb{T}^d)$$

when $p_0 < p < p_1$. Since $p_0 < 1$, this holds also for $p = 1$. Thus, by Theorem 2.4.15:

$$\sup_{\rho>0} \rho \, \lambda(|V_* f| > \rho) = \|V_* f\|_{1,\infty} \leq C \|f\|_{H_{1,\infty}^\square} \leq C \|f\|_1$$

for all $f \in L_1(\mathbb{T}^d)$. ■

Theorem 2.4.19 and Corollary 2.4.21 can be regarded also as an alternative tool to the Calderon-Zygmund decomposition lemma for proving weak type $(1, 1)$ inequalities. In many cases, this method can be applied better and more simply than the Calderon-Zygmund decomposition lemma.

2.5 Almost Everywhere Convergence of the ℓ_q-Summability Means

Since the kernels $K_n^{q,\alpha}$ and $K_n^{q,\alpha,\gamma}$ are integrable, the definition of the Fejér and Riesz means can be extended to distributions.

Definition 2.5.1 Let $f \in D(\mathbb{T}^d)$, $1 \leq q \leq \infty$, $n \in \mathbb{N}$ and $0 \leq \alpha, \gamma < \infty$. The nth ℓ_q-Cesàro means $\sigma_n^{q,\alpha} f$ and ℓ_q-Riesz means $\sigma_n^{q,\alpha,\gamma} f$ of the Fourier series of f are given by

$$\sigma_n^{q,\alpha} f := f * K_n^{q,\alpha}$$

and

$$\sigma_n^{q,\alpha,\gamma} f := f * K_n^{q,\alpha,\gamma},$$

respectively.

Definition 2.5.2 We define the maximal Cesàro and maximal Riesz operator by

$$\sigma_*^{q,\alpha} f := \sup_{n\in\mathbb{N}} |\sigma_n^{q,\alpha} f|$$

and

$$\sigma_*^{q,\alpha,\gamma} f := \sup_{n\in\mathbb{N}} |\sigma_n^{q,\alpha,\gamma} f|,$$

respectively.

If $\alpha = 1$, we obtain the maximal Fejér operator and write it simply as $\sigma_*^q f := \sigma_*^{q,1} f$. We will prove that the Cesàro and Riesz maximal operators, $\sigma_*^{q,\alpha} f$ and $\sigma_*^{q,\alpha,\gamma} f$ are bounded from the Hardy space $H_p^\square(\mathbb{T}^d)$ to the Lebesgue space $L_p(\mathbb{T}^d)$ when $q = 1, 2$ or $q = \infty$ and p is greater than a critical index $p_0 < 1$ which is depending on q, d and α. If p is equal to this critical index, then weak type inequality holds. As a consequence, we obtain the almost everywhere convergence of the ℓ_q-Cesàro and Riesz means to the original function. We start again with the triangular and cubic Cesàro summability.

2.5.1 Almost Everywhere Convergence for $q = 1$ and $q = \infty$

Proposition 2.5.3 *If $0 < \alpha \le 1$ and $q = 1$ or $q = \infty$, then*

$$\left\|\sigma_*^{q,\alpha} f\right\|_\infty \le C \left\|f\right\|_\infty \qquad (f \in L_\infty(\mathbb{T}^d)).$$

Proof The proof follows easily from the fact that the $L_1(\mathbb{T}^d)$-norms of the kernel functions are uniformly bounded (see Theorem 2.3.1) and from Lemma 2.2.8. ■

In what follows we use the notation $a \wedge b := \min(a, b)$.

Theorem 2.5.4 *Suppose that $q = 1, \infty$ and $0 < \alpha < \infty$. If*

$$p_0 := \frac{d}{d + \alpha \wedge 1} < p \le \infty,$$

then

$$\left\|\sigma_*^{q,\alpha} f\right\|_p \le C_p \left\|f\right\|_{H_p^\square} \qquad (f \in H_p^\square(\mathbb{T}^d)). \tag{2.5.1}$$

Corollary 2.5.5 *If $q = 1, \infty$, $0 < \alpha < \infty$ and $1 < p < \infty$, then*

$$\|\sigma_*^{q,\alpha} f\|_p \le C_p \|f\|_p \qquad (f \in L_p(\mathbb{T}^d)).$$

Proof This follows from Theorem 2.5.4 and from the fact that $H_p^\square(\mathbb{T}^d) \sim L_p(\mathbb{T}^d)$ for $1 < p \le \infty$. ■

***Proof of Theorem 2.5.4* for** $q = 1$. By Lemma 2.2.8, we may suppose again that $0 < \alpha \le 1$. It is enough to show that

$$\begin{aligned}\int_{\mathbb{T}^2} &\left|\sigma_*^{1,\alpha} a(x_1, x_2)\right|^p \, dx_1 \, dx_2 \\ &= \int_{\mathbb{T}^2} \sup_{n \ge 1} \left| \int_I a(t_1, t_2) K_n^{1,\alpha}(x_1 - t_1, x_2 - t_2) \, dt_1 \, dt_2 \right|^p \, dx_1 \, dx_2 \\ &\le C_p \end{aligned} \tag{2.5.2}$$

for every $H_p^\square$-atom a, where $2/(2+\alpha) < p < 1$ and I is the support of the atom. By Theorem 2.4.19 and Proposition 2.5.3, this will imply (2.5.1). Without loss of generality, we can suppose that a is a $H_p^\square$-atom with support $I = I_1 \times I_2$ and

$$[-2^{-K-2}, 2^{-K-2}] \subset I_j \subset [-2^{-K-1}, 2^{-K-1}] \qquad (j = 1, 2)$$

for some $K \in \mathbb{N}$. By symmetry, we may assume that $\pi > x_1 - t_1 > x_2 - t_2 > 0$, and so, instead of (2.5.2), it is enough to show that

$$\begin{aligned}\int_{\mathbb{T}^2} \sup_{n \ge 1} &\left| \int_I a(t_1, t_2) K_n^{1,\alpha}(x_1 - t_1, x_2 - t_2) 1_{A_i}(x_1 - t_1, x_2 - t_2) \, dt_1 \, dt_2 \right|^p \, dx_1 \, dx_2 \\ &\le C_p\end{aligned}$$

for all $i = 1, \ldots, 10$, where

$$\begin{aligned}
A_1 &:= \{(x_1, x_2) : 0 < x_1 \le 2^{-K+5}, 0 < x_2 < x_1 < \pi, x_2 \le \pi/2\}, \\
A_2 &:= \{(x_1, x_2) : 2^{-K+5} < x_1 < \pi, 0 < x_2 \le 2^{-K+2}, x_2 \le \pi/2\}, \\
A_3 &:= \{(x_1, x_2) : 2^{-K+5} < x_1 < \pi, 2^{-K+2} < x_2 \le x_1/2, x_2 \le \pi/2\}, \\
A_4 &:= \{(x_1, x_2) : 2^{-K+5} < x_1 < \pi, x_1/2 < x_2 \le x_1 - 2^{-K+2}, x_2 \le \pi/2\}, \\
A_5 &:= \{(x_1, x_2) : 2^{-K+5} < x_1 < \pi, x_1 - 2^{-K+2} < x_2 < x_1, x_2 \le \pi/2\} \\
A_6 &:= \{(x_1, x_2) : x_2 > \pi/2, \pi - 2^{-K+5} \le x_2 < \pi, 0 < x_2 < x_1 < \pi\}, \\
A_7 &:= \{(x_1, x_2) : \pi/2 < x_2 < \pi - 2^{-K+5}, \pi - 2^{-K+2} < x_1 < \pi\}, \\
A_8 &:= \{(x_1, x_2) : \pi/2 < x_2 < \pi - 2^{-K+5}, (\pi + x_2)/2 < x_1 \le \pi - 2^{-K+2}\}, \\
A_9 &:= \{(x_1, x_2) : \pi/2 < x_2 < \pi - 2^{-K+5}, x_2 + 2^{-K+2} < x_1 \le (\pi + x_2)/2\}, \\
A_{10} &:= \{(x_1, x_2) : \pi/2 < x_2 < \pi - 2^{-K+5}, x_2 < x_1 \le x_2 + 2^{-K+2}\}.
\end{aligned}$$

These sets are similar to those in Theorem 2.3.1 (see Fig. 2.11). If $0 < x_1 - t_1 \le 2^{-K+5}$, then $-2^{-K-1} < x_1 \le 2^{-K+6}$ and the same holds for x_2. If $\pi - 2^{-K+5} \le x_2 - t_2 < \pi$, then $\pi - 2^{-K+6} < x_2 \le \pi + 2^{-K-1}$ and the same is true for x_1. By the definition of the $H_p^\square$-atom and by Theorem 2.3.1,

$$\begin{aligned}
&\int_{\mathbb{T}^2} \sup_{n\ge 1} \left| \int_I a(t_1,t_2) K_n^{1,\alpha}(x_1-t_1,x_2-t_2) 1_{A_1}(x_1-t_1,x_2-t_2)\, dt_1\, dt_2 \right|^p dx_1\, dx_2 \\
&\le 2^{2K} \int_{\mathbb{T}^2} \sup_{n\ge 1} \left| \int_I \left|K_n^{1,\alpha}(x_1-t_1,x_2-t_2)\right| 1_{A_1}(x_1-t_1,x_2-t_2)\, dt_1\, dt_2 \right|^p dx_1\, dx_2 \\
&\le C_p 2^{2K} 2^{-2K} \le C_p
\end{aligned}$$

and

$$\begin{aligned}
&\int_{\mathbb{T}^2} \sup_{n\ge 1} \left| \int_I a(t_1,t_2) K_n^{1,\alpha}(x_1-t_1,x_2-t_2) 1_{A_6}(x_1-t_1,x_2-t_2)\, dt_1\, dt_2 \right|^p dx_1\, dx_2 \\
&\qquad \le C_p 2^{2K} 2^{-2K} \le C_p.
\end{aligned}$$

On the set A_2, we have $2^{-K+5} < x_1 - t_1 < \pi$ and $0 < x_2 - t_2 \le 2^{-K+2}$, thus

$$2^{-K+4} < x_1 < \pi + 2^{-K-1} \qquad \text{and} \qquad -2^{-K-1} < x_2 \le 2^{-K+3}.$$

Using (2.2.14), we conclude

$$\begin{aligned}
&\left| \int_I a(t_1,t_2) K_n^{1,\alpha}(x_1-t_1,x_2-t_2) 1_{A_2}(x_1-t_1,x_2-t_2)\, dt_1\, dt_2 \right| \\
&\quad \le C_p 2^{2K/p} \int_I (x_1-t_1-x_2+t_2)^{-3/2} (x_2-t_2)^{-1/2} \\
&\qquad 1_{A_2}(x_1-t_1,x_2-t_2)\, dt_1\, dt_2 \\
&\quad \le C_p 2^{2K/p} 1_{\{2^{-K+4}<x_1<\pi+2^{-K-1}\}} 1_{\{-2^{-K-1}<x_2\le 2^{-K+3}\}} \\
&\qquad \int_I (x_1-2^{-K+3})^{-3/2} (x_2-t_2)^{-1/2}\, dt_1\, dt_2 \\
&\quad \le C_p 2^{2K/p-3K/2} 1_{\{2^{-K+4}<x_1<\pi+2^{-K-1}\}} \\
&\qquad 1_{\{-2^{-K-1}<x_2\le 2^{-K+3}\}} (x_1-2^{-K+3})^{-3/2}
\end{aligned} \tag{2.5.3}$$

and

$$\begin{aligned}
&\int_{\mathbb{T}^2} \sup_{n\ge 1} \left| \int_I a(t_1,t_2) K_n^{1,\alpha}(x_1-t_1,x_2-t_2) 1_{A_2}(x_1-t_1,x_2-t_2)\, dt_1\, dt_2 \right|^p dx_1\, dx_2 \\
&\quad \le C_p 2^{2K-3Kp/2} \int_{2^{-K+4}}^{\pi+2^{-K-1}} \int_{-2^{-K-1}}^{2^{-K+3}} (x_1-2^{-K+3})^{-3p/2}\, dx_1\, dx_2 \\
&\quad \le C_p.
\end{aligned}$$

Here we have used that $p > 2/3$. Similarly, on A_7, $\pi/2 < x_2 - t_2 < \pi - 2^{-K+5}$ and $\pi - 2^{-K+2} < x_1 - t_1 < \pi$, thus

$$\pi/2 - 2^{-K-1} < x_2 < \pi - 2^{-K+4} \qquad \text{and} \qquad \pi - 2^{-K+3} < x_1 < \pi + 2^{-K-1}.$$

By (2.2.14),

$$\left| \int_I a(t_1, t_2) K_n^{1,\alpha}(x_1 - t_1, x_2 - t_2) 1_{A_7}(x_1 - t_1, x_2 - t_2)\, dt_1\, dt_2 \right|$$
$$\leq C_p 2^{2K/p} \int_I (x_1 - t_1 - x_2 + t_2)^{-3/2} (\pi - x_1 + t_1)^{-1/2}$$
$$1_{A_7}(x_1 - t_1, x_2 - t_2)\, dt_1\, dt_2$$
$$\leq C_p 2^{2K/p} 1_{\{\pi/2 - 2^{-K-1} < x_2 < \pi - 2^{-K+4}\}} 1_{\{\pi - 2^{-K+3} < x_1 < \pi + 2^{-K-1}\}}$$
$$\int_I (\pi - 2^{-K+3} - x_2)^{-3/2} (\pi - x_1 + t_1)^{-1/2}\, dt_1\, dt_2$$
$$\leq C_p 2^{2K/p - 3K/2} 1_{\{\pi/2 - 2^{-K-1} < x_2 < \pi - 2^{-K+4}\}}$$
$$1_{\{\pi - 2^{-K+3} < x_1 < \pi + 2^{-K-1}\}} (\pi - 2^{-K+3} - x_2)^{-3/2}$$

and

$$\int_{\mathbb{T}^2} \sup_{n \geq 1} \left| \int_I a(t_1, t_2) K_n^{1,\alpha}(x_1 - t_1, x_2 - t_2) 1_{A_7}(x_1 - t_1, x_2 - t_2)\, dt_1\, dt_2 \right|^p dx_1\, dx_2$$
$$\leq C_p 2^{2K - 3Kp/2} \int_{\pi/2 - 2^{-K-1}}^{\pi - 2^{-K+4}} \int_{\pi - 2^{-K+3}}^{\pi + 2^{-K-1}} (\pi - 2^{-K+3} - x_2)^{-3p/2}\, dx_2\, dx_1$$
$$\leq C_p.$$

We may suppose that the center of I is zero, in other words $I := (-\nu, \nu) \times (-\nu, \nu)$. Let

$$A_1(u, t_2) := \int_{-\nu}^{u} a(t_1, t_2)\, dt_1 \quad \text{and} \quad A_2(u, v) := \int_{-\nu}^{v} A_1(u, t_2)\, dt_2.$$

Observe that

$$|A_k(u, v)| \leq C_p 2^{K(2/p - k)} \qquad (k = 1, 2).$$

Integrating by parts, we can see that

$$\int_{I_1} a(t_1, t_2) K_n^{1,\alpha}(x_1 - t_1, x_2 - t_2) 1_{A_3 \cup A_8}(x_1 - t_1, x_2 - t_2)\, dt_1$$
$$= A_1(\nu, t_2) K_n^{1,\alpha}(x_1 - \nu, x_2 - t_2) 1_{A_3 \cup A_8}(x_1 - \nu, x_2 - t_2)$$
$$+ \int_{\nu}^{\nu} A_1(t_1, t_2) \partial_1 K_n^{1,\alpha}(x_1 - t_1, x_2 - t_2) 1_{A_3 \cup A_8}(x_1 - t_1, x_2 - t_2)\, dt_1,$$

because $A_1(-\nu, t_2) = 0$. Let us integrate the first term again by parts and use that

$$A_2(\nu, \nu) = \int_{I_1} \int_{I_2} a(t_1, t_2)\, dt_1\, dt_2 = 0$$

to obtain

$$\int_{I_1}\int_{I_2} a(t_1,t_2)K_n^{1,\alpha}(x_1-t_1,x_2-t_2)1_{A_3\cup A_8}(x_1-t_1,x_2-t_2)\,dt_1\,dt_2$$
$$=\int_{-\nu}^{\nu} A_2(\nu,t_2)\partial_2 K_n^{1,\alpha}(x_1-\nu,x_2-t_2)1_{A_3\cup A_8}(x_1-\nu,x_2-t_2)\,dt_2$$
$$+\int_{I_1}\int_{I_2} A_1(t_1,t_2)\partial_1 K_n^{1,\alpha}(x_1-t_1,x_2-t_2)1_{A_3\cup A_8}(x_1-t_1,x_2-t_2)\,dt_1\,dt_2.$$

Note that

$$x_1-t_1-x_2+t_2>(x_1-t_1)/2 \tag{2.5.4}$$

on the set A_3 and

$$x_1-t_1-x_2+t_2>(\pi-x_2+t_2)/2 \tag{2.5.5}$$

on the set A_8. If $n\le 2^K$, we get from Lemma 2.2.17 and (2.5.4) that

$$\begin{aligned}
&\left|\int_I a(t_1,t_2)K_n^{1,\alpha}(x_1-t_1,x_2-t_2)1_{A_3}(x_1-t_1,x_2-t_2)\,dt_1\,dt_2\right| \\
&\le C_p n^{1-\gamma}2^{2K/p-2K} \\
&\qquad \int_{I_2}(x_1-\nu)^{-1-\beta}(x_2-t_2)^{\beta-\gamma-1}1_{A_3}(x_1-\nu,x_2-t_2)\,dt_2 \\
&\qquad + C_p n^{1-\gamma}2^{2K/p-K} \\
&\qquad\quad \int_I (x_1-t_1)^{-1-\beta}(x_2-t_2)^{\beta-\gamma-1}1_{A_3}(x_1-t_1,x_2-t_2)\,dt_1dt_2 \\
&\le C_p 2^{2K/p-2K-K\gamma}1_{\{2^{-K+4}<x_1<\pi+2^{-K-1}\}}1_{\{2^{-K+1}<x_2\le x_1/2+2^{-K}\}} \\
&\qquad (x_1-2^{-K-1})^{-1-\beta}(x_2-2^{-K-1})^{\beta-\gamma-1},
\end{aligned} \tag{2.5.6}$$

where $0\le\beta\le 1$, $\gamma=\alpha$ or $\gamma=1$. On A_8, we use (2.5.5) to obtain

$$\begin{aligned}
&\left|\int_I a(t_1,t_2)K_n^{1,\alpha}(x_1-t_1,x_2-t_2)1_{A_8}(x_1-t_1,x_2-t_2)\,dt_1\,dt_2\right| \\
&\le C_p n^{1-\gamma}2^{2K/p-2K} \\
&\qquad \int_{I_2}(\pi-x_2+t_2)^{-1-\beta}(\pi-x_1+\nu)^{\beta-\gamma-1}1_{A_8}(x_1-\nu,x_2-t_2)\,dt_2 \\
&\qquad + C_p n^{1-\gamma}2^{2K/p-K} \\
&\qquad\quad \int_I (\pi-x_2+t_2)^{-1-\beta}(\pi-x_1+t_1)^{\beta-\gamma-1}1_{A_8}(x_1-t_1,x_2-t_2)\,dt_1dt_2 \\
&\le C_p 2^{2K/p-2K-K\gamma}1_{\{\pi/2-2^{-K-1}<x_2<\pi-2^{-K+4}\}}1_{\{(\pi+x_2)/2-2^{-K}<x_1<\pi-2^{-K+1}\}}
\end{aligned}$$

$$(\pi - x_2 - 2^{-K-1})^{-1-\beta}(\pi - x_1 - 2^{-K-1})^{\beta-\gamma-1}. \tag{2.5.7}$$

Similarly, if $n > 2^K$, then we get from (2.2.15) and (2.5.4) that

$$\begin{aligned}
&\left|\int_I a(t_1,t_2)K_n^{1,\alpha}(x_1-t_1,x_2-t_2)1_{A_3}(x_1-t_1,x_2-t_2)\,dt_1\,dt_2\right| \\
&\quad\le C_p n^{-\gamma}2^{2K/p}\int_I (x_1-t_1)^{-1-\beta}(x_2-t_2)^{\beta-\gamma-1}1_{A_3}(x_1-t_1,x_2-t_2)\,dt_1\,dt_2 \\
&\quad\le C_p 2^{2K/p-2K-K\gamma}1_{\{2^{-K+4}<x_1<\pi+2^{-K-1}\}}1_{\{2^{-K+1}<x_2\le x_1/2+2^{-K}\}} \\
&\qquad (x_1-2^{-K-1})^{-1-\beta}(x_2-2^{-K-1})^{\beta-\gamma-1}
\end{aligned} \tag{2.5.8}$$

and, by (2.5.5),

$$\begin{aligned}
&\left|\int_I a(t_1,t_2)K_n^{1,\alpha}(x_1-t_1,x_2-t_2)1_{A_8}(x_1-t_1,x_2-t_2)\,dt_1\,dt_2\right| \\
&\quad\le C_p n^{-\gamma}2^{2K/p} \\
&\qquad \int_I (\pi-x_2+t_2)^{-1-\beta}(\pi-x_1+t_1)^{\beta-\gamma-1}1_{A_8}(x_1-t_1,x_2-t_2)\,dt_1\,dt_2 \\
&\quad\le C_p 2^{2K/p-2K-K\gamma}1_{\{\pi/2-2^{-K-1}<x_2<\pi-2^{-K+4}\}}1_{\{(\pi+x_2)/2-2^{-K}<x_1<\pi-2^{-K+1}\}} \\
&\qquad (\pi-x_2-2^{-K-1})^{-1-\beta}(\pi-x_1-2^{-K-1})^{\beta-\gamma-1}.
\end{aligned} \tag{2.5.9}$$

Choosing $\beta = \gamma/2$, we conclude

$$\begin{aligned}
&\int_{\mathbb{T}^2}\sup_{n\ge 1}\left|\int_I a(t_1,t_2)K_n^{1,\alpha}(x_1-t_1,x_2-t_2)1_{A_3}(x_1-t_1,x_2-t_2)\,dt_1\,dt_2\right|^p\,dx_1dx_2 \\
&\quad\le C_p 2^{2K-2Kp-K\gamma p} \\
&\qquad \int_{2^{-K+4}}^{\pi+2^{-K-1}}\int_{2^{-K+1}}^{x_1/2+2^{-K}}(x_1-2^{-K-1})^{-p(1+\gamma/2)}(x_2-2^{-K-1})^{-p(1+\gamma/2)}\,dx_2dx_1 \\
&\quad\le C_p 2^{2K-2Kp-K\gamma p}2^{-K(1-p(1+\gamma/2))}2^{-K(1-p(1+\gamma/2))} \\
&\quad\le C_p
\end{aligned}$$

and

$$\begin{aligned}
&\int_{\mathbb{T}^2}\sup_{n\ge 1}\left|\int_I a(t_1,t_2)K_n^{1,\alpha}(x_1-t_1,x_2-t_2)1_{A_8}(x_1-t_1,x_2-t_2)\,dt_1\,dt_2\right|^p\,dx_1dx_2 \\
&\quad\le C_p 2^{2K-2Kp-K\gamma p}\int_{\pi/2-2^{-K-1}}^{\pi-2^{-K+4}}\int_{(\pi+x_2)/2-2^{-K}}^{\pi-2^{-K+1}} \\
&\qquad (\pi-x_2-2^{-K-1})^{-p(1+\gamma/2)}(\pi-x_1-2^{-K-1})^{-p(1+\gamma/2)}\,dx_1dx_2 \\
&\quad\le C_p 2^{2K-2Kp-K\gamma p}2^{-K(1-p(1+\gamma/2))}2^{-K(1-p(1+\gamma/2))}
\end{aligned}$$

$$\leq C_p,$$

whenever $p > 2/(2+\gamma)$. Recall that $\gamma = \alpha$ or $\gamma = 1$.

Since

$$x_2 - t_2 > (x_1 - t_1)/2 \qquad \text{on } A_4$$

and

$$\pi - x_1 + t_1 > (\pi - x_2 + t_2)/2 \qquad \text{on } A_9,$$

Lemma 2.2.17 implies

$$\begin{aligned}
&\left| \int_I a(t_1,t_2) K_n^{1,\alpha}(x_1-t_1, x_2-t_2) 1_{A_4}(x_1-t_1, x_2-t_2)\, dt_1\, dt_2 \right| \\
&\leq C_p n^{1-\gamma} 2^{2K/p-2K} \int_{I_2} (x_1-\nu-x_2+t_2)^{-1-\beta} \\
&\qquad (x_1-t_1)^{\beta-\gamma-1} 1_{A_4}(x_1-\nu, x_2-t_2)\, dt_2 \\
&\quad + C_p n^{1-\gamma} 2^{2K/p-K} \int_I (x_1-t_1-x_2+t_2)^{-1-\beta} \\
&\qquad (x_1-t_1)^{\beta-\gamma-1} 1_{A_4}(x_1-t_1, x_2-t_2)\, dt_1 dt_2 \\
&\leq C_p 2^{2K/p-2K-K\gamma} 1_{\{2^{-K+4}<x_1<\pi+2^{-K-1}\}} 1_{\{x_1/2-2^{-K}<x_2\leq x_1-2^{-K+1}\}} \\
&\qquad (x_1-x_2-2^{-K})^{-1-\beta} (x_1-2^{-K-1})^{\beta-\gamma-1}
\end{aligned} \tag{2.5.10}$$

and

$$\begin{aligned}
&\left| \int_I a(t_1,t_2) K_n^{1,\alpha}(x_1-t_1, x_2-t_2) 1_{A_9}(x_1-t_1, x_2-t_2)\, dt_1\, dt_2 \right| \\
&\leq C_p n^{1-\gamma} 2^{2K/p-2K} \int_{I_2} (x_1-\nu-x_2+t_2)^{-1-\beta} \\
&\qquad (\pi-x_2+\nu)^{\beta-\gamma-1} 1_{A_9}(x_1-\nu, x_2-t_2)\, dt_2 \\
&\quad + C_p n^{1-\gamma} 2^{2K/p-K} \int_I (x_1-t_1-x_2+t_2)^{-1-\beta} \\
&\qquad (\pi-x_2+t_2)^{\beta-\gamma-1} 1_{A_9}(x_1-t_1, x_2-t_2)\, dt_1 dt_2 \\
&\leq C_p 2^{2K/p-2K-K\gamma} 1_{\{\pi/2-2^{-K-1}<x_2<\pi-2^{-K+4}\}} 1_{\{x_2+2^{-K+1}<x_1<(\pi+x_2)/2+2^{-K}\}} \\
&\qquad (x_1-x_2-2^{-K})^{-1-\beta} (\pi-x_2-2^{-K-1})^{\beta-\gamma-1},
\end{aligned} \tag{2.5.11}$$

whenever $n \leq 2^K$. If $n > 2^K$, then by (2.2.15),

$$\begin{aligned}
&\left|\int_I a(t_1,t_2)K_n^{1,\alpha}(x_1-t_1,x_2-t_2)1_{A_4}(x_1-t_1,x_2-t_2)\,dt_1\,dt_2\right| \\
&\quad \le C_p n^{-\gamma}2^{2K/p}\int_I (x_1-t_1-x_2+t_2)^{-1-\beta} \\
&\qquad\qquad (x_1-t_1)^{\beta-\gamma-1}1_{A_4}(x_1-t_1,x_2-t_2)\,dt_1dt_2 \\
&\quad \le C_p 2^{2K/p-2K-K\gamma}1_{\{2^{-K+4}<x_1<\pi+2^{-K-1}\}}1_{\{x_1/2-2^{-K}<x_2\le x_1-2^{-K+1}\}} \\
&\qquad (x_1-x_2-2^{-K})^{-1-\beta}(x_1-2^{-K-1})^{\beta-\gamma-1}
\end{aligned} \tag{2.5.12}$$

and

$$\begin{aligned}
&\left|\int_I a(t_1,t_2)K_n^{1,\alpha}(x_1-t_1,x_2-t_2)1_{A_9}(x_1-t_1,x_2-t_2)\,dt_1\,dt_2\right| \\
&\quad \le C_p n^{-\gamma}2^{2K/p}\int_I (x_1-t_1-x_2+t_2)^{-1-\beta} \\
&\qquad\qquad (\pi-x_2+t_2)^{\beta-\gamma-1}1_{A_9}(x_1-t_1,x_2-t_2)\,dt_1dt_2 \\
&\quad \le C_p 2^{2K/p-2K-K\gamma}1_{\{\pi/2-2^{-K-1}<x_2<\pi-2^{-K+4}\}}1_{\{x_2+2^{-K+1}<x_1<(\pi+x_2)/2+2^{-K}\}} \\
&\qquad (x_1-x_2-2^{-K})^{-1-\beta}(\pi-x_2-2^{-K-1})^{\beta-\gamma-1}.
\end{aligned} \tag{2.5.13}$$

Choosing again $\beta=\gamma/2$, we obtain

$$\begin{aligned}
&\int_{\mathbb{T}^2}\sup_{n\ge 1}\left|\int_I a(t_1,t_2)K_n^{1,\alpha}(x_1-t_1,x_2-t_2)1_{A_4}(x_1-t_1,x_2-t_2)\,dt_1\,dt_2\right|^p dx_1dx_2 \\
&\quad \le C_p 2^{2K-2Kp-K\gamma p} \\
&\qquad \int_{2^{-K+4}}^{\pi+2^{-K-1}}\int_{x_1/2-2^{-K}}^{x_1-2^{-K+1}} (x_1-x_2-2^{-K})^{-p(1+\gamma/2)}(x_1-2^{-K-1})^{-p(1+\gamma/2)}\,dx_2dx_1 \\
&\quad \le C_p 2^{2K-2Kp-K\gamma p}2^{-K(1-p(1+\gamma/2))}2^{-K(1-p(1+\gamma/2))} \\
&\quad \le C_p
\end{aligned}$$

and

$$\begin{aligned}
&\int_{\mathbb{T}^2}\sup_{n\ge 1}\left|\int_I a(t_1,t_2)K_n^{1,\alpha}(x_1-t_1,x_2-t_2)1_{A_9}(x_1-t_1,x_2-t_2)\,dt_1\,dt_2\right|^p dx_1dx_2 \\
&\quad \le C_p 2^{2K-2Kp-K\gamma p}\int_{\pi/2-2^{-K-1}}^{\pi-2^{-K+4}}\int_{x_2+2^{-K+1}}^{(\pi+x_2)/2+2^{-K}} \\
&\qquad (x_1-x_2-2^{-K})^{-p(1+\gamma/2)}(\pi-x_2-2^{-K-1})^{-p(1+\gamma/2)}\,dx_1dx_2 \\
&\quad \le C_p 2^{2K-2Kp-K\gamma p}2^{-K(1-p(1+\gamma/2))}2^{-K(1-p(1+\gamma/2))} \\
&\quad \le C_p,
\end{aligned}$$

whenever $p>2/(2+\gamma)$.

Finally, inequality (2.2.19) imply

$$\left|\int_I a(t_1,t_2)K_n^{1,\alpha}(x_1-t_1,x_2-t_2)1_{A_5}(x_1-t_1,x_2-t_2)\,dt_1\,dt_2\right|$$
$$\leq C_p 2^{2K/p}\int_I (x_1-t_1-x_2+t_2)^{\gamma-1}$$
$$(x_2-t_2)^{-\gamma-1}1_{A_5}(x_1-t_1,x_2-t_2)\,dt_1\,dt_2$$
$$\leq C_p 2^{2K/p-K\gamma}1_{\{2^{-K+4}<x_1<\pi+2^{-K-1}\}}$$
$$1_{\{x_1-2^{-K+3}<x_2\leq x_1+2^{-K}\}}\int_{I_2}(x_2-t_2)^{-\gamma-1}\,dt_2$$
$$\leq C_p 2^{2K/p-K\gamma-K}1_{\{2^{-K+4}<x_1<\pi+2^{-K-1}\}}$$
$$1_{\{x_1-2^{-K+3}<x_2\leq x_1+2^{-K}\}}(x_2-2^{-K-1})^{-\gamma-1}$$

and

$$\left|\int_I a(t_1,t_2)K_n^{1,\alpha}(x_1-t_1,x_2-t_2)1_{A_{10}}(x_1-t_1,x_2-t_2)\,dt_1\,dt_2\right|$$
$$\leq C_p 2^{2K/p}\int_I (x_1-t_1-x_2+t_2)^{\gamma-1}$$
$$(\pi-x_1+t_1)^{-\gamma-1}1_{A_5}(x_1-t_1,x_2-t_2)\,dt_1\,dt_2$$
$$\leq C_p 2^{2K/p-K\gamma}1_{\{\pi/2-2^{-K-1}<x_2<\pi-2^{-K+4}\}}$$
$$1_{\{x_2-2^{-K}<x_1<x_2+2^{-K+3}\}}\int_{I_2}(\pi-x_1+t_1)^{-\gamma-1}\,dt_2$$
$$\leq C_p 2^{2K/p-K\gamma-K}1_{\{\pi/2-2^{-K-1}<x_2<\pi-2^{-K+4}\}}$$
$$1_{\{x_2-2^{-K}<x_1<x_2+2^{-K+3}\}}(\pi-x_1-2^{-K-1})^{-\gamma-1}.$$

Hence

$$\int_{\mathbb{T}^2}\sup_{n\geq 1}\left|\int_I a(t_1,t_2)K_n^{1,\alpha}(x_1-t_1,x_2-t_2)1_{A_5}(x_1-t_1,x_2-t_2)\,dt_1\,dt_2\right|^p\,dx_1dx_2$$
$$\leq C_p 2^{2K-K\gamma p-Kp}\int_{2^{-K+4}}^{\pi+2^{-K-1}}\int_{x_1-2^{-K+3}}^{x_1+2^{-K}}(x_2-2^{-K-1})^{-p(\gamma+1)}\,dx_2dx_1$$
$$\leq C_p 2^{2K-K\gamma p-Kp}\int_{2^{-K+3}}^{\pi+2^{-K+5}}\int_{x_2-2^{-K}}^{x_2+2^{-K+3}}(x_2-2^{-K-1})^{-p(\gamma+1)}\,dx_1dx_2$$
$$\leq C_p$$

and

$$
\begin{aligned}
&\int_{\mathbb{T}^2} \sup_{n\ge 1} \left| \int_I a(t_1,t_2) K_n^{1,\alpha}(x_1-t_1,x_2-t_2) 1_{A_{10}}(x_1-t_1,x_2-t_2)\,dt_1\,dt_2 \right|^p dx_1 dx_2 \\
&\le C_p 2^{2K-K\gamma p-Kp} \int_{\pi/2-2^{-K-1}}^{\pi-2^{-K+4}} \int_{x_2-2^{-K}}^{x_2+2^{-K+3}} (\pi-x_1-2^{-K-1})^{-p(\gamma+1)}\,dx_1 dx_2 \\
&\le C_p 2^{2K-K\gamma p-Kp} \int_{\pi/2-2^{-K+1}}^{\pi-2^{-K+3}} \int_{x_1+2^{-K}}^{x_1-2^{-K+3}} (\pi-x_1-2^{-K-1})^{-p(\gamma+1)}\,dx_2 dx_1 \\
&\le C_p,
\end{aligned}
$$

whenever $p > 1/(1+\gamma)$, which finishes the proof of the theorem. ■

Proof of Theorem 2.5.4 **for** $q=\infty$. We assume again that $\alpha \le 1$ and a is a cube $H_p^\square$-atom with support $I = I_1 \times I_2$,

$$
[-2^{-K-2}, 2^{-K-2}] \subset I_j \subset [-2^{-K-1}, 2^{-K-1}] \qquad (j=1,2)
$$

for some $K \in \mathbb{Z}$. As before, it is enough to show that

$$
\begin{aligned}
&\int_{\mathbb{T}^2} \sup_{n\ge 1} \left| \int_I a(t_1,t_2) K_n^{\infty,\alpha}(x_1-t_1,x_2-t_2) 1_{A_i}(x_1-t_1,x_2-t_2)\,dt_1\,dt_2 \right|^p dx_1\,dx_2 \\
&\quad \le C_p
\end{aligned}
$$

for all $i = 1,2,3,4,5$, where $\pi > x_1 - t_1 > x_2 - t_2 > 0$ and

$$
\begin{aligned}
A_1 &:= \{(x_1,x_2) : 0 < x_1 \le 2^{-K+5}, 0 < x_2 < x_1 < \pi\}, \\
A_2 &:= \{(x_1,x_2) : 2^{-K+5} < x_1 < \pi, 0 < x_2 \le 2^{-K+2}\}, \\
A_3 &:= \{(x_1,x_2) : 2^{-K+5} < x_1 < \pi, 2^{-K+2} < x_2 \le x_1/2\}, \\
A_4 &:= \{(x_1,x_2) : 2^{-K+5} < x_1 < \pi, x_1/2 < x_2 \le x_1 - 2^{-K+2}\}, \\
A_5 &:= \{(x_1,x_2) : 2^{-K+5} < x_1 < \pi, x_1 - 2^{-K+2} < x_2 < x_1\}
\end{aligned}
$$

(see Fig. 2.12). The estimation on the set A_1 is the same as before in the proof for $q=1$. Inequality (2.2.28) implies

$$
\begin{aligned}
&\left| \int_I a(t_1,t_2) K_n^{\infty,\alpha}(x_1-t_1,x_2-t_2) 1_{A_2}(x_1-t_1,x_2-t_2)\,dt_1\,dt_2 \right| \\
&\le C_p 2^{2K/p} \int_I (x_1-t_1-x_2+t_2)^{-1-\gamma}(x_2-t_2)^{\gamma-1} \\
&\qquad 1_{A_2}(x_1-t_1,x_2-t_2)\,dt_1\,dt_2 \\
&\le C_p 2^{2K/p} 1_{\{2^{-K+4}<x_1<\pi+2^{-K-1}\}} 1_{\{-2^{-K-1}<x_2\le 2^{-K+3}\}} \\
&\qquad \int_I (x_1-2^{-K+3})^{-1-\gamma}(x_2-t_2)^{\gamma-1}\,dt_1\,dt_2
\end{aligned}
$$

$$\le C_p 2^{2K/p-K-K\gamma} 1_{\{2^{-K+4}<x_1<\pi+2^{-K-1}\}}$$
$$1_{\{-2^{-K-1}<x_2\le 2^{-K+3}\}}(x_1-2^{-K+3})^{-1-\gamma},$$

where $\gamma=\alpha$ or $\gamma=1$ in the whole proof. Furthermore,

$$\int_{\mathbb{T}^2}\sup_{n\ge 1}\left|\int_I a(t_1,t_2)K_n^{\infty,\alpha}(x_1-t_1,x_2-t_2)1_{A_2}(x_1-t_1,x_2-t_2)\,dt_1\,dt_2\right|^p\,dx_1\,dx_2$$
$$\le C_p 2^{2K-Kp-K\gamma p}\int_{2^{-K+4}}^{\pi+2^{-K-1}}\int_{-2^{-K-1}}^{2^{-K+3}}(x_1-2^{-K+3})^{-p(1+\gamma)}\,dx_2dx_1$$
$$\le C_p,$$

provided that $p>1/(1+\gamma)$. For any $0<\beta<\alpha$, we get from (2.3.1) that

$$\left|\int_I a(t_1,t_2)K_n^{\infty,\alpha}(x_1-t_1,x_2-t_2)1_{A_3}(x_1-t_1,x_2-t_2)\,dt_1\,dt_2\right|$$
$$\le C_p 2^{2K/p}n^{-\gamma}\int_I (x_1-t_1)^{-1-\gamma+\beta}(x_2-t_2)^{-1-\beta}$$
$$1_{A_3}(x_1-t_1,x_2-t_2)\,dt_1\,dt_2$$
$$\le C_p 2^{2K/p-2K-K\gamma} 1_{\{2^{-K+4}<x_1<\pi+2^{-K-1}\}}1_{\{2^{-K+1}<x_2\le x_1/2+2^{-K}\}}$$
$$(x_1-2^{-K-1})^{-1-\gamma+\beta}(x_2-2^{-K-1})^{-1-\beta},$$

whenever $n>2^K$. Lemma 2.2.21 and (2.5.4) imply that

$$\left|\partial_j K_n^{\infty,\alpha}(x_1,x_2)\right|\le Cn^{1-\alpha}x_1^{-1-\alpha+\beta}x_2^{-1-\beta}$$

on A_3, where $j=1,2$. Similar to the proof for $q=1$, we get by integration by parts that

$$\left|\int_I a(t_1,t_2)K_n^{\infty,\alpha}(x_1-t_1,x_2-t_2)1_{A_3}(x_1-t_1,x_2-t_2)\,dt_1\,dt_2\right|$$
$$\le\left|\int_{-\nu}^{\nu}A_2(\nu,t_2)\partial_2 K_n^{\infty,\alpha}(x_1-\nu,x_2-t_2)1_{A_3}(x_1-\nu,x_2-t_2)\,dt_2\right|$$
$$+\left|\int_{I_1}\int_{I_2}A_1(t_1,t_2)\partial_1 K_n^{\infty,\alpha}(x_1-t_1,x_2-t_2)1_{A_3}(x_1-t_1,x_2-t_2)\,dt_1\,dt_2\right|$$
$$\le C_p n^{1-\alpha}2^{2K/p-2K}\int_{I_2}(x_1-\nu)^{-1-\alpha+\beta}(x_2-t_2)^{-1-\beta}1_{A_3}(x_1-t_1,x_2-t_2)\,dt_2$$
$$+C_p n^{1-\alpha}2^{2K/p-K}\int_I (x_1-t_1)^{-1-\alpha+\beta}(x_2-t_2)^{-1-\beta}$$
$$1_{A_3}(x_1-t_1,x_2-t_2)\,dt_1\,dt_2$$
$$\le C_p 2^{2K/p-2K-K\alpha}1_{\{2^{-K+4}<x_1<\pi+2^{-K-1}\}}1_{\{2^{-K+1}<x_2\le x_1/2+2^{-K}\}}$$

$$(x_1 - 2^{-K-1})^{-1-\alpha+\beta}(x_2 - 2^{-K-1})^{-1-\beta}$$

if $n \leq 2^K$. Thus

$$\begin{aligned}
&\int_{\mathbb{T}^2} \sup_{n\geq 1} \left| \int_I a(t_1, t_2) K_n^{\infty,\alpha}(x_1 - t_1, x_2 - t_2) 1_{A_3}(x_1 - t_1, x_2 - t_2)\, dt_1\, dt_2 \right|^p dx_1\, dx_2 \\
&\leq C_p 2^{2K-2Kp-K\gamma p} \\
&\quad \int_{2^{-K+4}}^{\pi+2^{-K-1}} \int_{2^{-K+1}}^{x_1/2+2^{-K}} (x_1 - 2^{-K-1})^{(-1-\gamma+\beta)p}(x_2 - 2^{-K-1})^{-(1+\beta)p}\, dx_2 dx_1 \\
&\leq C_p 2^{2K-2Kp-K\gamma p} 2^{-K(1-(\gamma-\beta+1)p)} 2^{-K(1-(1+\beta)p)} \\
&\leq C_p,
\end{aligned}$$

whenever $p > 1/(1+\beta)$ and $p > 1/(\gamma - \beta + 1)$. $\beta = \gamma/2$ implies $p > \frac{2}{2+\gamma}$.

Using (2.3.2), we see that

$$\begin{aligned}
&\left| \int_I a(t_1, t_2) K_n^{\infty,\alpha}(x_1 - t_1, x_2 - t_2) 1_{A_4}(x_1 - t_1, x_2 - t_2)\, dt_1\, dt_2 \right| \\
&\quad \leq C_p n^{-\gamma} 2^{2K/p} \int_I (x_1 - t_1 - x_2 + t_2)^{-1-\gamma+\beta}(x_1 - t_1)^{-1-\beta} \\
&\qquad 1_{A_4}(x_1 - t_1, x_2 - t_2)\, dt_1\, dt_2 \\
&\quad \leq C_p 2^{2K/p-2K-K\gamma} 1_{\{2^{-K+4} < x_1 < \pi + 2^{-K-1}\}} 1_{\{x_1/2 - 2^{-K} < x_2 \leq x_1 - 2^{-K+1}\}} \\
&\qquad (x_1 - x_2 - 2^{-K})^{-1-\gamma+\beta}(x_1 - 2^{-K-1})^{-1-\beta},
\end{aligned} \tag{2.5.14}$$

where $n > 2^K$ and $0 < \beta < \alpha$. Since $x_2 > x_1/2$ and $x_2 > x_1 - x_2$ on A_4, Lemma 2.2.21 implies

$$\left| \partial_j K_n^{\infty,\alpha}(x_1, x_2) \right| \leq C n^{1-\alpha} x_1^{-1-\beta}(x_1 - x_2)^{-1-\alpha+\beta},$$

where $j = 1, 2$. For $n \leq 2^K$, we get by integration by parts that

$$\begin{aligned}
&\left| \int_I a(t_1, t_2) K_n^{\infty,\alpha}(x_1 - t_1, x_2 - t_2) 1_{A_4}(x_1 - t_1, x_2 - t_2)\, dt_1\, dt_2 \right| \\
&\leq \left| \int_{-\nu}^{\nu} A_2(\nu, t_2) \partial_2 K_n^{\infty,\alpha}(x_1 - \nu, x_2 - t_2) 1_{A_4}(x_1 - \nu, x_2 - t_2)\, dt_2 \right| \\
&\quad + \left| \int_{I_1} \int_{I_2} A_1(t_1, t_2) \partial_1 K_n^{\infty,\alpha}(x_1 - t_1, x_2 - t_2) 1_{A_4}(x_1 - t_1, x_2 - t_2)\, dt_1\, dt_2 \right| \\
&\leq C_p n^{1-\alpha} 2^{2K/p-2K} \int_{I_2} (x_1 - \nu)^{-1-\beta}(x_1 - t_1 - x_2 + t_2)^{-1-\alpha+\beta} \\
&\qquad 1_{A_4}(x_1 - t_1, x_2 - t_2)\, dt_2
\end{aligned}$$

$$
\begin{aligned}
&+ C_p n^{1-\alpha} 2^{2K/p-K} \int_I (x_1-t_1)^{-1-\beta}(x_1-t_1-x_2+t_2)^{-1-\alpha+\beta} \\
&\qquad 1_{A_4}(x_1-t_1,x_2-t_2)\,dt_1\,dt_2 \\
&\le C_p 2^{2K/p-2K-K\alpha} 1_{\{2^{-K+4}<x_1<\pi+2^{-K-1}\}} 1_{\{x_1/2-2^{-K}<x_2\le x_1-2^{-K+1}\}} \\
&\qquad (x_1-x_2-2^{-K})^{-1-\alpha+\beta}(x_1-2^{-K-1})^{-1-\beta}. \qquad (2.5.15)
\end{aligned}
$$

From this it follows that

$$
\begin{aligned}
&\int_{\mathbb{T}^2} \sup_{n\ge 1} \left| \int_I a(t_1,t_2) K_n^{\infty,\alpha}(x_1-t_1,x_2-t_2) 1_{A_4}(x_1-t_1,x_2-t_2)\,dt_1\,dt_2 \right|^p dx_1\,dx_2 \\
&\le C_p 2^{2K-2Kp-K\gamma p} \int_{2^{-K+4}}^{\pi+2^{-K-1}} \int_{x_1/2-2^{-K}}^{x_1-2^{-K+1}} \\
&\qquad (x_1-x_2-2^{-K})^{(-1-\gamma+\beta)p}(x_1-2^{-K-1})^{-(1+\beta)p}\,dx_2 dx_1 \\
&\le C_p 2^{2K-2Kp-K\gamma p} 2^{-K(1-(\gamma-\beta+1)p)} 2^{-K(1-(1+\beta)p)} \\
&\le C_p,
\end{aligned}
$$

whenever $\beta=\gamma/2$ and $p>\frac{2}{2+\gamma}$.

Finally, since $x_2>x_1/2$ also on A_5,

$$
\begin{aligned}
&\left| \int_I a(t_1,t_2) K_n^{\infty,\alpha}(x_1-t_1,x_2-t_2) 1_{A_5}(x_1-t_1,x_2-t_2)\,dt_1\,dt_2 \right| \\
&\le C_p 2^{2K/p} \int_I (x_1-t_1)^{-2} 1_{A_5}(x_1-t_1,x_2-t_2)\,dt_1\,dt_2 \\
&\le C_p 2^{2K/p-2K} 1_{\{2^{-K+4}<x_1<\pi+2^{-K-1}\}} 1_{\{x_1-2^{-K+3}<x_2\le x_1+2^{-K}\}} (x_1-2^{-K-1})^{-2}
\end{aligned}
$$

and so

$$
\begin{aligned}
&\int_{\mathbb{T}^2} \sup_{n\ge 1} \left| \int_I a(t_1,t_2) K_n^{\infty,\alpha}(x_1-t_1,x_2-t_2) 1_{A_4}(x_1-t_1,x_2-t_2)\,dt_1\,dt_2 \right|^p dx_1\,dx_2 \\
&\le C_p 2^{2K-2Kp} \int_{2^{-K+4}}^{\pi+2^{-K-1}} \int_{x_1-2^{-K+3}}^{x_1+2^{-K}} (x_1-2^{-K-1})^{-2p}\,dx_2 dx_1 \\
&\le C_p.
\end{aligned}
$$

This completes the proof. ■

If p is smaller than or equal to the critical index, then this theorem is not true (see Oswald [253] and Stein, Taibleson and Weiss [292]). More exactly, we have

Theorem 2.5.6 *If $q=\infty$ and $\alpha=1$, then the operator $\sigma_*^{q,\alpha}$ is not bounded from $H_p^\square(\mathbb{T}^d)$ to $L_p(\mathbb{T}^d)$ if p is smaller than or equal to the critical index $d/(d+1)$.*

However, if p is equal to the critical index, then we can verify a weak type inequality.

Theorem 2.5.7 *Suppose that $q = 1, \infty$ and $0 < \alpha < \infty$. If*

$$p_0 := \frac{d}{d + \alpha \wedge 1}$$

and $f \in H_{p_0}^{\square}(\mathbb{T}^d)$, then

$$\left\| \sigma_*^{q,\alpha} f \right\|_{p_0,\infty} = \sup_{\rho>0} \rho \lambda(\sigma_*^{q,\alpha} f > \rho)^{1/p_0} \le C \left\| f \right\|_{H_{p_0}^{\square}}. \tag{2.5.16}$$

***Proof of Theorem 2.5.7* for $q = 1$.** We may suppose again that $0 < \alpha \le 1$. We use Theorem 2.4.20 and prove that

$$\sup_{\rho>0} \rho^{2/(2+\alpha)} \lambda(\sigma_*^{1,\alpha} a > \rho) \le C$$

for all $H_{2/(2+\alpha)}^{\square}$-atoms a. In other words, we have to show that

$$\lambda\left(\sup_{n\ge1} \left| \int_I a(t_1,t_2) K_n^{1,\alpha}(x_1-t_1, x_2-t_2) 1_{A_i}(x_1-t_1, x_2-t_2)\, dt_1\, dt_2 \right| > \rho \right)$$
$$\le C\rho^{-2/(2+\alpha)}$$

for $i = 1, \ldots, 10$ and $\rho > 0$. Since

$$\rho^{2/(2+\alpha)} \lambda(|g| > \rho) \le \int_{\mathbb{T}^d} |g|^{2/(2+\alpha)}, \tag{2.5.17}$$

the desired inequality follows from the proof of Theorem 2.5.4 for $i = 1, 6, 5, 10$. The same holds for $i = 2, 7$ if $\alpha < 1$. So for $i = 2, 7$, we suppose that $\alpha = 1$.

For $i = 2$ and $p = 2/3$, we have seen in (2.5.3) that

$$\left| \int_I a(t_1,t_2) K_n^{1,\alpha}(x_1-t_1, x_2-t_2) 1_{A_2}(x_1-t_1, x_2-t_2)\, dt_1\, dt_2 \right|$$
$$\le C 2^{3K/2} 1_{\{2^{-K+4} < x_1 < \pi + 2^{-K-1}\}} 1_{\{-2^{-K-1} < x_2 \le 2^{-K+3}\}} (x_1 - 2^{-K+3})^{-3/2}.$$

If this is greater than ρ, then

$$1_{\{2^{-K+4} < x_1 < \pi + 2^{-K-1}\}} (x_1 - 2^{-K+3}) < C\rho^{-2/3} 2^K 1_{\{-2^{-K-1} < x_2 \le 2^{-K+3}\}}$$

and

$$2^{-K+4} < x_1 < C\rho^{-2/3} 2^K + 2^{-K+4}.$$

Consequently,

$$\lambda\left(\sup_{n\ge 1}\left|\int_I a(t_1,t_2)K_n^{1,\alpha}(x_1-t_1,x_2-t_2)1_{A_2}(x_1-t_1,x_2-t_2)\,dt_1\,dt_2\right|>\rho\right)$$
$$\le \int_{\mathbb{T}^2} 1_{\left\{1_{\{2^{-K+4}<x_1<\pi+2^{-K-1}\}}(x_1-2^{-K+3})<C\rho^{-2/3}2^K 1_{\{-2^{-K-1}<x_2\le 2^{-K+3}\}}\right\}}\,dx_1\,dx_2$$
$$\le C\rho^{-2/3}2^K\int_{\mathbb{T}} 1_{\{-2^{-K-1}<x_2\le 2^{-K+3}\}}\,dx_2$$
$$\le C\rho^{-2/3}.$$

Similarly,

$$\left|\int_I a(t_1,t_2)K_n^{1,\alpha}(x_1-t_1,x_2-t_2)1_{A_7}(x_1-t_1,x_2-t_2)\,dt_1\,dt_2\right|$$
$$\le C2^{3K/2}1_{\{\pi/2-2^{-K-1}<x_2<\pi-2^{-K+4}\}}$$
$$1_{\{\pi-2^{-K+3}<x_1<\pi+2^{-K-1}\}}(\pi-2^{-K+3}-x_2)^{-3/2}.$$

If this is greater than ρ, then

$$1_{\{\pi/2-2^{-K-1}<x_2<\pi-2^{-K+4}\}}(\pi-2^{-K+3}-x_2)$$
$$< C\rho^{-2/3}2^K 1_{\{\pi-2^{-K+3}<x_1<\pi+2^{-K-1}\}}.$$

Let us denote the set of (x_1,x_2) for which the preceding inequality holds by H_7. If $(x_1,x_2)\in H_7$, then

$$\pi-2^{-K+3}-C\rho^{-2/3}2^K<x_2<\pi-2^{-K+3}.$$

Furthermore,

$$\lambda\left(\sup_{n\ge 1}\left|\int_I a(t_1,t_2)K_n^{1,\alpha}(x_1-t_1,x_2-t_2)1_{A_7}(x_1-t_1,x_2-t_2)\,dt_1\,dt_2\right|>\rho\right)$$
$$\le \int_{\mathbb{T}^2} 1_{H_7}(x_1,x_2)\,dx_2\,dx_1$$
$$\le C\rho^{-2/3}2^K\int_{\mathbb{T}} 1_{\{\pi-2^{-K+3}<x_1<\pi+2^{-K-1}\}}\,dx_1$$
$$\le C\rho^{-2/3}.$$

For $i=3,8,4,9$, we may suppose that $\gamma=\alpha$ and $p=2/(2+\alpha)$. We get by (2.5.6) and (2.5.8) that

$$\left|\int_I a(t_1,t_2)K_n^{1,\alpha}(x_1-t_1,x_2-t_2)1_{A_3}(x_1-t_1,x_2-t_2)\,dt_1\,dt_2\right|$$
$$\leq C1_{\{2^{-K+4}<x_1<\pi+2^{-K-1}\}}1_{\{2^{-K+1}<x_2\leq x_1/2+2^{-K}\}}$$
$$(x_1-2^{-K-1})^{-1-\beta}(x_2-2^{-K-1})^{\beta-\alpha-1}.$$

If this is greater than ρ, then

$$1_{\{2^{-K+1}<x_2\leq x_1/2+2^{-K}\}}(x_2-2^{-K-1})$$
$$<C\rho^{-\frac{1}{1+\alpha-\beta}}1_{\{2^{-K+4}<x_1<\pi+2^{-K-1}\}}(x_1-2^{-K-1})^{-\frac{1+\beta}{1+\alpha-\beta}}.$$

Note that $x_1/2+2^{-K}<x_1$. Choosing β such that $-\frac{1+\beta}{1+\alpha-\beta}+1<0$, i.e., $\alpha/2<\beta\leq 1$, we obtain

$$\lambda\left(\sup_{n\geq 1}\left|\int_I a(t_1,t_2)K_n^{1,\alpha}(x_1-t_1,x_2-t_2)1_{A_3}(x_1-t_1,x_2-t_2)\,dt_1\,dt_2\right|>\rho\right)$$
$$\leq\int_{2^{-K+4}}^{\rho^{-1/(2+\alpha)}+2^{-K-1}}\frac{x_1}{2}+2^{-K}\,dx_1$$
$$+C\rho^{-\frac{1}{1+\alpha-\beta}}\int_{\rho^{-1/(2+\alpha)}+2^{-K-1}}^{\pi}(x_1-2^{-K-1})^{-\frac{1+\beta}{1+\alpha-\beta}}\,dx_1$$
$$\leq C\rho^{-2/(2+\alpha)}+C\rho^{-\frac{1}{1+\alpha-\beta}}\rho^{\frac{-1}{2+\alpha}(-\frac{1+\beta}{1+\alpha-\beta}+1)}$$
$$=C\rho^{-2/(2+\alpha)}.$$

Similarly, by (2.5.7) and (2.5.9),

$$\left|\int_I a(t_1,t_2)K_n^{1,\alpha}(x_1-t_1,x_2-t_2)1_{A_8}(x_1-t_1,x_2-t_2)\,dt_1\,dt_2\right|$$
$$\leq C1_{\{\pi/2-2^{-K-1}<x_2<\pi-2^{-K+4}\}}1_{\{(\pi+x_2)/2-2^{-K}<x_1<\pi-2^{-K+1}\}}$$
$$(\pi-x_2-2^{-K-1})^{-1-\beta}(\pi-x_1-2^{-K-1})^{\beta-\alpha-1}.$$

If this is greater than ρ, then

$$1_{\{(\pi+x_2)/2-2^{-K}<x_1<\pi-2^{-K+1}\}}(\pi-x_1-2^{-K-1})$$
$$<C\rho^{-\frac{1}{1+\alpha-\beta}}1_{\{\pi/2-2^{-K-1}<x_2<\pi-2^{-K+4}\}}(\pi-x_2-2^{-K-1})^{-\frac{1+\beta}{1+\alpha-\beta}}.$$

Here $(\pi-x_2)/2+2^{-K}<\pi-x_2$. Choosing β as before, we obtain

$$\lambda\left(\sup_{n\geq 1}\left|\int_I a(t_1,t_2)K_n^{1,\alpha}(x_1-t_1,x_2-t_2)1_{A_8}(x_1-t_1,x_2-t_2)\,dt_1\,dt_2\right|>\rho\right)$$
$$\leq\int_{\pi-\rho^{-1/(2+\alpha)}-2^{-K-1}}^{\pi-2^{-K+4}}\frac{\pi-x_2}{2}+2^{-K}\,dx_2$$

$$
\begin{aligned}
&+ C\rho^{-\frac{1}{1+\alpha-\beta}} \int_{-\pi}^{\pi-\rho^{-1/(2+\alpha)}-2^{-K-1}} (\pi - x_2 - 2^{-K-1})^{-\frac{1+\beta}{1+\alpha-\beta}} \, dx_2 \\
&\leq C\rho^{-2/(2+\alpha)} + C\rho^{-\frac{1}{1+\alpha-\beta}} \rho^{\frac{-1}{2+\alpha}(-\frac{1+\beta}{1+\alpha-\beta}+1)} \\
&= C\rho^{-2/(2+\alpha)}.
\end{aligned}
$$

For A_4, we get from (2.5.10) and (2.5.12) that

$$
\begin{aligned}
&\left| \int_I a(t_1, t_2) K_n^{1,\alpha}(x_1 - t_1, x_2 - t_2) 1_{A_4}(x_1 - t_1, x_2 - t_2) \, dt_1 \, dt_2 \right| \\
&\leq C 1_{\{2^{-K+4} < x_1 < \pi + 2^{-K-1}\}} 1_{\{x_1/2 - 2^{-K} < x_2 \leq x_1 - 2^{-K+1}\}} \\
&\quad (x_1 - x_2 - 2^{-K})^{-1-\beta} (x_1 - 2^{-K-1})^{\beta-\alpha-1}.
\end{aligned}
$$

If this is greater than ρ, then

$$
\begin{aligned}
&1_{\{x_1/2 - 2^{-K} < x_2 \leq x_1 - 2^{-K+1}\}} (x_1 - x_2 - 2^{-K}) \\
&\quad < C\rho^{-\frac{1}{1+\beta}} 1_{\{2^{-K+4} < x_1 < \pi + 2^{-K-1}\}} (x_1 - 2^{-K-1})^{\frac{\beta-\alpha-1}{1+\beta}}.
\end{aligned}
$$

Hence

$$
\begin{aligned}
&\lambda\left(\sup_{n \geq 1} \left| \int_I a(t_1, t_2) K_n^{1,\alpha}(x_1 - t_1, x_2 - t_2) 1_{A_4}(x_1 - t_1, x_2 - t_2) \, dt_1 \, dt_2 \right| > \rho \right) \\
&\leq \int_{2^{-K+4}}^{\rho^{-1/(2+\alpha)}+2^{-K-1}} x_1 \, dx_1 \\
&\quad + C\rho^{-\frac{1}{1+\beta}} \int_{\rho^{-1/(2+\alpha)}+2^{-K-1}}^{\pi} (x_1 - 2^{-K-1})^{\frac{\beta-\alpha-1}{1+\beta}} \, dx_1 \\
&\leq C\rho^{-2/(2+\alpha)} + C\rho^{-\frac{1}{1+\beta}} \rho^{\frac{-1}{2+\alpha}(\frac{\beta-\alpha-1}{1+\beta}+1)} \\
&= C\rho^{-2/(2+\alpha)}.
\end{aligned}
$$

Here we have chosen β such that $\frac{\beta-\alpha-1}{1+\beta} + 1 < 0$, i.e., $0 < \beta < \alpha/2$.

Finally, by (2.5.11) and (2.5.13),

$$
\begin{aligned}
&\left| \int_I a(t_1, t_2) K_n^{1,\alpha}(x_1 - t_1, x_2 - t_2) 1_{A_9}(x_1 - t_1, x_2 - t_2) \, dt_1 \, dt_2 \right| \\
&\leq C 1_{\{\pi/2 - 2^{-K-1} < x_2 < \pi - 2^{-K+4}\}} 1_{\{x_2 + 2^{-K+1} < x_1 < (\pi + x_2)/2 + 2^{-K}\}} \\
&\quad (x_1 - x_2 - 2^{-K})^{-1-\beta} (\pi - x_2 - 2^{-K-1})^{\beta-\alpha-1}
\end{aligned}
$$

and

$$1_{\{x_2+2^{-K+1}<x_1<(\pi+x_2)/2+2^{-K}\}}(x_1 - x_2 - 2^{-K})$$
$$< C\rho^{-\frac{1}{1+\beta}} 1_{\{\pi/2-2^{-K-1}<x_2<\pi-2^{-K+4}\}}(\pi - x_2 - 2^{-K-1})^{\frac{\beta-\alpha-1}{1+\beta}}.$$

This implies that

$$x_2 + 2^{-K+1} < x_1 < (\pi - x_2 - 2^{-K-1})^{\frac{\beta-\alpha-1}{1+\beta}} + x_2 + 2^{-K+1}$$

and so

$$\lambda\left(\sup_{n\geq 1}\left|\int_I a(t_1,t_2)K_n^{1,\alpha}(x_1-t_1,x_2-t_2)1_{A_9}(x_1-t_1,x_2-t_2)\,dt_1\,dt_2\right| > \rho\right)$$
$$\leq \int_{\pi-\rho^{-1/(2+\alpha)}-2^{-K-1}}^{\pi-2^{-K+4}} \frac{\pi - x_2}{2} + 2^{-K}\,dx_2$$
$$+ C\rho^{-\frac{1}{1+\beta}}\int_{-\pi}^{\pi-\rho^{-1/(2+\alpha)}-2^{-K-1}} (\pi - x_2 - 2^{-K-1})^{\frac{\beta-\alpha-1}{1+\beta}}\,dx_2$$
$$\leq C\rho^{-2/(2+\alpha)} + C\rho^{-\frac{1}{1+\beta}}\rho^{\frac{-1}{2+\alpha}(\frac{\beta-\alpha-1}{1+\beta}+1)}$$
$$= C\rho^{-2/(2+\alpha)}$$

with the same β as for A_4, i.e., $0 < \beta < \alpha/2$. The proof of the theorem is complete. ■

***Proof of Theorem 2.5.7* for** $q = \infty$. Similar to the proof for $q = 1$, we have to show that

$$\lambda\left(\sup_{n>1}\left|\int_I a(t_1,t_2)K_n^{\infty,\alpha}(x_1-t_1,x_2-t_2)1_{A_i}(x_1-t_1,x_2-t_2)\,dt_1\,dt_2\right| > \rho\right)$$
$$\leq C\rho^{-2/(2+\alpha)}$$

for $\alpha \leq 1$, $i = 1,\ldots,5$, for all $H^{\square}_{2/(2+\alpha)}$-atoms a and $\rho > 0$. For $i = 1, 2, 5$, this inequality follows from (2.5.17) and the proof of Theorem 2.5.4. For $i = 3, 4$, we may suppose that $\gamma = \alpha$ and $p = 2/(2+\alpha)$. We have seen in (2.5.6) and (2.5.8) that

$$\left|\int_I a(t_1,t_2)K_n^{\infty,\alpha}(x_1-t_1,x_2-t_2)1_{A_3}(x_1-t_1,x_2-t_2)\,dt_1\,dt_2\right|$$
$$\leq C 1_{\{2^{-K+4}<x_1<\pi+2^{-K-1}\}} 1_{\{2^{-K+1}<x_2\leq x_1/2+2^{-K}\}}$$
$$(x_1 - 2^{-K-1})^{-1-\alpha+\beta}(x_2 - 2^{-K-1})^{-1-\beta}.$$

If this is greater than ρ, then

$$
\begin{aligned}
&1_{\{2^{-K+1}<x_2\le x_1/2+2^{-K}\}}(x_2-2^{-K-1})\\
&\qquad < C\rho^{-\frac{1}{1+\beta}} 1_{\{2^{-K+4}<x_1<\pi+2^{-K-1}\}}(x_1-2^{-K-1})^{-\frac{1+\alpha-\beta}{1+\beta}}.
\end{aligned}
$$

Since $x_1/2+2^{-K}<x_1$ and β can be chosen such that $-\frac{1+\alpha-\beta}{1+\beta}+1<0$, i.e., $0<\beta<\alpha/2$, we obtain

$$
\begin{aligned}
&\lambda\left(\sup_{n\ge1}\left|\int_I a(t_1,t_2)K_n^{\infty,\alpha}(x_1-t_1,x_2-t_2)1_{A_3}(x_1-t_1,x_2-t_2)\,dt_1\,dt_2\right|>\rho\right)\\
&\le \int_{2^{-K+4}}^{\rho^{-1/(2+\alpha)}+2^{-K-1}} \frac{x_1}{2}+2^{-K}\,dx_1\\
&\quad + C\rho^{-\frac{1}{1+\beta}}\int_{\rho^{-1/(2+\alpha)}+2^{-K-1}}^{\pi}(x_1-2^{-K-1})^{-\frac{1+\alpha-\beta}{1+\beta}}\,dx_1\\
&\le C\rho^{-2/(2+\alpha)}+C\rho^{-\frac{1}{1+\beta}}\rho^{\frac{-1}{2+\alpha}(-\frac{1+\alpha-\beta}{1+\beta}+1)}\\
&= C\rho^{-2/(2+\alpha)}.
\end{aligned}
$$

Similarly, by (2.5.14) and (2.5.15),

$$
\begin{aligned}
&\left|\int_I a(t_1,t_2)K_n^{\infty,\alpha}(x_1-t_1,x_2-t_2)1_{A_4}(x_1-t_1,x_2-t_2)\,dt_1\,dt_2\right|\\
&\qquad\le C_p2^{2K/p-2K-K\gamma}1_{\{2^{-K+4}<x_1<\pi+2^{-K-1}\}}1_{\{x_1/2-2^{-K}<x_2\le x_1-2^{-K+1}\}}\\
&\qquad\quad (x_1-x_2-2^{-K})^{-1-\alpha+\beta}(x_1-2^{-K-1})^{-1-\beta},
\end{aligned}
$$

which implies that

$$
\begin{aligned}
&1_{\{x_1/2-2^{-K}<x_2\le x_1-2^{-K+1}\}}(x_1-x_2-2^{-K})\\
&\qquad < C\rho^{-\frac{1}{1+\alpha-\beta}}1_{\{2^{-K+4}<x_1<\pi+2^{-K-1}\}}(x_1-2^{-K-1})^{-\frac{1+\beta}{1+\alpha-\beta}}.
\end{aligned}
$$

Hence

$$
\begin{aligned}
&\lambda\left(\sup_{n\ge1}\left|\int_I a(t_1,t_2)K_n^{\infty,\alpha}(x_1-t_1,x_2-t_2)1_{A_4}(x_1-t_1,x_2-t_2)\,dt_1\,dt_2\right|>\rho\right)\\
&\le \int_{2^{-K+4}}^{\rho^{-1/(2+\alpha)}+2^{-K-1}} x_1\,dx_1\\
&\quad + C\rho^{-\frac{1}{1+\alpha-\beta}}\int_{\rho^{-1/(2+\alpha)}+2^{-K-1}}^{\pi}(x_1-2^{-K-1})^{-\frac{1+\beta}{1+\alpha-\beta}}\,dx_1\\
&\le C\rho^{-2/(2+\alpha)}+C\rho^{-\frac{1}{1+\alpha-\beta}}\rho^{\frac{-1}{2+\alpha}(-\frac{1+\beta}{1+\alpha-\beta}+1)}\\
&= C\rho^{-2/(2+\alpha)},
\end{aligned}
$$

where $-\frac{1+\beta}{1+\alpha-\beta}+1<0$, i.e., $\alpha/2<\beta<\alpha$. This completes the proof of the theorem. ■

Of course, (2.5.16) cannot be true for $p<p_0$, i.e., $\sigma_*^{q,\alpha}$ is not bounded from $H_p^\square(\mathbb{T}^d)$ to the weak $L_{p,\infty}(\mathbb{T}^d)$ space for $p<p_0$. If the operator was bounded, then by interpolation (2.5.1) would hold for $p=p_0$, which contradicts Theorem 2.5.6.

Oswald [253] proved a similar theorem to Theorem 2.5.4 for the Riesz means of the Fourier transforms and for $q=\infty$. Theorems 2.5.4 and 2.5.7 can be found in Weisz [330, 339]. For a detailed proof of the multi-dimensional version, see [337, 338, 341, 344].

Marcinkiewicz [233] verified for two-dimensional Fourier series that the cubic (i.e., $q=\infty$) Fejér means of a function $f\in L\log L(\mathbb{T}^2)$ converge almost everywhere to f as $n\to\infty$. Later Zhizhiashvili [364, 366] extended this result to all $f\in L_1(\mathbb{T}^2)$ and to Cesàro means and Berens, Li and Xu [30] to $q=1$. The general convergence result can be found in [330, 337–339, 341].

The next corollary follows easily from Theorem 2.5.4.

Corollary 2.5.8 *Suppose that $q=1,\infty$ and $0<\alpha<\infty$. If $f\in L_1(\mathbb{T}^d)$, then*

$$\sup_{\rho>0}\rho\,\lambda(\sigma_*^{q,\alpha}f>\rho)\le C\|f\|_1.$$

The density argument of Marcinkiewicz and Zygmund implies

Corollary 2.5.9 *Suppose that $q=1,\infty$ and $0<\alpha<\infty$. If $f\in L_1(\mathbb{T}^d)$, then*

$$\lim_{n\to\infty}\sigma_n^{q,\alpha}f=f \qquad a.e.$$

Proof Since the trigonometric polynomials are dense in $L_1(\mathbb{T}^d)$, the corollary follows from Theorem 1.3.6 and Corollary 2.5.8. ■

2.5.2 Almost Everywhere Convergence for *q* = 2

Theorem 2.5.10 *Suppose that $q=2$, $(d-1)/2<\alpha<\infty$ and $\gamma\in\mathbb{P}$. If*

$$p_0:=\frac{d}{d/2+\alpha+1/2}<p<\infty,$$

and $f\in H_p^\square(\mathbb{T}^d)$, then

$$\left\|\sigma_*^{q,\alpha,\gamma}f\right\|_p\le C_p\|f\|_{H_p^\square}.$$

Proof Let us choose $N\in\mathbb{N}$ such that $N<\alpha-(d-1)/2\le N+1$. As we mentioned in Sect. 2.4, we may suppose that the support of an atom a is a ball B with radius β, $2^{-K-1} < \beta \le 2^{-K}$ ($K \in \mathbb{N}$). Moreover, we may suppose that the center of B is zero, i.e., $B = B(0, \beta)$. Obviously,

$$\begin{aligned}
&\int_{\mathbb{T}^d \setminus (rB)} |\sigma_*^{2,\alpha,\gamma} a(x)|^p \, dx \\
&\le \sum_{i=4\lfloor d^{1/2}\rfloor -1}^{\lfloor d^{1/2} 2^K \pi \rfloor -1} \int_{B(0,(i+2)2^{-K}) \setminus B(0,(i+1)2^{-K}) \cap \mathbb{T}^d} \sup_{n \ge d^{1/2} 2^{K+1}} |\sigma_n^{2,\alpha,\gamma} a(x)|^p \, dx \\
&+ \sum_{i=4\lfloor d^{1/2}\rfloor -1}^{\lfloor d^{1/2} 2^K \pi \rfloor -1} \int_{B(0,(i+2)2^{-K}) \setminus B(0,(i+1)2^{-K}) \cap \mathbb{T}^d} \sup_{n < d^{1/2} 2^{K+1}} |\sigma_n^{2,\alpha,\gamma} a(x)|^p \, dx \\
&=: (A) + (B),
\end{aligned}$$

where $r = 8d^{1/2}$. Note that if $K \le 3$, then the integral is equal to 0.

We use Taylor's formula for $g_k(t) = \widehat{\theta}_0(n(x - 2k\pi - t))$:

$$g_k(t) = \sum_{l=0}^{N-1} \sum_{\|i\|_1 = l} \partial_1^{i_1} \dots \partial_d^{i_d} g_k(0) \prod_{j=1}^{d} \frac{t_j^{i_j}}{i_j!} + \sum_{\|i\|_1 = N} \partial_1^{i_1} \dots \partial_d^{i_d} g_k(\nu t) \prod_{j=1}^{d} \frac{t_j^{i_j}}{i_j!}$$

for some $0 < \nu < 1$. Here

$$\partial_1^{i_1} \dots \partial_d^{i_d} g_k(t) = (-1)^{\|i\|_1} n^{\|i\|_1} \partial_1^{i_1} \dots \partial_d^{i_d} \widehat{\theta}_0(n(x - 2k\pi - t)).$$

Using this with $t - 2k\pi$ instead of t, Theorem 2.2.30 and the definition of the atom, we obtain

$$\begin{aligned}
\sigma_n^{2,\alpha,\gamma} a(x) &= \frac{1}{(2\pi)^d} \sum_{k \in \mathbb{Z}^d} n^d \int_{B+2k\pi} a(t) \widehat{\theta}_0(n(x-t)) \, dt \\
&= \frac{1}{(2\pi)^d} \sum_{k \in \mathbb{Z}^d} n^d \int_{B+2k\pi} a(t) \\
&\qquad \left(\widehat{\theta}_0(n(x-t)) - \sum_{l=0}^{N-1} \sum_{\|i\|_1 = l} \partial_1^{i_1} \dots \partial_d^{i_d} g_k(0) \prod_{j=1}^{d} \frac{(t_j - 2k_j\pi)^{i_j}}{i_j!} \right) dt \\
&= \frac{1}{(2\pi)^d} \sum_{k \in \mathbb{Z}^d} n^d \sum_{\|i\|_1 = N} (-1)^{\|i\|_1} n^{\|i\|_1} \int_{B+2k\pi} a(t) \\
&\qquad \partial_1^{i_1} \cdots \partial_d^{i_d} \widehat{\theta}_0 \Big(n(x - 2k\pi) - n v_k (t - 2k\pi) \Big) \prod_{j=1}^{d} \frac{(t_j - 2k_j\pi)^{i_j}}{i_j!} \, dt,
\end{aligned}$$

where $0 < v_k < 1$. Then, by Corollary 2.2.28,

$$\left|\sigma_n^{2,\alpha,\gamma} a(x)\right| \le C_p \sum_{k\in\mathbb{Z}^d} n^{(d-1)/2+N-\alpha} 2^{Kd/p} 2^{-KN} \int_{B+2k\pi} \|x-2k\pi - v_k(t-2k\pi)\|_2^{-d/2-\alpha-1/2}\,dt. \tag{2.5.18}$$

Moreover,

$$\begin{aligned}\sup_{n\ge d^{1/2}2^{K+1}} \left|\sigma_n^{2,\alpha,\gamma} a(x)\right| &\le C_p \sum_{k\in\mathbb{Z}^d} 2^{K((d-1)/2-\alpha)} 2^{Kd/p} \int_{B+2k\pi} \|x-2k\pi - v_k(t-2k\pi)\|_2^{-d/2-\alpha-1/2}\,dt \\ &=: A_1(x)+A_2(x),\end{aligned}$$

where

$$A_1(x) := 2^{K((d-1)/2-\alpha)} 2^{Kd/p} \int_{B+2k\pi} \|x-v_0 t\|_2^{-d/2-\alpha-1/2}\,dt$$

and

$$A_2(x) := \sum_{k\in\mathbb{Z}^d, k\ne 0} 2^{K((d-1)/2-\alpha)} 2^{Kd/p} \int_{B+2k\pi} \|x-2k\pi - v_k(t-2k\pi)\|_2^{-d/2-\alpha-1/2}\,dt.$$

If $k=0$, $u\in B$ and $x\in B(0,(i+2)2^{-K})\setminus B(0,(i+1)2^{-K})\cap\mathbb{T}^d$ for some $i=4\lfloor d^{1/2}\rfloor-1,\ldots,\lfloor d^{1/2}2^K\pi\rfloor-1$, then

$$\|x-u\|_2 \ge \|x\|_2 - \|u\|_2 \ge i2^{-K}.$$

In case $k\ne 0$, $u\in B+2k\pi$ and $x\in B(0,(i+2)2^{-K})\setminus B(0,(i+1)2^{-K})\cap\mathbb{T}^d$, one can see that

$$\|x-u\|_2 \ge \|k\|_2/4.$$

Then

$$\begin{aligned}A_1(x) &\le C_p 2^{K((d-1)/2-\alpha)} 2^{Kd/p} \int_B (i2^{-K})^{-d/2-\alpha-1/2}\,dt \\ &\le C_p 2^{Kd/p} i^{-d/2-\alpha-1/2}\end{aligned}$$

and

$$
\begin{aligned}
A_2(x) &\le C_p \sum_{k\in\mathbb{Z}^d, k\neq 0} 2^{K((d-1)/2-\alpha)} 2^{Kd/p} \int_{B+2k\pi} \|k\|_2^{-d/2-\alpha-1/2}\, dt \\
&\le C_p \sum_{k\in\mathbb{Z}^d, k\neq 0} 2^{K(-d/2-1/2-\alpha)} 2^{Kd/p} \|k\|_2^{-d/2-1/2-\alpha} \\
&\le C_p \sum_{j=1}^{\infty} 2^{K(-d/2-1/2-\alpha)} 2^{Kd/p} j^{(-d/2-1/2-\alpha)} j^{d-1} \\
&\le C_p
\end{aligned}
$$

for $p \ge d/(d/2+\alpha+1/2)$. Hence,

$$
(A) \le C_p \sum_{i=4\lfloor d^{1/2}\rfloor -1}^{\lfloor d^{1/2}2^K\pi\rfloor -1} 2^{-Kd} i^{d-1} 2^{Kd} i^{p(-d/2-\alpha-1/2)} + C_p \sum_{i=4\lfloor d^{1/2}\rfloor -1}^{\lfloor d^{1/2}2^K\pi\rfloor -1} 2^{-Kd} i^{d-1} \le C_p
$$

if $p > d/(d/2+\alpha+1/2)$.

Applying Taylor's formula for $N+1$ instead of N, we get similar to (2.5.18) that

$$
\begin{aligned}
\left|\sigma_n^{2,\alpha,\gamma} a(x)\right| \le C_p \sum_{k\in\mathbb{Z}^d} & n^{(d-1)/2+(N+1)-\alpha} 2^{Kd/p} 2^{-K(N+1)} \\
& \int_{B+2k\pi} \|x-2k\pi - v_k(t-2k\pi)\|_2^{-d/2-\alpha-1/2}\, dt
\end{aligned}
$$

and

$$
\begin{aligned}
\sup_{n<d^{1/2}2^{K+1}} \left|\sigma_n^{2,\alpha,\gamma} a(x)\right| \le C_p \sum_{k\in\mathbb{Z}^d} & 2^{K((d-1)/2-\alpha)} 2^{Kd/p} \\
& \int_{B+2k\pi} \|x-2k\pi - v_k(t-2k\pi)\|_2^{-d/2-\alpha-1/2}\, dt.
\end{aligned}
$$

The inequality

$$
(B) \le C_p
$$

can be shown as above. ■

Corollary 2.5.11 *Suppose that $q=2$, $(d-1)/2<\alpha<\infty$ and $\gamma\in\mathbb{P}$. If $1<p<\infty$, then*

$$
\|\sigma_*^{q,\alpha,\gamma} f\|_p \le C_p \|f\|_p \qquad (f\in L_p(\mathbb{T}^d)).
$$

Theorem 2.5.10 was proved by Stein, Taibleson and Weiss [292] and Lu [224]. The author generalized it for other summability methods in Weisz [332, 334]. The theorem is not true if p is smaller than or equal to the critical index $d/(d/2+\alpha+1/2)$ (see Stein, Taibleson and Weiss [292]).

Theorem 2.5.12 *If $q = 2$, $(d-1)/2 < \alpha < \infty$ and $\gamma \in \mathbb{P}$, then the operator $\sigma_*^{q,\alpha,\gamma}$ is not bounded from $H_p^\square(\mathbb{T}^d)$ to $L_p(\mathbb{T}^d)$ if p is smaller than or equal to the critical index $d/(d/2+\alpha+1/2)$.*

If p is equal to the critical index, then we have again a weak type inequality.

Theorem 2.5.13 *Suppose that $q = 2$, $(d-1)/2 < \alpha < \infty$ and $\gamma \in \mathbb{P}$. If*

$$p_0 := \frac{d}{d/2+\alpha+1/2}$$

and $f \in H_{p_0}^\square(\mathbb{T}^d)$, then

$$\left\|\sigma_*^{q,\alpha,\gamma} f\right\|_{p_0,\infty} = \sup_{\rho>0} \rho\lambda(\sigma_*^{q,\alpha,\gamma} f > \rho)^{1/p_0} \le C \left\|f\right\|_{H_{p_0}^\square}.$$

Proof We will use Theorem 2.4.20. Let us introduce the set

$$E_\rho := \left\{i \ge 4\lfloor d^{1/2}\rfloor - 1 : i^{-d/2-\alpha-1/2} > C^{-1}\rho 2^{-Kd/p}\right\},$$

where $p = d/(d/2+\alpha+1/2)$. Observe that

$$\rho^p \lambda\Big(\{A_1 > \rho\} \cap \{\mathbb{T}^d \setminus (rB)\}\Big) \le C\rho^p \sum_{i\in E_\rho} i^{d-1} 2^{-Kd}.$$

If k is the largest integer for which $k^{-d/2-\alpha-1/2} > C^{-1}\rho 2^{-Kd/p}$, then

$$\rho^p \lambda\Big(\{A_1 > \rho\} \cap \{\mathbb{T}^d \setminus (rB)\}\Big) \le \rho^p 2^{-Kd} k^d \le C.$$

The same inequality for (A_2) is trivial. We can estimate $\sup_{n<d^{1/2}2^{K+1}} |\sigma_n^{2,\alpha,\gamma} a(x)|$ similarly, which shows the theorem. ∎

Corollary 2.5.14 *Suppose that $q = 2$, $(d-1)/2 < \alpha < \infty$ and $\gamma \in \mathbb{P}$. If $f \in L_1(\mathbb{T}^d)$, then*

$$\sup_{\rho>0} \rho\, \lambda(\sigma_*^{q,\alpha,\gamma} f > \rho) \le C\|f\|_1.$$

As in the previous subsection, this implies

Corollary 2.5.15 *Suppose that $q = 2$, $(d-1)/2 < \alpha < \infty$ and $\gamma \in \mathbb{P}$. If $f \in L_1(\mathbb{T}^d)$, then*

$$\lim_{n\to\infty} \sigma_n^{q,\alpha,\gamma} f = f \qquad a.e.$$

2.6 ℓ_q-Summability Defined by a Function θ

Now we generalize the ℓ_q-Fejér and Riesz means investigated above. We introduce a general summability method, the so-called θ-summability generated by a given one-dimensional function θ.

We suppose that $\theta : \mathbb{R} \to \mathbb{R}$ and

$$\sum_{k\in\mathbb{Z}^d} \left| \theta\left(\frac{\|k\|_q}{n}\right) \right| < \infty \tag{2.6.1}$$

for all $n \in \mathbb{N}$. If θ has compact support, then this holds obviously. As we will see in Sect. 2.6.1, (2.6.2) implies (2.6.1).

Definition 2.6.1 Suppose that θ satisfies (2.6.1). For $f \in L_1(\mathbb{T}^d)$, $1 \le q \le \infty$ and $n \in \mathbb{N}$, the nth ℓ_q-θ-means $\sigma_n^{q,\theta} f$ of the Fourier series of f and the nth ℓ_q-θ kernel $K_n^{q,\theta}$ are defined by

$$\sigma_n^{q,\theta} f(x) := \sum_{k\in\mathbb{Z}^d} \theta\left(\frac{\|k\|_q}{n}\right) \widehat{f}(k) e^{\imath k\cdot x}$$

and

$$K_n^{q,\theta}(t) := \sum_{k\in\mathbb{Z}^d} \theta\left(\frac{\|k\|_q}{n}\right) e^{\imath k\cdot t},$$

respectively.

Lemma 2.6.2 *Suppose that θ satisfies* (2.6.1). *For $f \in L_1(\mathbb{T}^d)$ and $n \in \mathbb{N}$,*

$$\sigma_n^{q,\theta} f(x) = \frac{1}{(2\pi)^d} \int_{\mathbb{T}^d} f(x-t) K_n^{q,\theta}(t)\, dt.$$

The definition of the ℓ_q-θ-means can be extended to distributions as usual.

Definition 2.6.3 Suppose that θ satisfies (2.6.1). For $f \in D(\mathbb{T}^d)$, $1 \le q \le \infty$ and $n \in \mathbb{N}$, the nth ℓ_q-θ-means $\sigma_n^{q,\theta} f$ of the Fourier series of f are given by

$$\sigma_n^{q,\theta} f := f * K_n^{q,\theta}.$$

Definition 2.6.4 We define the maximal θ-operator by

$$\sigma_*^{q,\theta} f := \sup_{n\in\mathbb{N}} \left|\sigma_n^{q,\theta} f\right|.$$

Note that $K_n^{q,\theta}$ is bounded and integrable. If $\theta(t) = \max((1-|t|^\gamma)^\alpha, 0)$, then we get back the Riesz (or in special case $\alpha = \gamma = 1$, the Fejér) means. θ-summability was considered in many papers and books, such as Butzer and Nessel [47], Trigub and Belinsky [319], Natanson and Zuk [244], Bokor, Schipp, Szili and Vértesi [38, 272, 274, 300, 301], and Feichtinger and Weisz [103, 104, 332, 337, 338, 342, 346].

2.6.1 Triangular and Cubic Summability

For $q = 1$ or ∞, instead of (2.6.1), we suppose that

$$\begin{cases} \text{the support of } \theta \text{ is } [-c, c] \ (0 < c \leq \infty), \\ \theta \text{ is even and continuous, } \theta(0) = 1, \\ \sum_{k=0}^{\infty} k^d \left| \Delta_1 \theta(\frac{k}{n}) \right| < \infty, \\ \lim_{t \to \infty} t^d \theta(t) = 0, \end{cases} \tag{2.6.2}$$

where

$$\Delta_1 \theta \left(\frac{k}{n} \right) := \theta \left(\frac{k}{n} \right) - \theta \left(\frac{k+1}{n} \right)$$

is the first difference. If the support of θ is not compact, then we say that $c = \infty$. Abel rearrangement implies

$$\sum_{j \in \mathbb{Z}^d} \left| \theta \left(\frac{\|j\|_q}{n} \right) \right| \leq C \sum_{k=0}^{\infty} k^{d-1} \left| \theta \left(\frac{k}{n} \right) \right| \leq C \sum_{k=0}^{\infty} k^d \left| \Delta_1 \theta \left(\frac{k}{n} \right) \right| < \infty,$$

thus (2.6.1) holds.

Lemma 2.6.5 *Suppose that θ satisfies* (2.6.2). *For $f \in L_1(\mathbb{T}^d)$, $q = 1, \infty$ and $n \in \mathbb{N}$, we have*

$$\sigma_n^{q,\theta} f(x) = \sum_{j=0}^{\infty} \Delta_1 \theta \left(\frac{j}{n} \right) s_j^q f(x).$$

and

$$K_n^{q,\theta}(t) = \sum_{j=0}^{\infty} \Delta_1 \theta \left(\frac{j}{n} \right) D_j^q(t)$$

Proof The proof follows from

$$K_n^{q,\theta}(t) = \sum_{k \in \mathbb{Z}^d} \sum_{j \geq \|k\|_q} \Delta_1 \theta \left(\frac{j}{n} \right) e^{\imath k \cdot t} = \sum_{j=0}^{\infty} \Delta_1 \theta \left(\frac{j}{n} \right) D_j^q(t)$$

■

We need also the following condition:

$$\begin{cases} \theta \text{ is twice continuously differentiable on}(0,c), \\ \theta'' \neq 0 \text{ except at finitely many points and finitely many intervals}, \\ \lim_{t\to 0+0} t\theta'(t) \text{ is finite}, \\ \lim_{t\to c-0} t\theta'(t) \text{ is finite}, \\ \lim_{t\to\infty} t\theta'(t) = 0. \end{cases} \tag{2.6.3}$$

The norm convergence follows easily from Theorem 2.6.7.

Theorem 2.6.6 *Assume that $q = 1$ or $q = \infty$ and* (2.6.2) *and* (2.6.3) *are satisfied. If $1 \le p < \infty$, then*

$$\sup_{n\in\mathbb{N}} \left\|\sigma_n^{q,\theta} f\right\|_p \le C\|f\|_p$$

and

$$\lim_{n\to\infty} \sigma_n^{q,\theta} f = f \qquad \text{in the } L_p(\mathbb{T}^d)\text{-norm for all } f \in L_p(\mathbb{T}^d).$$

For the almost everywhere convergence, we introduce some notations. Let $\mathbb{X}$ and $\mathbb{Y}$ be two complete quasi-normed spaces of measurable functions, $L_\infty(\mathbb{T}^d)$ be continuously embedded into $\mathbb{X}$ and $L_\infty(\mathbb{T}^d)$ be dense in $\mathbb{X}$. Suppose that if $0 \le f \le g$, $f, g \in \mathbb{Y}$, then $\|f\|_{\mathbb{Y}} \le \|g\|_{\mathbb{Y}}$. If $f_n, f \in \mathbb{Y}$, $f_n \ge 0$ $(n \in \mathbb{N})$ and $f_n \nearrow f$ a.e. as $n \to \infty$, then assume that $\|f - f_n\|_{\mathbb{Y}} \to 0$. Recall that σ_*^q denotes the maximal Fejér operator.

Theorem 2.6.7 *Assume that $q = 1$ or $q = \infty$ and* (2.6.2) *and* (2.6.3) *are satisfied. If $\sigma_*^q : \mathbb{X} \to \mathbb{Y}$ is bounded, i.e.,*

$$\|\sigma_*^q f\|_{\mathbb{Y}} \le C\|f\|_{\mathbb{X}} \qquad (f \in \mathbb{X} \cap L_\infty(\mathbb{T}^d)),$$

then $\sigma_*^{q,\theta}$ is also bounded,

$$\|\sigma_*^{q,\theta} f\|_{\mathbb{Y}} \le C\|f\|_{\mathbb{X}} \qquad (f \in \mathbb{X}).$$

Proof By Abel rearrangement,

$$\sum_{k=0}^{m} \Delta_1\theta\left(\frac{k}{n}\right) D_k^q(x) = \sum_{k=0}^{m-1} \Delta_2\theta\left(\frac{k}{n}\right) k K_k^q(x) + \Delta_1\theta\left(\frac{m}{n}\right) m K_m^q(x),$$

where

$$\Delta_2\theta\left(\frac{k}{n}\right) := \Delta_1\theta\left(\frac{k}{n}\right) - \Delta_1\theta\left(\frac{k+1}{n}\right)$$

is the second difference and K_m^q denotes the Fejér kernel. Observe that for a fixed x, we have that $K_m^q(x)$ is uniformly bounded in m. By Lagrange's mean value theorem there exists $m < \xi(m) < m + 1$, such that

$$m\Delta_1\theta\left(\frac{m}{n}\right) = -\frac{m}{n}\theta'\left(\frac{\xi(m)}{n}\right)$$

and this converges to zero if $m \to \infty$. Thus,

$$K_n^{q,\theta}(x) = \sum_{k=0}^{\infty} k\,\Delta_2\theta\left(\frac{k}{n}\right) K_k^q(x).$$

Now we prove that

$$\sup_{n\ge 1}\sum_{k=0}^{\infty} k\left|\Delta_2\theta\left(\frac{k}{n}\right)\right| \le C < \infty. \tag{2.6.4}$$

If $\theta'' \ge 0$ on the interval $(i/n, (j+2)/n)$, then θ is convex on this interval and this yields that

$$\Delta_2\theta\left(\frac{k}{n}\right) \ge 0 \quad \text{for} \quad i \le k \le j.$$

Hence

$$\begin{aligned}\sum_{k=i}^{j} k\left|\Delta_2\theta\left(\frac{k}{n}\right)\right| &= \sum_{k=i}^{j} k\,\Delta_2\theta\left(\frac{k}{n}\right)\\ &= \theta\left(\frac{i}{n}\right) + (i-1)\,\Delta_1\theta\left(\frac{i}{n}\right) -\\ &\qquad j\,\Delta_1\theta\left(\frac{j+1}{n}\right) - \theta\left(\frac{j+1}{n}\right).\end{aligned}$$

Applying again Lagrange's mean value theorem, we have

$$(i-1)\left|\Delta_1\theta\left(\frac{i}{n}\right)\right| = \frac{i-1}{n}\left|\theta'\left(\frac{\xi(i)}{n}\right)\right| = \frac{i-1}{\xi(i)}\left|\frac{\xi(i)}{n}\theta'\left(\frac{\xi(i)}{n}\right)\right| \le C,$$

where $i < \xi(i) < i+1$. Here, we used the fact that the function $x \mapsto |x\theta'(x)|$ is bounded, which follows from (2.6.3). If $\theta'' = 0$ at an isolated point u or if θ'' is not twice continuously differentiable at u, $u \in (k/n, (k+1)/n)$, then the boundedness of $k\left|\Delta_2\theta\left(\frac{k}{n}\right)\right|$ can be seen in the same way. Since there are only finitely many intervals and isolated points satisfying the above properties, we have shown (2.6.4).

Hence

$$\begin{aligned}\sigma_n^{q,\theta} f(x) &= \int_{\mathbb{T}^d} f(t) K_n^{q,\theta}(x-t)\,dt\\ &= \sum_{k=0}^{\infty}\int_{\mathbb{T}^d} k\,\Delta_2\theta\left(\frac{k}{n}\right) f(t) K_k^q(x-t)\,dt\end{aligned}$$

for all $f \in L_\infty(\mathbb{T}^d)$. Thus

$$\sigma_*^{q,\theta} f \le C\sigma_*^q f \qquad (f \in L_\infty(\mathbb{T}^d))$$

and so

$$\left\|\sigma_*^{q,\theta} f\right\|_{\mathbb{Y}} \le C \left\|f\right\|_{\mathbb{X}} \qquad (f \in \mathbb{X} \cap L_\infty(\mathbb{T}^d)).$$

By a usual density argument, we finish the proof of the theorem. ■

It is easy to see that $\mathbb{X}$ can be chosen to be the Hardy space $H_p^\square(\mathbb{T}^d)$ and $\mathbb{Y}$ to be the space $L_p(\mathbb{T}^d)$ or $L_{p,\infty}(\mathbb{T}^d)$ $(0 < p \le \infty)$. Theorems 2.6.7 and 2.5.4 imply

Theorem 2.6.8 *Assume that* $q = 1$ *or* $q = \infty$ *and* (2.6.2) *and* (2.6.3) *are satisfied. If*

$$\frac{d}{d+1} < p \le \infty,$$

then

$$\left\|\sigma_*^{q,\theta} f\right\|_p \le C_p \left\|f\right\|_{H_p^\square} \qquad (f \in H_p^\square(\mathbb{T}^d))$$

and, for $f \in H_{d/(d+1)}^\square(\mathbb{T}^d)$,

$$\left\|\sigma_*^{q,\theta} f\right\|_{d/(d+1),\infty} = \sup_{\rho>0} \rho\lambda(\sigma_*^{q,\theta} f > \rho)^{(d+1)/d} \le C \left\|f\right\|_{H_{d/(d+1)}^\square}.$$

Moreover,

$$\sup_{\rho>0} \rho\,\lambda(\sigma_*^{q,\theta} f > \rho) \le C\|f\|_1 \qquad (f \in L_1(\mathbb{T}^d)).$$

Corollary 2.6.9 *Assume that* $q = 1$ *or* $q = \infty$ *and* (2.6.2) *and* (2.6.3) *are satisfied. If* $f \in L_1(\mathbb{T}^d)$, *then*

$$\lim_{n\to\infty} \sigma_n^{q,\theta} f = f \qquad a.e.$$

2.6.2 *Circular Summability*

If $q = 2$, then we have to assume other additional conditions instead of (2.6.2) and (2.6.3). Recall that

$$\theta_0(x) = \theta(\|x\|_2).$$

Let

$$\theta_0 \in L_1(\mathbb{R}^d) \qquad \text{and} \qquad \widehat{\theta_0} \in L_1(\mathbb{R}^d). \tag{2.6.5}$$

Assume that $\widehat{\theta_0}$ is $(N+1)$-times differentiable $(N \ge 0)$ and there exists

$$d+N-1<\beta\le d+N$$

such that

$$\left|\partial_1^{i_1}\cdots\partial_d^{i_d}\widehat{\theta_0}(x)\right|\le C\|x\|_2^{-\beta-1}\qquad (x\neq 0), \tag{2.6.6}$$

whenever $i_1+\cdots+i_d=N$ or $i_1+\cdots+i_d=N+1$. If $\beta=d+N$, then it is enough to suppose (2.6.6) for $i_1+\cdots+i_d=N+1$.

We recall that the Riesz summability, i.e., if $\theta(t)=\max((1-|t|^\gamma)^\alpha,0)$, satisfy (2.6.5) and (2.6.6) with $\beta=d/2+\alpha-1/2$ (see Corollary 2.2.28).

The norm convergence can be proved as Theorem 2.3.2.

Theorem 2.6.10 *Assume that* $q=2$, $\theta(0)=1$ *and* (2.6.1) *and* (2.6.5) *are satisfied. If* $1\le p<\infty$, *then*

$$\sup_{n\in\mathbb{N}}\left\|\sigma_n^{q,\theta}f\right\|_p\le C\|f\|_p$$

and

$$\lim_{n\to\infty}\sigma_n^{q,\theta}f=f\qquad \text{in the } L_p(\mathbb{T}^d)\text{-norm for all } f\in L_p(\mathbb{T}^d).$$

We can prove the next theorem similar to Theorem 2.5.10. The details are left to the reader.

Theorem 2.6.11 *Assume that* $q=2$ *and* (2.6.1), (2.6.5) *and* (2.6.6) *are satisfied. If*

$$\frac{d}{\beta+1}<p\le\infty,$$

then

$$\left\|\sigma_*^{q,\theta}f\right\|_p\le C_p\|f\|_{H_p^\square}\qquad (f\in H_p^\square(\mathbb{T}^d))$$

and, for $f\in H_{d/(\beta+1)}^\square(\mathbb{T}^d)$,

$$\left\|\sigma_*^{q,\theta}f\right\|_{d/(\beta+1),\infty}=\sup_{\rho>0}\rho\lambda(\sigma_*^{q,\theta}f>\rho)^{(\beta+1)/d}\le C\|f\|_{H_{d/(\beta+1)}^\square}.$$

Moreover,

$$\sup_{\rho>0}\rho\,\lambda(\sigma_*^{q,\theta}f>\rho)\le C\|f\|_1\qquad (f\in L_1(\mathbb{T}^d)).$$

Corollary 2.6.12 *Assume that* $q=2$, $\theta(0)=1$ *and* (2.6.1), (2.6.5) *and* (2.6.6) *are satisfied. If* $f\in L_1(\mathbb{T}^d)$, *then*

$$\lim_{n\to\infty}\sigma_n^{q,\theta}f=f\qquad a.e.$$

We note again, that (2.6.2) implies (2.6.1).

2.6.3 Some Summability Methods

Now we give some examples for the θ-summation.

Example 2.6.13 (Fejér summation). Let

$$\theta(t) = \begin{cases} 1 - |t| & \text{if } |t| \le 1; \\ 0 & \text{if } |t| > 1. \end{cases}$$

Example 2.6.14 (de La Vallée-Poussin summation). Let

$$\theta(t) = \begin{cases} 1 & \text{if if } |t| \le 1/2; \\ -2|t| + 2 & \text{if } 1/2 < |t| \le 1; \\ 0 & \text{if } |t| > 1. \end{cases}$$

Example 2.6.15 (Jackson-de La Vallée-Poussin summation). Let

$$\theta(t) = \begin{cases} 1 - 3t^2/2 + 3|t|^3/4 & \text{if } |t| \le 1; \\ (2 - |t|)^3/4 & \text{if } 1 < |t| \le 2; \\ 0 & \text{if } |t| > 2. \end{cases}$$

Example 2.6.16 Let $0 = \alpha_0 < \alpha_1 < \cdots < \alpha_m$ and $\beta_0, \ldots, \beta_m$ $(m \in \mathbb{N})$ be real numbers, $\beta_0 = 1$, $\beta_m = 0$. Suppose that θ is even, $\theta(\alpha_j) = \beta_j$ $(j = 0, 1, \ldots, m)$, $\theta(t) = 0$ for $t \ge \alpha_m$, θ is a polynomial on the interval $[\alpha_{j-1}, \alpha_j]$ $(j = 1, \ldots, m)$.

Example 2.6.17 (Rogosinski summation). Let

$$\theta(t) = \begin{cases} \cos \pi t/2 & \text{if } |t| \le 1 + 2j; \\ 0 & \text{if } |t| > 1 + 2j \end{cases} \qquad \text{for some } j \in \mathbb{N}.$$

Example 2.6.18 (Weierstrass summation). Let

$$\theta(t) = e^{-|t|^\gamma} \quad \text{for some} \quad 1 \le \gamma < \infty.$$

Note that if $\gamma = 1$, then we obtain the Abel means.

Example 2.6.19 Let

$$\theta(t) = e^{-(1+|t|^q)^\gamma} \quad \text{for some} \quad 1 \le q < \infty, 0 < \gamma < \infty.$$

Example 2.6.20 (Picard and Bessel summations). Let

$$\theta(t) = (1 + |t|^\gamma)^{-\alpha} \quad \text{for some} \quad 0 < \alpha < \infty, 1 \le \gamma < \infty, \alpha\gamma > d.$$

Example 2.6.21 (Riesz summation). Let

$$\theta(t) = \begin{cases} (1-|t|^\gamma)^\alpha & \text{if } |t| \le 1; \\ 0 & \text{if } |t| > 1 \end{cases}$$

for some $0 < \alpha, \gamma < \infty$.

It is easy to see that all of these examples satisfy (2.6.2) and (2.6.3).

Theorem 2.6.22 *Suppose that θ is one of the Examples 2.6.13–2.6.21. Then Theorems 2.6.6, 2.6.8 and Corollary 2.6.9 hold.*

One can show [334, 343] that Example 2.6.21 with $\alpha > (d-1)/2$, $\gamma \in \mathbb{P}$ and $\beta = d/2 + \alpha - 1/2$, Example 2.6.18 with $0 < \gamma < \infty$ and $\beta = d + N$, Example 2.6.19 with $0 < \gamma, q < \infty$ and $\beta = d + N$ and Example 2.6.20 with $\beta = d + N$ satisfy (2.6.2), (2.6.5) and (2.6.6).

Theorem 2.6.23 *Suppose that θ is one of the Examples 2.6.18, 2.6.19, 2.6.20 or 2.6.21 with the parameter β just defined. Then Theorems 2.6.10, 2.6.11 and Corollary 2.6.12 hold.*

Chapter 3
Rectangular Summability of Higher Dimensional Fourier Series

In this chapter, we investigate the rectangular summability of d-dimensional Fourier series. We consider two types of convergence, the so-called restricted and unrestricted convergence. In the first case, $n \in \mathbb{N}^d$ is in a cone or a cone-like set and $n \to \infty$ while in the second case, we have $n \in \mathbb{N}^d$ and $\min(n_1, \ldots, n_d) \to \infty$, which is called Pringsheim's convergence. Similarly, we consider two types of maximal operators, the restricted one defined on a cone or cone-like set and the unrestricted one defined on $\mathbb{N}^d$. We prove similar results as for the ℓ_q-summability. In the restricted case, we use the Hardy space $H_p^{\square}(\mathbb{T}^d)$ and in the unrestricted case a new Hardy space $H_p(\mathbb{T}^d)$.

In the first section, we present the basic definitions for the rectangular summability and verify some estimations for the kernel functions. In the next section, we can find the $L_p(\mathbb{T}^d)$ convergence of the rectangular Cesàro and Riesz means. In Sect. 3.3, we investigate the restricted maximal operators of the rectangular Cesàro and Riesz means by taking the supremum over a cone. We show that these operators are bounded from the Hardy space $H_p^{\square}(\mathbb{T}^d)$ to $L_p(\mathbb{T}^d)$ for any $p > p_0$, where $p_0 < 1$ is depending again on the summation and on the dimension. As a consequence, we obtain the restricted almost everywhere convergence of the summability means. Similar results are also shown for cone-like sets.

We introduce the product Hardy spaces $H_p(\mathbb{T}^d)$ and present the atomic decomposition and a boundedness result for these spaces. Moreover, we show that the unrestricted maximal operator of the rectangular Cesàro and Riesz means is bounded from $H_p(\mathbb{T}^d)$ to $L_p(\mathbb{T}^d)$ for any $p > p_0$. This implies the almost everywhere convergence of the summability means in Pringsheim's sense. In the last section, we consider the rectangular θ-summability and prove similar results as mentioned above. We give a sufficient and necessary condition for the uniform and $L_1(\mathbb{T}^d)$ convergence of the rectangular θ-means.

F. Weisz, *Lebesgue Points and Summability of Higher Dimensional Fourier Series*,
https://doi.org/10.1007/978-3-030-74636-0_3

3.1 Summability Kernels

Definition 3.1.1 For $f \in L_1(\mathbb{T}^d)$ and $n \in \mathbb{N}^d$, the nth rectangular Fejér means $\sigma_n f$ of the Fourier series of f and the nth rectangular Fejér kernel K_n are introduced by

$$\sigma_n f(x) = \sum_{|k_1|\le n_1} \cdots \sum_{|k_d|\le n_d} \prod_{i=1}^{d} \left(1 - \frac{|k_i|}{n_i}\right) \widehat{f}(k) e^{\imath k\cdot x}$$

and

$$K_n(t) := \sum_{|k_1|\le n_1} \cdots \sum_{|k_d|\le n_d} \prod_{i=1}^{d} \left(1 - \frac{|k_i|}{n_i}\right) e^{\imath k\cdot t},$$

respectively.

Again, we generalize this definition as follows.

Definition 3.1.2 Let $f \in L_1(\mathbb{T}^d)$, $n \in \mathbb{N}^d$ and $\alpha \ge 0$. The nth rectangular Cesàro means $\sigma_n^\alpha f$ of the Fourier series of f and the nth rectangular Cesàro kernel K_n^α are introduced by

$$\sigma_n^\alpha f(x) := \frac{1}{\prod_{i=1}^d A_{n_i-1}^\alpha} \sum_{|k_1|\le n_1} \cdots \sum_{|k_d|\le n_d} \prod_{i=1}^{d} A_{n_i-1-|k_i|}^\alpha \widehat{f}(k) e^{\imath k\cdot x}$$

and

$$K_n^\alpha(t) := \frac{1}{\prod_{i=1}^d A_{n_i-1}^\alpha} \sum_{|k_1|\le n_1} \cdots \sum_{|k_d|\le n_d} \prod_{i=1}^{d} A_{n_i-1-|k_i|}^\alpha e^{\imath k\cdot t},$$

respectively.

The Cesàro means are also called rectangular (C, α)-means. If $\alpha = 1$, then these are the rectangular Fejér means and if $\alpha = 0$, then the rectangular partial sums (see Fig. 3.1).

Definition 3.1.3 For $f \in L_1(\mathbb{T}^d)$, $n \in \mathbb{N}^d$ and $0 < \alpha, \gamma < \infty$, the nth rectangular Riesz means $\sigma_n^{\alpha,\gamma} f$ of the Fourier series of f and the nth rectangular Riesz kernel $K_n^{\alpha,\gamma}$ are given by

$$\sigma_n^{\alpha,\gamma} f(x) := \sum_{|k_1|\le n_1} \cdots \sum_{|k_d|\le n_d} \prod_{i=1}^{d} \left(1 - \left(\frac{|k_i|}{n_i}\right)^\gamma\right)^\alpha \widehat{f}(k) e^{\imath k\cdot x}$$

and

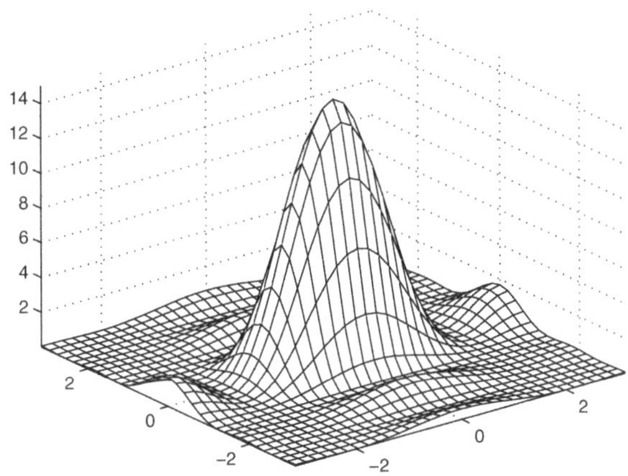

Fig. 3.1 The rectangular Fejér kernel K_n with $d = 2$, $n_1 = 3, n_2 = 5$

$$K_n^{\alpha,\gamma}(t) := \sum_{|k_1| \le n_1} \cdots \sum_{|k_d| \le n_d} \prod_{i=1}^{d} \left(1 - \left(\frac{|k_i|}{n_i}\right)^{\gamma}\right)^{\alpha} e^{\imath k \cdot t},$$

respectively.

For $\alpha = \gamma = 1$, we get back the rectangular Fejér means. The next results follow from

$$K_n^{\alpha} = K_{n_1}^{\alpha} \otimes \cdots \otimes K_{n_d}^{\alpha} \tag{3.1.1}$$

and

$$K_n^{\alpha,\gamma} = K_{n_1}^{\alpha,\gamma} \otimes \cdots \otimes K_{n_d}^{\alpha,\gamma}, \tag{3.1.2}$$

where $K_{n_j}^{\alpha}$ and $K_{n_j}^{\alpha,\gamma}$ are the corresponding one dimensional kernels.

Lemma 3.1.4 *If $0 \le \alpha, \gamma < \infty$ and $n \in \mathbb{N}^d$, then*

$$\frac{1}{(2\pi)^d} \int_{\mathbb{T}^d} K_n^{\alpha}(t)\, dt = 1$$

and

$$\frac{1}{(2\pi)^d} \int_{\mathbb{T}^d} K_n^{\alpha,\gamma}(t)\, dt = 1.$$

Lemma 3.1.5 *If $0 \le \alpha, \gamma < \infty$ and $n \in \mathbb{N}^d$, then*

$$|K_n^{\alpha}(t)| \le C \prod_{i=1}^{d} n_i \quad \text{and} \quad |K_n^{\alpha,\gamma}(t)| \le C \prod_{i=1}^{d} n_i \quad (t \in \mathbb{T}^d).$$

Lemma 3.1.6 *For $f \in L_1(\mathbb{T}^d)$, $n \in \mathbb{N}^d$ and $0 < \alpha, \gamma < \infty$,*

$$\sigma_n^\alpha f(x) = \frac{1}{(2\pi)^d} \int_{\mathbb{T}^d} f(x-t) K_n^\alpha(t)\, dt$$

and

$$\sigma_n^{\alpha,\gamma} f(x) = \frac{1}{(2\pi)^d} \int_{\mathbb{T}^d} f(x-t) K_n^{\alpha,\gamma}(t)\, dt.$$

The rectangular Cesàro means are the weighted arithmetic means of the rectangular partial sums.

Lemma 3.1.7 *For $f \in L_1(\mathbb{T}^d)$, $\alpha > 0$ and $n \in \mathbb{N}^d$, we have*

$$\sigma_n f(x) = \frac{1}{\prod_{i=1}^d n_i} \sum_{k_1=1}^{n_1-1} \cdots \sum_{k_d=1}^{n_d-1} s_k f(x),$$

$$\sigma_n^\alpha f(x) = \frac{1}{\prod_{i=1}^d A_{n_i-1}^\alpha} \sum_{k_1=0}^{n_1-1} \cdots \sum_{k_d=0}^{n_d-1} \prod_{i=1}^d A_{n_i-1-k_i}^{\alpha-1} s_k f(x)$$

and

$$K_n^\alpha(t) = \frac{1}{\prod_{i=1}^d A_{n_i-1}^\alpha} \sum_{k_1=0}^{n_1-1} \cdots \sum_{k_d=0}^{n_d-1} \prod_{i=1}^d A_{n_i-1-k_i}^{\alpha-1} D_k(t).$$

We will use the next estimation of the derivatives of the one-dimensional kernel functions.

Theorem 3.1.8 *For $0 < \alpha \le r+1$, $n \in \mathbb{P}$ and $t \in \mathbb{T}$, $t \neq 0$,*

$$\left|\left(K_n^\alpha\right)^{(r)}(t)\right| \le C n^{r+1} \qquad \text{and} \qquad \left|\left(K_n^\alpha\right)^{(r)}(t)\right| \le \frac{C}{n^{\alpha-r}|t|^{\alpha+1}}.$$

Proof Similar to Lemma 1.2.4 and Theorem 1.4.16, we have

$$|D_k^{(r)}| \le C k^{r+1} \qquad (k \in \mathbb{P}),$$

which implies the first inequality.

We have seen in Theorem 1.4.16 and Lemma 1.4.14 that

$$\begin{aligned}
K_n^\alpha(t) &= \frac{1}{A_{n-1}^\alpha}\sum_{k=0}^{n-1} A_{n-1-k}^{\alpha-1}\frac{\sin((k+1/2)t)}{\sin(t/2)} \\
&= \frac{1}{A_{n-1}^\alpha \sin(t/2)}\Im\left(\sum_{k=0}^{n-1} A_{n-1-k}^{\alpha-1} e^{\imath(k+1/2)t}\right) \\
&= \frac{1}{A_{n-1}^\alpha \sin(t/2)}\Im\left(e^{\imath(n-1/2)t}\sum_{j=0}^{n-1} A_j^{\alpha-1} e^{-\imath jt}\right).
\end{aligned}$$

In this proof, we use the notation

$$u(\beta) := \sum_{k=0}^{n-1} A_k^\beta e^{-\imath kt}.$$

Abel rearrangement and Lemma 1.4.8 imply

$$\begin{aligned}
u(\beta) &= \sum_{k=0}^{n-2}\left(A_k^\beta - A_{k+1}^\beta\right) S_k + A_{n-1}^\beta S_{n-1} \\
&= -\sum_{k=0}^{n-2} A_{k+1}^{\beta-1} S_k + A_{n-1}^\beta S_{n-1} \\
&= -\sum_{k=1}^{n-1} A_k^{\beta-1} S_{k-1} + A_{n-1}^\beta S_{n-1},
\end{aligned}$$

where

$$S_k := \sum_{j=0}^{k} e^{-\imath jt} = \frac{1-e^{-\imath(k+1)t}}{1-e^{-\imath t}}.$$

Then

$$\begin{aligned}
u(\beta) &= -\sum_{k=1}^{n-1} A_k^{\beta-1}\frac{1-e^{-\imath kt}}{1-e^{-\imath t}} + A_{n-1}^\beta \frac{1-e^{-\imath nt}}{1-e^{-\imath t}} \\
&= \left(1-e^{-\imath t}\right)^{-1}\left(\sum_{k=1}^{n-1} A_k^{\beta-1} e^{-\imath kt} - \sum_{k=1}^{n-1} A_k^{\beta-1} + A_{n-1}^\beta - A_{n-1}^\beta e^{-\imath nt}\right) \\
&= \left(1-e^{-\imath t}\right)^{-1} u(\beta-1) - \left(1-e^{-\imath t}\right)^{-1} A_{n-1}^\beta e^{-\imath nt}.
\end{aligned}$$

Iterating this result s-times ($s \in \mathbb{N}$),

$$\begin{aligned} u(\beta) &= \left(1-e^{-\imath t}\right)^{-2} u(\beta-2) - \left(1-e^{-\imath t}\right)^{-2} A_{n-1}^{\beta-1} e^{-\imath nt} \\ &\quad - \left(1-e^{-\imath t}\right)^{-1} A_{n-1}^{\beta} e^{-\imath nt} \\ &= \dots \\ &= \left(1-e^{-\imath t}\right)^{-s} u(\beta-s) - e^{-\imath nt} \sum_{j=1}^{s} A_{n-1}^{\beta-j+1} \left(1-e^{-\imath t}\right)^{-j}. \end{aligned}$$

Writing $\beta = \alpha - 1$ and using (1.4.11), we conclude

$$\begin{aligned} K_n^{\alpha}(t) &= \frac{1}{A_{n-1}^{\alpha} \sin(t/2)} \Im\left(e^{\imath(n-1/2)t} u(\alpha-1)\right) \\ &= \frac{1}{A_{n-1}^{\alpha} \sin(t/2)} \Im\Bigg(e^{\imath(n-1/2)t} \left(1-e^{-\imath t}\right)^{-s} u(\alpha-1-s) \\ &\quad - e^{-\imath t/2} \sum_{j=1}^{s} A_{n-1}^{\alpha-j} \left(1-e^{-\imath t}\right)^{-j}\Bigg) \\ &= \frac{1}{A_{n-1}^{\alpha} \sin(t/2)} \Im\Bigg(e^{\imath(n-1/2)t} \left(1-e^{-\imath t}\right)^{-s} \sum_{k=0}^{n-1} A_k^{\alpha-1-s} e^{-\imath kt} \\ &\quad - e^{-\imath t/2} \sum_{j=1}^{s} A_{n-1}^{\alpha-j} \left(1-e^{-\imath t}\right)^{-j}\Bigg). \end{aligned}$$

The equality

$$\begin{aligned} K_n^{\alpha}(t) &= \frac{1}{A_{n-1}^{\alpha} \sin(t/2)} \Im\Bigg(e^{\imath(n-1/2)t} \left(1-e^{-\imath t}\right)^{-\alpha} \\ &\quad - \left(1-e^{-\imath t}\right)^{-s} \sum_{k=n}^{\infty} A_k^{\alpha-1-s} e^{-\imath(k-n+1/2)t} - e^{-\imath t/2} \sum_{j=1}^{s} A_{n-1}^{\alpha-j} \left(1-e^{-\imath t}\right)^{-j}\Bigg) \\ &=: I_1(t) + I_2(t) + I_3(t) \end{aligned}$$

follows from (1.4.5). Suppose that $|t| \geq 1/n$. The rth derivative of I_1 can be estimated as

$$\begin{aligned} \left|I_1^{(r)}(t)\right| &\leq \frac{C}{A_{n-1}^{\alpha}} \sum_{l=0}^{r} \frac{n^l}{|t|^{1+\alpha+r-l}} \\ &\leq C|t|^{-r-1} \sum_{l=0}^{r} (n|t|)^{l-\alpha} \\ &\leq C|t|^{-r-1} (n|t|)^{r-\alpha} = C n^{r-\alpha} |t|^{-\alpha-1}. \end{aligned}$$

To estimate the second term, we choose $s > \alpha + r$. Then the r times termwise differentiated series in I_2 is absolutely convergent. Thus

$$\begin{aligned}
\left|I_2^{(r)}(t)\right| &\le \frac{C}{A_{n-1}^{\alpha}} \sum_{l=0}^{r} \sum_{k=n}^{\infty} A_k^{\alpha-1-s} \frac{(k-n+1/2)^l}{|t|^{1+s+r-l}} \\
&\le \frac{C}{A_{n-1}^{\alpha}} \sum_{l=0}^{r} |t|^{-1-s-r+l} \sum_{k=n}^{\infty} k^{\alpha-1-s+l} \\
&\le C \sum_{l=0}^{r} |t|^{-1-s-r+l} n^{l-s} \\
&\le C|t|^{-r-1} \sum_{l=0}^{r} (n|t|)^{l-s} \\
&\le C|t|^{-r-1}(n|t|)^{r-s} \le C|t|^{-r-1}(n|t|)^{r-\alpha} = Cn^{r-\alpha}|t|^{-\alpha-1}.
\end{aligned}$$

Similarly,

$$\begin{aligned}
\left|I_3^{(r)}(t)\right| &\le \frac{C}{A_{n-1}^{\alpha}} \sum_{j=1}^{s} A_{n-1}^{\alpha-j} \frac{1}{|t|^{1+j+r}} \\
&\le C|t|^{-r-1} \sum_{j=1}^{s} (n|t|)^{-j} \\
&\le C|t|^{-r-1}(n|t|)^{-1} \le C|t|^{-r-1}(n|t|)^{r-\alpha} = Cn^{r-\alpha}|t|^{-\alpha-1},
\end{aligned}$$

because $0 < \alpha \le r+1$. Finally, if $|t| < 1/n$, then the first inequality of our theorem implies the second one. ∎

The next lemma can be proved as Lemma 1.4.13.

Lemma 3.1.9 *For $\alpha > -1$ and $h > 0$, we have*

$$\sigma_n^{\alpha+h} f = \frac{1}{\prod_{i=1}^{d} A_{n_i-1}^{\alpha+h}} \sum_{k_1=1}^{n_1} \cdots \sum_{k_d=1}^{n_d} \prod_{i=1}^{d} A_{n_i-k_i}^{h-1} A_{k_i-1}^{\alpha} \sigma_k^{\alpha} f.$$

The same results hold if we choose different exponents α_i and γ_i in the products.

3.2 Norm Convergence of Rectangular Summability Means

The next results follow from (3.1.1), (3.1.2), Theorem 2.3.3 and from the one-dimensional theorems.

Theorem 3.2.1 *If* $0 < \alpha \le 1$, *then*

$$\sup_{n \in \mathbb{N}^d} \int_{\mathbb{T}^d} \left| K_n^{\alpha}(x) \right| dx \le C.$$

If $0 < \alpha < \infty$ *and* $\gamma \in \mathbb{P}$, *then*

$$\sup_{n \in \mathbb{N}^d} \int_{\mathbb{T}^d} \left| K_n^{\alpha,\gamma}(x) \right| dx \le C.$$

Theorem 3.2.2 *If* $1 \le p < \infty$, $0 < \alpha < \infty$ *and* $\gamma \in \mathbb{P}$, *then*

$$\sup_{n \in \mathbb{N}^d} \left\| \sigma_n^{\alpha} f \right\|_p \le C \| f \|_p$$

and

$$\sup_{n \in \mathbb{N}^d} \left\| \sigma_n^{\alpha,\gamma} f \right\|_p \le C \| f \|_p.$$

Moreover, for all $f \in L_p(\mathbb{T}^d)$,

$$\lim_{n \to \infty} \sigma_n^{\alpha} f = f \qquad \text{in the } L_p(\mathbb{T}^d)\text{-norm}$$

and

$$\lim_{n \to \infty} \sigma_n^{\alpha,\gamma} f = f \qquad \text{in the } L_p(\mathbb{T}^d)\text{-norm}.$$

Here, the convergence is understood in Pringsheim's sense as in Theorem 2.1.8.

3.3 Almost Everywhere Restricted Summability over a Cone

In this section, we investigate the convergence of the rectangular Cesàro and Riesz summability means taken in a cone. For a given $\tau \ge 1$, we define a cone by

$$\mathbb{R}_\tau^d := \{x \in \mathbb{R}_+^d : \tau^{-1} \le x_i/x_j \le \tau, i, j = 1, \ldots, d\}. \tag{3.3.1}$$

The choice $\tau = 1$ yields the diagonal. The definition of the Cesàro and Riesz means can be extended to distributions as follows.

Definition 3.3.1 Let $f \in D(\mathbb{T}^d)$, $n \in \mathbb{N}^d$ and $0 \le \alpha, \gamma < \infty$. The nth rectangular Cesàro means $\sigma_n^{\alpha} f$ and rectangular Riesz means $\sigma_n^{\alpha,\gamma} f$ of the Fourier series of f are given by

$$\sigma_n^{\alpha} f := f * K_n^{\alpha}$$

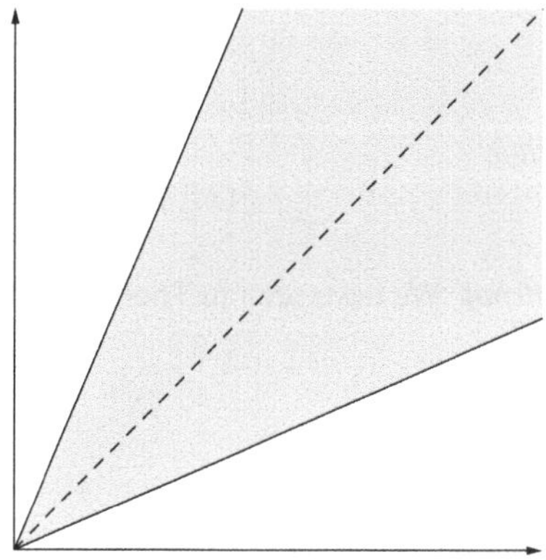

Fig. 3.2 The cone for $d = 2$

and

$$\sigma_n^{\alpha,\gamma} f := f * K_n^{\alpha,\gamma},$$

respectively.

Definition 3.3.2 We define the restricted maximal Cesàro and restricted maximal Riesz operator by

$$\sigma_{\square}^{\alpha} f := \sup_{n \in \mathbb{R}_\tau^d} |\sigma_n^{\alpha} f|$$

and

$$\sigma_{\square}^{\alpha,\gamma} f := \sup_{n \in \mathbb{R}_\tau^d} |\sigma_n^{\alpha,\gamma} f|,$$

respectively.

If $\alpha - 1$, we obtain the restricted maximal Fejér operator $\sigma_{\square} f$. As we can see on Fig. 3.2, in the restricted maximal operator the supremum is taken on a cone only. Marcinkiewicz and Zygmund [234] were the first who considered the restricted convergence. We show that the restricted maximal operator is bounded from $H_p^{\square}(\mathbb{T}^d)$ to $L_p(\mathbb{T}^d)$.

The next result follows easily from Theorem 3.2.1.

Theorem 3.3.3 *If* $0 < \alpha \leq 1$, *then*

$$\left\| \sigma_{\square}^{\alpha} f \right\|_\infty \leq C \left\| f \right\|_\infty \qquad (f \in L_\infty(\mathbb{T}^d)).$$

If $0 < \alpha < \infty$ *and* $\gamma \in \mathbb{P}$, *then*

$$\left\| \sigma_{\square}^{\alpha,\gamma} f \right\|_\infty \leq C \left\| f \right\|_\infty \qquad (f \in L_\infty(\mathbb{T}^d)).$$

Theorem 3.3.4 *If* $0 < \alpha \leq 1$ *and*

$$\max\left\{\frac{d}{d+1},\frac{1}{\alpha+1}\right\} < p \le \infty,$$

then

$$\left\|\sigma_{\square}^{\alpha} f\right\|_p \le C_p \|f\|_{H_p^{\square}} \qquad (f \in H_p^{\square}(\mathbb{T}^d)).$$

Proof We have seen in Theorem 3.1.8 that

$$\left|K_{n_j}^{\alpha}(t)\right| \le \frac{C}{n_j^{\alpha}|t|^{\alpha+1}} \qquad (t \ne 0) \tag{3.3.2}$$

and

$$\left|(K_{n_j}^{\alpha})'(t)\right| \le \frac{C}{n_j^{\alpha-1}|t|^{\alpha+1}} \qquad (t \ne 0). \tag{3.3.3}$$

Let a be an arbitrary $H_p^{\square}$-atom with support $I = I_1 \times I_2$ and

$$2^{-K-1} < |I_1|/\pi = |I_2|/\pi \le 2^{-K} \qquad (K \in \mathbb{N}).$$

We can suppose again that the center of I is zero. In this case,

$$[-\pi 2^{-K-2}, \pi 2^{-K-2}] \subset I_1, I_2 \subset [-\pi 2^{-K-1}, \pi 2^{-K-1}].$$

Choose $s \in \mathbb{N}$ such that $2^{s-1} < \tau \le 2^s$. It is easy to see that if $n_1 \ge k$ or $n_2 \ge k$, then we have $n_1, n_2 \ge k2^{-s}$. Indeed, since (n_1, n_2) is in a cone, $n_1 \ge k$ implies $n_2 \ge \tau^{-1} n_1 \ge k2^{-s}$. By Theorem 2.4.19, it is enough to prove that

$$\int_{\mathbb{T}^2 \setminus 4(I_1 \times I_2)} \left|\sigma_{\square}^{\alpha} a(x_1, x_2)\right|^p \, dx_1 \, dx_2 \le C_p. \tag{3.3.4}$$

First we integrate over $(\mathbb{T} \setminus 4I_1) \times 4I_2$. Obviously,

$$\begin{aligned}
&\int_{\mathbb{T}\setminus 4I_1} \int_{4I_2} \left|\sigma_{\square}^{\alpha} a(x_1, x_2)\right|^p \, dx_1 \, dx_2 \\
&\le \sum_{|i|=1}^{2^K-1} \int_{\pi i 2^{-K}}^{\pi(i+1)2^{-K}} \int_{4I_2} \sup_{n_1,n_2 \ge 2^{K-s}} \left|\sigma_{n_1,n_2}^{\alpha} a(x_1, x_2)\right|^p \, dx_1 \, dx_2 \\
&\quad + \sum_{|i|=1}^{2^K-1} \int_{\pi i 2^{-K}}^{\pi(i+1)2^{-K}} \int_{4I_2} \sup_{n_1,n_2 < 2^K} \left|\sigma_{n_1,n_2}^{\alpha} a(x_1, x_2)\right|^p \, dx_1 \, dx_2 \\
&=: (A) + (B).
\end{aligned}$$

We can suppose that $i > 0$. Using that

$$\int_{\mathbb{T}} \left|K_{n_2}^{\alpha}(x_2)\right| dx_2 \le C \qquad (n_2 \in \mathbb{N})$$

(see Corollary 1.5.3), (3.3.2) and the definition of the atom, we conclude

$$\begin{aligned}\left|\sigma_{n_1,n_2}^{\alpha} a(x_1, x_2)\right| &= \left|\int_{I_1}\int_{I_2} a(t_1, t_2) K_n^{\alpha}(x_1 - t_1) K_{n_2}^{\alpha}(x_2 - t_2)\, dt_1\, dt_2\right| \\ &\le C_p 2^{2K/p} \int_{I_1} \frac{1}{n_1^{\alpha}|x_1 - t_1|^{\alpha+1}}\, dt_1.\end{aligned}$$

For $x_1 \in [\pi i 2^{-K}, \pi(i+1)2^{-K})$ $(i \ge 1)$ and $t_1 \in I_1$, we have

$$\frac{1}{|x_1 - t_1|^{\nu}} \le \frac{1}{(\pi i 2^{-K} - \pi 2^{-K-1})^{\nu}} \le \frac{C2^{K\nu}}{i^{\nu}} \qquad (\nu > 0). \tag{3.3.5}$$

From this, it follows that

$$\left|\sigma_{n_1,n_2}^{\alpha} a(x_1, x_2)\right| \le C_p 2^{2K/p + K\alpha} \frac{1}{n_1^{\alpha} i^{\alpha+1}}.$$

Since $n_1 \ge 2^K 2^{-s}$, we obtain

$$(A) \le C_p \sum_{i=1}^{2^K-1} 2^{-2K} 2^{2K+K\alpha p} \frac{1}{2^{K\alpha p} i^{(\alpha+1)p}} \le C_p \sum_{i=1}^{2^K-1} \frac{1}{i^{(\alpha+1)p}},$$

which is a convergent series if $p > 1/(\alpha+1)$.

To consider (B), let $I_1 = I_2 = (-\mu, \mu)$ and

$$A_1(x_1, v) := \int_{-\pi}^{x_1} a(t_1, v)\, dt_1, \qquad A_2(x_1, x_2) := \int_{-\pi}^{x_2} A_1(x_1, t_2)\, dt_2. \tag{3.3.6}$$

Then

$$|A_k(x_1, x_2)| \le C_p 2^{K(2/p-k)}. \tag{3.3.7}$$

Integrating by parts, we get that

$$\begin{aligned}&\int_{I_1} a(t_1, t_2) K_{n_1}^{\alpha}(x_1 - t_1)\, dt_1 \\ &\quad = A_1(\mu, t_2) K_{n_1}^{\alpha}(x_1 - \mu) - \int_{I_1} A_1(t_1, t_2)(K_{n_1}^{\alpha})'(x_1 - t_1)\, dt_1.\end{aligned} \tag{3.3.8}$$

Recall that the one-dimensional kernel $K_{n_2}^{\alpha}$ satisfies

$$\left|K_{n_2}^{\alpha}\right| \le C n_2 \qquad (n_2 \in \mathbb{N}).$$

For $x_1 \in [\pi i 2^{-K}, \pi(i+1)2^{-K})$, the inequalities (3.3.2), (3.3.5) and (3.3.7) imply

$$\begin{aligned}
&\left| \int_{I_2} A_1(\mu, t_2) K_{n_1}^\alpha(x_1 - \mu) K_{n_2}^\alpha(x_2 - t_2)\, dt_2 \right| \\
&\quad \le C_p 2^{2K/p-K} 2^{-K} \frac{1}{n_1^\alpha |x_1 - \mu|^{\alpha+1}} n_2 \\
&\quad \le C_p 2^{2K/p+K\alpha-K} n_1^{1-\alpha} \frac{1}{i^{\alpha+1}}.
\end{aligned}$$

Moreover, by (3.3.3), (3.3.5) and (3.3.7),

$$\begin{aligned}
&\left| \int_{I_2} \int_{I_1} A_1(t_1, t_2) (K_{n_1}^\alpha)'(x_1 - t_1) K_{n_2}^\alpha(x_2 - t_2)\, dt_2\, dt_1 \right| \\
&\quad \le C_p 2^{2K/p-K} \int_{I_1} \frac{1}{n^{\alpha-1} |x_1 - t_1|^{\alpha+1}}\, dt_1 \\
&\quad \le C_p 2^{2K/p+K\alpha-K} n_1^{1-\alpha} \frac{1}{i^{\alpha+1}}.
\end{aligned}$$

Consequently,

$$(B) \le C_p \sum_{i=1}^{2^K-1} 2^{-2K} 2^{2K+K\alpha p-Kp} 2^{K(1-\alpha)p} \frac{1}{i^{(\alpha+1)p}} \le C_p \sum_{i=1}^{2^K-1} \frac{1}{i^{(\alpha+1)p}} < \infty,$$

because $p > 1/(\alpha+1)$. Hence, we have proved that in this case

$$\int_{\mathbb{T}\setminus 4I_1} \int_{4I_2} \left|\sigma_\square^\alpha a(x_1, x_2)\right|^p\, dx_1\, dx_2 \le C_p.$$

Next, we integrate over $(\mathbb{T} \setminus 4I_1) \times (\mathbb{T} \setminus 4I_2)$:

$$\begin{aligned}
&\int_{\mathbb{T}\setminus 4I_1} \int_{\mathbb{T}\setminus 4I_2} \left|\sigma_\square^\alpha a(x_1, x_2)\right|^p\, dx_1\, dx_2 \\
&\quad \le \sum_{|i|=1}^\infty \sum_{|j|=1}^\infty \int_{\pi i 2^{-K}}^{\pi(i+1)2^{-K}} \int_{\pi j 2^{-K}}^{\pi(j+1)2^{-K}} \sup_{n_1, n_2 \ge 2^{K-s}} \left|\sigma_{n_1,n_2}^\alpha a(x_1, x_2)\right|^p\, dx_1\, dx_2 \\
&\quad + \sum_{|i|=1}^\infty \sum_{|j|=1}^\infty \int_{\pi i 2^{-K}}^{\pi(i+1)2^{-K}} \int_{\pi j 2^{-K}}^{\pi(j+1)2^{-K}} \sup_{n_1, n_2 < 2^K} \left|\sigma_{n_1,n_2}^\alpha a(x_1, x_2)\right|^p\, dx_1\, dx_2 \\
&\quad =: (C) + (D).
\end{aligned}$$

We may suppose again that $i, j > 0$. For $x_1 \in [\pi i 2^{-K}, \pi(i+1)2^{-K})$ and $x_2 \in [\pi j 2^{-K}, \pi(j+1)2^{-K})$, we have by (3.3.2) and (3.3.5) that

$$\begin{aligned}\left|\sigma_{n_1,n_2}^{\alpha} a(x_1, x_2)\right| &\le C_p 2^{2K/p} \int_{I_1} \frac{1}{n_1^{\alpha}|x_1 - t_1|^{\alpha+1}}\, dt_1 \int_{I_2} \frac{1}{n_2^{\alpha}|x_2 - t_2|^{\alpha+1}}\, dt_2 \\ &\le C_p \frac{2^{2K/p+K\alpha+K\alpha}}{n_1^{\alpha} n_2^{\alpha} i^{\alpha+1} j^{\alpha+1}}.\end{aligned}$$

This implies that

$$\begin{aligned}(C) &\le C_p \sum_{i=1}^{2^K-1} \sum_{j=1}^{2^K-1} 2^{-2K} \frac{2^{2K+K\alpha p+K\alpha p}}{2^{K\alpha p+K\alpha p} i^{(\alpha+1)p} j^{(\alpha+1)p}} \\ &\le C_p \sum_{i=1}^{\infty} \sum_{j=1}^{\infty} \frac{1}{i^{(\alpha+1)p} j^{(\alpha+1)p}} < \infty.\end{aligned}$$

Using (3.3.8) and integrating by parts in both variables, we get that

$$\begin{aligned}&\int_{I_1} \int_{I_2} a(t_1, t_2) K_{n_1}^{\alpha}(x_1 - t_1) K_{n_2}^{\alpha}(x_2 - t_2)\, dt_1\, dt_2 \\ &= -\int_{I_2} A_2(\mu, t_2) K_{n_1}^{\alpha}(x_1 - \mu)(K_{n_2}^{\alpha})'(x_2 - t_2)\, dt_2 \\ &\quad + \int_{I_1} A_2(t_1, \mu)(K_{n_1}^{\alpha})'(x_1 - t_1) K_{n_2}^{\alpha}(x_2 - \mu)\, dt_1 \\ &\quad - \int_{I_1} \int_{I_2} A_2(t_1, t_2)(K_{n_1}^{\alpha})'(x_1 - t_1)(K_{n_2}^{\alpha})'(x_2 - t_2)\, dt_1\, dt_2 \\ &=: D_{n_1,n_2}^1(x_1, x_2) + D_{n_1,n_2}^2(x_1, x_2) + D_{n_1,n_2}^3(x_1, x_2).\end{aligned} \tag{3.3.9}$$

Note that $A(\mu, -\mu) = A(\mu, \mu) = 0$. Since $|K_{n_1}^{\alpha}| \le Cn_1$ and (3.3.2) holds as well, we obtain

$$|K_{n_1}^{\alpha}(x_1)| \le C \frac{n_1^{\eta+\alpha(\eta-1)}}{|x_1|^{(\alpha+1)(1-\eta)}}$$

for all $0 \le \eta \le 1$. Moreover, the inequality

$$|(K_{n_2}^{\alpha})'| \le Cn_2^2 \qquad (n_2 \in \mathbb{N})$$

and (3.3.3) imply

$$|(K_{n_2}^{\alpha})'(x_2)| \le C \frac{n_2^{2\zeta+(\alpha-1)(\zeta-1)}}{|x_2|^{(\alpha+1)(1-\zeta)}} = C \frac{n_2^{\zeta+1+\alpha(\zeta-1)}}{|x_2|^{(\alpha+1)(1-\zeta)}} \tag{3.3.10}$$

for all $0 \le \zeta \le 1$. We use inequalities (3.3.5) and (3.3.7) to obtain

$$\begin{aligned}\left|D^1_{n_1,n_2}(x_1,x_2)\right| &\le C_p 2^{2K/p-2K} \frac{n_1^{\eta+\alpha(\eta-1)}}{|x_1-\mu|^{(\alpha+1)(1-\eta)}} \int_{I_2} \frac{n_2^{\zeta+1+\alpha(\zeta-1)}}{|x_2-t_2|^{(\alpha+1)(1-\zeta)}}\,dt_2 \\ &\le C_p 2^{2K/p-3K} n_1^{\eta+\alpha(\eta-1)} \left(\frac{2^K}{i}\right)^{(\alpha+1)(1-\eta)} \\ &\quad n_2^{\zeta+1+\alpha(\zeta-1)} \left(\frac{2^K}{j}\right)^{(\alpha+1)(1-\zeta)}, \end{aligned} \tag{3.3.11}$$

whenever $x_1 \in [\pi i 2^{-K}, \pi(i+1)2^{-K})$, $x_2 \in [\pi j 2^{-K}, \pi(j+1)2^{-K})$ and $0 \le \eta, \zeta \le 1$. If

$$\eta + \alpha(\eta-1) + \zeta + 1 + \alpha(\zeta-1) \ge 0,$$

then

$$\sup_{n_1,n_2<2^K} \left|D^1_{n_1,n_2}(x_1,x_2)\right| \le C_p 2^{2K/p} \frac{1}{i^{(\alpha+1)(1-\eta)}} \frac{1}{j^{(\alpha+1)(1-\mu)}}$$

because (n_1, n_2) is in a cone. Choosing

$$\eta := \zeta := \max\left\{\frac{2\alpha-1}{2(\alpha+1)}, 0\right\},$$

we can see that

$$\begin{aligned}&\int_{\mathbb{T}\setminus 4I_1} \int_{\mathbb{T}\setminus 4I_2} \sup_{n_1,n_2<2^K} \left|D^1_{n_1,n_2}(x_1,x_2)\right|^p dx_1\,dx_2 \\ &\quad \le C_p \sum_{i=1}^{\infty}\sum_{j=1}^{\infty} 2^{-2K} 2^{2K} \frac{1}{i^{3p/2\wedge(\alpha+1)p}} \frac{1}{j^{3p/2\wedge(\alpha+1)p}},\end{aligned}$$

which is a convergent series. The analogous estimate for $\left|D^2_{n_1,n_2}(x_1,x_2)\right|$ can be similarly proved.

For $x_1 \in [\pi i 2^{-K}, \pi(i+1)2^{-K})$ and $x_2 \in [\pi j 2^{-K}, \pi(j+1)2^{-K})$, we conclude that

$$\begin{aligned}\left|D^3_{n_1,n_2}(x_1,x_2)\right| &\le C_p 2^{2K/p-2K} \int_{I_1} \frac{1}{n_1^{\alpha-1}|x_1-t_1|^{\alpha+1}}\,dt_1 \int_{I_2} \frac{1}{n_2^{\alpha-1}|x_2-t_2|^{\alpha+1}}\,dt_2 \\ &\le C_p \frac{2^{2K/p-2K+K\alpha+K\alpha} n_1^{1-\alpha} n_2^{1-\alpha}}{i^{\alpha+1} j^{\alpha+1}}.\end{aligned}$$

So

$$\int_{\mathbb{T}\setminus 4I_1}\int_{\mathbb{T}\setminus 4I_2}\sup_{n_1,n_2<2^K}\left|D^3_{n_1,n_2}(x_1,x_2)\right|^p\,dx_1\,dx_2$$
$$\le C_p\sum_{i=1}^{2^K-1}\sum_{j=1}^{2^K-1}2^{-2K}\frac{2^{2K-2Kp+K\alpha p+K\alpha p}2^{K(2-\alpha-\alpha)p}}{i^{(\alpha+1)p}j^{(\alpha+1)p}}$$
$$\le C_p\sum_{i=1}^{\infty}\sum_{j=1}^{\infty}\frac{1}{i^{(\alpha+1)p}}\frac{1}{j^{(\alpha+1)p}}<\infty$$

by the hypothesis. The integration over $4I_1\times(\mathbb{T}\setminus 4I_2)$ can be done as above. This finishes the proof of (3.3.4) as well as the theorem. ■

Remark 3.3.5 In the d-dimensional case, the constant $d/(d+1)$ appears if we investigate the corresponding term to D^1_n. More exactly, if we integrate the term

$$\int_{I_d}A(\mu,\cdots,\mu,t_d)K^\alpha_{n_1}(x_1-\mu)\cdots K^\alpha_{n_{d-1}}(x_{d-1}-\mu)(K^\alpha_{n_d})'(x_d-t_d)\,dt_d$$

over $(\mathbb{T}\setminus 4I_1)\times\cdots\times(\mathbb{T}\setminus 4I_d)$ similar to (3.3.11), then we get that $p>d/(d+1)$.

Corollary 3.3.6 *If* $0<\alpha\le 1$ *and* $1<p<\infty$, *then*

$$\left\|\sigma^\alpha_\square f\right\|_p\le C_p\|f\|_p\qquad (f\in L_p(\mathbb{T}^d)).$$

Let us turn to the Riesz means.

Theorem 3.3.7 *If* $0<\alpha<\infty$, $\gamma\in\mathbb{P}$ *and*

$$\max\left\{\frac{d}{d+1},\frac{1}{\alpha\wedge 1+1}\right\}<p\le\infty,$$

then

$$\left\|\sigma^{\alpha,\gamma}_\square f\right\|_p\le C_p\|f\|_{H^\square_p}\qquad (f\in H^\square_p(\mathbb{T}^d)).$$

Proof Let

$$\theta(s):=\begin{cases}(1-|s|^\gamma)^\alpha & \text{if } |s|\le 1;\\ 0, & \text{if } |s|>1\end{cases}\qquad (s\in\mathbb{R}).$$

By the one-dimensional version of Corollary 2.2.28,

$$\left|\widehat{\theta}(t)\right|,\left|(\widehat{\theta})'(t)\right|\le C|t|^{-\alpha-1}\qquad (t\ne 0).$$

Taking into account (2.2.34), we conclude that

$$\left|K^{\alpha,\gamma}_{n_j}(t)\right|\le\frac{C}{n_j^\alpha|t|^{\alpha+1}}\qquad (t\ne 0)\tag{3.3.12}$$

and

$$\left|(K_{n_j}^{\alpha,\gamma})'(t)\right| \le \frac{C}{n_j^{\alpha-1}|t|^{\alpha+1}} \qquad (t \neq 0). \tag{3.3.13}$$

For $0 < \alpha \le 1$, the inequality can be proved as in Theorem 3.3.4. Now let $\alpha > 1$. Since

$$\left|\widehat{\theta}(t)\right|, \left|(\widehat{\theta})'(t)\right| \le C$$

trivially and since $|t|^{-\alpha-1} \le |t|^{-2}$ if $|t| \ge 1$, we conclude that

$$\left|\widehat{\theta}(t)\right|, \left|(\widehat{\theta})'(t)\right| \le C|t|^{-2} \qquad (t \neq 0).$$

Hence

$$\left|K_{n_j}^{\alpha,\gamma}(t)\right| \le \frac{C}{n_j|t|^2}, \qquad \left|(K_{n_j}^{\alpha,\gamma})'(t)\right| \le \frac{C}{|t|^2} \qquad (t \neq 0)$$

and the theorem can be proved as above. ∎

Corollary 3.3.8 *Suppose that* $0 < \alpha < \infty$ *and* $\gamma \in \mathbb{P}$. *If* $1 < p < \infty$, *then*

$$\left\|\sigma_{\square}^{\alpha,\gamma} f\right\|_p \le C_p \|f\|_p \qquad (f \in L_p(\mathbb{T}^d)).$$

As we have seen in Theorems 2.5.6 and 2.5.12, in the one-dimensional case, the operators $\sigma_{\square}^{\alpha}$ and $\sigma_{\square}^{\alpha,\gamma}$ are not bounded from $H_p^{\square}(\mathbb{T})$ to $L_p(\mathbb{T})$ if $0 < p \le 1/(\alpha+1)$ and $\alpha = 1$. Using interpolation, we obtain the weak type (1, 1) inequality.

Corollary 3.3.9 *If* $0 < \alpha \le 1$, *then*

$$\sup_{\rho>0} \rho\lambda(\sigma_{\square}^{\alpha} f > \rho) \le C\|f\|_1 \qquad (f \in L_1(\mathbb{T}^d)).$$

If $0 < \alpha < \infty$ *and* $\gamma \in \mathbb{P}$, *then*

$$\sup_{\rho>0} \rho\lambda(\sigma_{\square}^{\alpha,\gamma} f > \rho) \le C\|f\|_1 \qquad (f \in L_1(\mathbb{T}^d)).$$

The density argument of Marcinkiewicz and Zygmund (Theorem 1.3.6) implies

Corollary 3.3.10 *Suppose that* $f \in L_1(\mathbb{T}^d)$. *If* $0 < \alpha \le 1$, *then*

$$\lim_{n\to\infty,\, n\in\mathbb{R}_\tau^d} \sigma_n^{\alpha} f = f \qquad a.e.$$

If $0 < \alpha < \infty$ *and* $\gamma \in \mathbb{P}$, *then*

$$\lim_{n\to\infty,\, n\in\mathbb{R}_\tau^d} \sigma_n^{\alpha,\gamma} f = f \qquad a.e.$$

This result was proved by Marcinkiewicz and Zygmund [234] for the two-dimensional Fejér means. The general version of Corollary 3.3.10 is due to the author [328, 329].

3.4 Almost Everywhere Restricted Summability over a Cone-Like Set

Now we generalize the results of Sect. 3.3 to so-called cone-like sets (see Fig. 3.3). Suppose that for all $j = 2, \ldots, d$, $\kappa_j : \mathbb{R}_+ \to \mathbb{R}_+$ are strictly increasing and continuous functions such that

$$\lim_{j\to\infty} \kappa_j = \infty \qquad \text{and} \qquad \lim_{j\to+0} \kappa_j = 0.$$

Moreover, suppose that there exist $c_{j,1}, c_{j,2}, \xi > 1$ such that

$$c_{j,1}\kappa_j(x) \le \kappa_j(\xi x) \le c_{j,2}\kappa_j(x) \qquad (x > 0). \tag{3.4.1}$$

Note that this is satisfied if κ_j is a power function. Let us define the numbers $\omega_{j,1}$ and $\omega_{j,2}$ via the formula

$$c_{j,1} = \xi^{\omega_{j,1}} \quad \text{and} \quad c_{j,2} = \xi^{\omega_{j,2}} \qquad (j = 2, \ldots, d). \tag{3.4.2}$$

For convenience, we extend the notations for $j = 1$ by $\kappa_1 := \mathcal{I}$, $c_{1,1} = c_{1,2} = \xi$. Here $\mathcal{I}$ denotes the identity function $\mathcal{I}(x) = x$. Let $\kappa = (\kappa_1, \ldots, \kappa_d)$ and $\tau = (\tau_1, \ldots, \tau_d)$ with $\tau_1 = 1$ and fixed $\tau_j \ge 1$ $(j = 2, \ldots, d)$. We define the cone-like set (with respect to the first dimension) by

$$\mathbb{R}^d_{\kappa,\tau} := \{x \in \mathbb{R}^d_+ : \tau_j^{-1}\kappa_j(n_1) \le n_j \le \tau_j\kappa_j(n_1), j = 2, \ldots, d\}.$$

Figure 3.3 shows a cone-like set for $d = 2$.

If $\kappa_j = \mathcal{I}$ for all $j = 2, \ldots, d$, then we get a cone investigated above. The condition on κ_j seems to be natural, because Gát [119] proved in the two-dimensional case that to each cone-like set with respect to the first dimension there exists a larger cone-like set with respect to the second dimension and conversely, if and only if (3.4.1) holds.

Here we have to consider a new Hardy space. We modify slightly the definition of $H_p^\square(\mathbb{T}^d)$. Fix $\psi \in S(\mathbb{R})$ such that $\int_{\mathbb{R}} \psi(x)dx \ne 0$. For $f \in D(\mathbb{T}^d)$, let

$$\psi_+^\kappa(f)(x) := \sup_{t\in(0,\infty)} \left| f * (\psi_t \otimes \psi_{\kappa_2(t)} \otimes \cdots \otimes \psi_{\kappa_d(t)})(x) \right|.$$

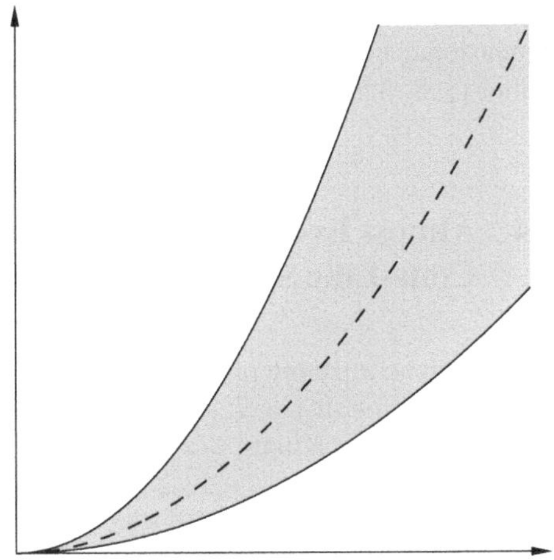

Fig. 3.3 Cone-like set for $d = 2$

Definition 3.4.1 For $0 < p < \infty$ the Hardy spaces $H_p^\kappa(\mathbb{T}^d)$ and weak Hardy spaces $H_{p,\infty}^\kappa(\mathbb{T}^d)$ consist of all distributions $f \in D(\mathbb{T}^d)$ for which

$$\|f\|_{H_p^\kappa} := \left\|\psi_+^\kappa(f)\right\|_p < \infty \qquad \text{and} \qquad \|f\|_{H_{p,\infty}^\kappa} := \left\|\psi_+^\kappa(f)\right\|_{p,\infty} < \infty.$$

We can prove all the theorems of Sect. 2.4 for $H_p^\kappa(\mathbb{T}^d)$. Among others,

$$\|f\|_{H_p^\kappa} \sim \left\|P_+^\kappa(f)\right\|_p \qquad (0 < p < \infty),$$

where P_t is the one-dimensional Poisson kernel and

$$P_+^\kappa(f)(x) := \sup_{t\in(0,\infty)} \left|f * (P_t \otimes P_{\kappa_2(t)} \otimes \cdots \otimes P_{\kappa_d(t)})(x)\right|.$$

If each $\kappa_j = \mathcal{I}$, we get back the Hardy spaces $H_p^\square(\mathbb{T}^d)$. We have to modify slightly the definition of atoms, too.

Definition 3.4.2 A bounded function a is an H_p^κ-atom if there exists a rectangle $I := I_1 \times \cdots \times I_d \subset \mathbb{T}^d$ with $|I_j| = \kappa_j(|I_1|^{-1})^{-1}$ such that

(i) $\operatorname{supp} a \subset I$,
(ii) $\|a\|_\infty \le |I|^{-1/p}$,
(iii) $\int_I a(x)x^k\,dx = 0$ for all multi-indices k with $|k| \le \lfloor d(1/p-1)\rfloor$.

The following two results can be proved as Theorems 2.4.18 and 2.4.19.

Theorem 3.4.3 *A distribution $f \in D(\mathbb{T}^d)$ is in $H_p^\kappa(\mathbb{T}^d)$ $(0 < p \le 1)$ if and only if there exist a sequence $(a_k, k \in \mathbb{N})$ of H_p^κ-atoms and a sequence $(\mu_k, k \in \mathbb{N})$ of real numbers such that*

$$\sum_{k=0}^{\infty} |\mu_k|^p < \infty \quad \text{and} \quad \sum_{k=0}^{\infty} \mu_k a_k = f \quad \text{in } D(\mathbb{T}^d). \tag{3.4.3}$$

Moreover,

$$\|f\|_{H_p^\kappa} \sim \inf \left(\sum_{k=0}^{\infty} |\mu_k|^p \right)^{1/p},$$

where the infimum is taken over all decompositions of f of the form (3.4.3).

Theorem 3.4.4 *For each $n \in \mathbb{N}^d$, let $K_n \in L_1(\mathbb{T}^d)$ and $V_n f := f * K_n$. Suppose that*

$$\int_{\mathbb{T}^d \setminus rI} |V_* a|^{p_0} \, d\lambda \le C_{p_0}$$

for all $H_{p_0}^\kappa$-atoms a and for some fixed $r \in \mathbb{N}$ and $0 < p_0 \le 1$, where the rectangle I is the support of the atom. If V_* is bounded from $L_{p_1}(\mathbb{T}^d)$ to $L_{p_1}(\mathbb{T}^d)$ for some $1 < p_1 \le \infty$, then

$$\|V_* f\|_p \le C_p \|f\|_{H_p^\kappa} \qquad (f \in H_p^\kappa(\mathbb{T}^d))$$

for all $p_0 \le p \le p_1$.

Definition 3.4.5 For given κ, τ satisfying the above conditions, we define the restricted maximal Cesàro and restricted maximal Riesz operator by

$$\sigma_\kappa^\alpha f := \sup_{n \in \mathbb{R}_{\kappa,\tau}^d} |\sigma_n^\alpha f|$$

and

$$\sigma_\kappa^{\alpha,\gamma} f := \sup_{n \in \mathbb{R}_{\kappa,\tau}^d} |\sigma_n^{\alpha,\gamma} f|,$$

respectively.

The next theorem holds obviously.

Theorem 3.4.6 *If $0 < \alpha \le 1$, then*

$$\left\|\sigma_\kappa^\alpha f\right\|_\infty \le C \, \|f\|_\infty \qquad (f \in L_\infty(\mathbb{T}^d)).$$

If $0 < \alpha < \infty$ and $\gamma \in \mathbb{P}$, then

$$\left\|\sigma_\kappa^{\alpha,\gamma} f\right\|_\infty \le C \, \|f\|_\infty \qquad (f \in L_\infty(\mathbb{T}^d)).$$

Let H be an arbitrary subset of $\{1, \ldots, d\}$, $H \ne \emptyset$, $H \ne \{1, \ldots, d\}$ and $H^c := \{1, \ldots, d\} \setminus H$. Define

$$p_1 := \sup_{H \subset \{1,\ldots,d\}} \frac{\sum_{j\in H} \omega_{j,2} + \sum_{j \in H^c} \omega_{j,1}}{\sum_{j\in H} \omega_{j,2} + 2\sum_{j \in H^c} \omega_{j,1}}, \tag{3.4.4}$$

where the numbers $\omega_{j,1}$ and $\omega_{j,2}$ are defined in (3.4.2).

Theorem 3.4.7 *If* $0 < \alpha \leq 1$ *and*

$$\max\left\{p_1, \frac{1}{\alpha+1}\right\} < p \leq \infty,$$

then

$$\left\|\sigma_\kappa^\alpha f\right\|_p \leq C_p \|f\|_{H_p^\kappa} \qquad (f \in H_p^\kappa(\mathbb{T}^d)).$$

Proof Since we will prove the result for $d = 2$, we simplify the notation. Instead of $c_{2,1}, c_{2,2}$ and $\omega_{2,1}, \omega_{2,2}$, we will write c_1, c_2 and ω_1, ω_2, respectively. Let a be an arbitrary H_p^κ-atom with support $I = I_1 \times I_2$, $|I_2|^{-1} = \kappa(|I_1|^{-1})$ and

$$2^{-K-1} < |I_1|/\pi \leq 2^{-K}, \qquad \kappa(2^{K+1})^{-1} < |I_2|/\pi \leq \kappa(2^K)^{-1}$$

for some $K \in \mathbb{N}$. We can suppose that the center of I is zero. In this case

$$[-\pi 2^{-K-2}, \pi 2^{-K-2}] \subset I_1 \subset [-\pi 2^{-K-1}, \pi 2^{-K-1}]$$

and

$$[-\pi\kappa(2^{K+1})^{-1}/2, \pi\kappa(2^{K+1})^{-1}/2] \subset I_2 \subset [-\pi\kappa(2^K)^{-1}/2, \pi\kappa(2^K)^{-1}/2].$$

To prove

$$\int_{\mathbb{T}^2 \setminus 4(I_1 \times I_2)} \left|\sigma_\kappa^\alpha a(x_1, x_2)\right|^p \, dx_1 \, dx_2 \leq C_p,$$

first we integrate over $(\mathbb{T} \setminus 4I_1) \times 4I_2$:

$$\begin{aligned}
&\int_{\mathbb{T}\setminus 4I_1} \int_{4I_2} |\sigma_\kappa^\alpha a(x_1, x_2)|^p \, dx_1 \, dx_2 \\
&\leq \int_{\mathbb{T}\setminus 4I_1} \int_{4I_2} \sup_{n_1 \geq 2^K, (n_1,n_2) \in \mathbb{R}^d_{\kappa,\tau}} \left|\sigma_{n_1,n_2}^\alpha a(x_1, x_2)\right|^p \, dx_1 \, dx_2 \\
&\quad + \int_{\mathbb{T}\setminus 4I_1} \int_{4I_2} \sup_{n_1 < 2^K, (n_1,n_2) \in \mathbb{R}^d_{\kappa,\tau}} \left|\sigma_{n_1,n_2}^\alpha a(x_1, x_2)\right|^p \, dx_1 \, dx_2 \\
&=: (A) + (B).
\end{aligned}$$

If $n_1 \geq 2^K$ and $x \in [\pi i 2^{-K}, \pi(i+1)2^{-K})$ $(i \geq 1)$, then by (3.3.5),

$$
\begin{aligned}
\left|\sigma^{\alpha}_{n_1,n_2}a(x_1,x_2)\right| &= \left|\int_{I_1}\int_{I_2} a(t_1,t_2)K^{\alpha}_{n_1}(x_1-t_1)K^{\alpha}_{n_2}(x_2-t_2)\,dt_1\,dt_2\right| \\
&\leq C_p 2^{K/p}\kappa(2^K)^{1/p}\int_{I_1}\frac{1}{n_1^{\alpha}|x_1-t_1|^{\alpha+1}}\,dt_1 \\
&\leq C_p 2^{K/p+K\alpha}\kappa(2^K)^{1/p}\frac{1}{n_1^{\alpha}i^{\alpha+1}} \\
&\leq C_p 2^{K/p}\kappa(2^K)^{1/p}\frac{1}{i^{\alpha+1}}.
\end{aligned}
$$

From this, it follows that

$$
\begin{aligned}
(A) &\leq \sum_{i=1}^{2^K-1}\int_{\pi i2^{-K}}^{\pi(i+1)2^{-K}}\int_{4I_2}\sup_{n_1\geq 2^K}\left|\sigma^{\alpha}_{n_1,n_2}a(x_1,x_2)\right|^p\,dx_1\,dx_2 \\
&\leq C_p\sum_{i=1}^{2^K-1}2^{-K}\kappa(2^K)^{-1}2^K\kappa(2^K)\frac{1}{i^{(\alpha+1)p}} \\
&= C_p\sum_{i=1}^{2^K-1}\frac{1}{i^{(\alpha+1)p}},
\end{aligned}
$$

which is a convergent series if $p>1/(1+\alpha)$.

We estimate (B) by

$$
\begin{aligned}
(B) &\leq \sum_{k=0}^{\infty}\int_{\mathbb{T}\setminus 4I_1}\int_{4I_2}\sup_{\frac{2^K}{\xi^{k+1}}\leq n_1<\frac{2^K}{\xi^k},(n_1,n_2)\in\mathbb{R}^d_{\kappa,\tau}}\left|\sigma^{\alpha}_{n_1,n_2}a(x_1,x_2)\right|^p\,dx_1\,dx_2 \\
&\leq \sum_{k=0}^{\infty}\left(\int_{\mathbb{T}\setminus\left[-\frac{\pi\xi^k}{2^K},\frac{\pi\xi^k}{2^K}\right]}\int_{4I_2}+\int_{\left[-\frac{\pi\xi^k}{2^K},\frac{\pi\xi^k}{2^K}\right]}\int_{4I_2}\right) \\
&\qquad \sup_{\frac{2^K}{\xi^{k+1}}\leq n_1<\frac{2^K}{\xi^k},(n_1,n_2)\in\mathbb{R}^d_{\kappa,\tau}}\left|\sigma^{\alpha}_{n_1,n_2}a(x_1,x_2)\right|^p\,dx_1\,dx_2 \\
&=: (B_1)+(B_2).
\end{aligned}
$$

If $(n_1,n_2)\in\mathbb{R}^d_{\kappa,\tau}$ and $\frac{2^K}{\xi^{k+1}}\leq n<\frac{2^K}{\xi^k}$, then $n_2<\tau\kappa(\frac{2^K}{\xi^k})$. The inequality $|K^{\alpha}_{n_2}|\leq Cn_2$ and (3.3.2) imply

$$
\begin{aligned}
&\left|\sigma^{\alpha}_{n_1,n_2}a(x_1,x_2)\right| \\
&\quad\leq C_p 2^{K/p}\kappa(2^K)^{1/p-1}n_2\int_{I_1}\frac{1}{n_1^{\alpha}|x_1-t_1|^{\alpha+1}}\,dt_1 \\
&\quad\leq C_p 2^{K/p-K}\kappa(2^K)^{1/p-1}\kappa\left(\frac{2^K}{\xi^k}\right)\left(\frac{2^K}{\xi^{k+1}}\right)^{-\alpha}|x_1-\pi 2^{-K-1}|^{-\alpha-1}.
\end{aligned}
$$

Hence

$$
\begin{aligned}
(B_1) &\le C_p \sum_{k=0}^{\infty} 2^{K(1-p-\alpha p)} \kappa(2^K)^{-p} \kappa\left(\frac{2^K}{\xi^k}\right)^p \xi^{k\alpha p} \\
&\qquad \int_{\mathbb{T}\setminus\left[-\frac{\pi\xi^k}{2^K}, \frac{\pi\xi^k}{2^K}\right]} |x_1 - \pi 2^{-K-1}|^{-(\alpha+1)p} \, dx_1 \\
&\le C_p \sum_{k=0}^{\infty} 2^{K(1-p-\alpha p)} \kappa(2^K)^{-p} \kappa\left(\frac{2^K}{\xi^k}\right)^p \xi^{k\alpha p} (\xi^k 2^{-K})^{-(\alpha+1)p+1}.
\end{aligned}
$$

Since $\kappa(x) \le c_1^{-1}\kappa(\xi x)$ by (3.4.1), we conclude

$$
(B_1) \le C_p \sum_{k=0}^{\infty} \kappa(2^K)^{-p} \kappa(2^K)^p c_1^{-kp} \xi^{k(1-p)} = C_p \sum_{k=0}^{\infty} \xi^{k(1-p-\omega_1 p)},
$$

which is convergent if $p > 1/(1+\omega_1)$. Note that

$$
\frac{1}{1+\omega_1} < \frac{1+\omega_1}{1+2\omega_1} \le p_1 < p.
$$

For (B_2), we obtain similarly that

$$
\begin{aligned}
\left|\sigma_{n_1,n_2}^{\alpha} a(x_1, x_2)\right| &\le C_p 2^{K/p-K} \kappa(2^K)^{1/p-1} n_1 n_2 \\
&\le C_p 2^{K/p-K} \kappa(2^K)^{1/p-1} \frac{2^K}{\xi^k} \kappa\left(\frac{2^K}{\xi^k}\right)
\end{aligned} \tag{3.4.5}
$$

and, moreover,

$$
\begin{aligned}
(B_2) &\le C_p \sum_{k=0}^{\infty} \frac{\xi^k}{2^K} \kappa(2^K)^{-1} 2^K \kappa(2^K)^{1-p} \xi^{-kp} \kappa\left(\frac{2^K}{\xi^k}\right)^p \\
&\le C_p \sum_{k=0}^{\infty} \xi^{k(1-p)} c_1^{-kp},
\end{aligned}
$$

which was just considered. Hence, we have proved that

$$
\int_{\mathbb{T}\setminus 4I_1} \int_{4I_2} |\sigma_{\kappa}^{\alpha} a(x_1, x_2)|^p \, dx_1 \, dx_2 \le C_p \qquad (p_1 < p \le 1).
$$

The integral over $4I_1 \times (\mathbb{T} \setminus 4I_2)$ can be handled with a similar idea. Indeed, let us denote the terms corresponding to (A), (B), (B_1), (B_2) by (A'), (B'), (B_1'), (B_2'). If we take the integrals in (A') over

$$4I_1 \times \left[\pi j \kappa(2^K)^{-1}, \pi(j+1)\kappa(2^K)^{-1}\right] \qquad (j = 1, \ldots, \kappa(2^K)/2 - 1),$$

then we get in the same way that (A') is bounded if $p > 1/(1+\alpha)$. For (B_1'), we can see that

$$\begin{aligned}
(B_1') &= \sum_{k=0}^{\infty} \int_{4I_1} \int_{\mathbb{T}\setminus\left[-\pi\kappa\left(\frac{2^K}{\xi^k}\right)^{-1}, \pi\kappa\left(\frac{2^K}{\xi^k}\right)^{-1}\right]} \\
&\qquad \sup_{\frac{2^K}{\xi^{k+1}} \le n_1 < \frac{2^K}{\xi^k}, (n_1,n_2)\in\mathbb{R}^d_{\kappa,\tau}} \left|\sigma^{\alpha}_{n_1,n_2} a(x_1,x_2)\right|^p \, dx_1 \, dx_2 \\
&\le C_p \sum_{k=0}^{\infty} 2^K \kappa(2^K) 2^{-K} 2^{-Kp} \int_{\mathbb{T}\setminus\left[-\pi\kappa\left(\frac{2^K}{\xi^k}\right)^{-1}, \pi\kappa\left(\frac{2^K}{\xi^k}\right)^{-1}\right]} \\
&\qquad \sup_{\frac{2^K}{\xi^{k+1}} \le n_1 < \frac{2^K}{\xi^k}, (n_1,n_2)\in\mathbb{R}^d_{\kappa,\tau}} \left(n_1 \int_{I_2} \frac{1}{n_2^{\alpha}|x_2 - t_2|^{\alpha+1}} \, dt_2\right)^p \, dx_2.
\end{aligned}$$

Thus

$$\begin{aligned}
(B_1') &\le C_p \sum_{k=0}^{\infty} \xi^{-kp} \kappa(2^K)^{1-p} \kappa\left(\frac{2^K}{\xi^{k+1}}\right)^{-\alpha p} \\
&\qquad \int_{\mathbb{T}\setminus\left[-\pi\kappa\left(\frac{2^K}{\xi^k}\right)^{-1}, \pi\kappa\left(\frac{2^K}{\xi^k}\right)^{-1}\right]} |x_2 - \pi\kappa(2^K)^{-1}/2|^{-(\alpha+1)p} \, dx_2 \\
&\le C_p \sum_{k=0}^{\infty} \xi^{-kp} \kappa(2^K)^{1-p} \kappa\left(\frac{2^K}{\xi^k}\right)^{p-1} \\
&\le C_p \sum_{k=0}^{\infty} \xi^{-kp} c_2^{k(1-p)} \\
&= C_p \sum_{k=0}^{\infty} \xi^{k(\omega_2 - \omega_2 p - p)}
\end{aligned}$$

and this converges if $p > \omega_2/(1+\omega_2)$, which is less than

$$\frac{1+\omega_2}{2+\omega_2} \le p_1.$$

Using (3.4.5), we establish that

$$
\begin{aligned}
(B_2') &= \sum_{k=0}^{\infty} \int_{4I_1} \int_{\left[-\kappa\left(\frac{2^K}{\xi^k}\right)^{-1}, \kappa\left(\frac{2^K}{\xi^k}\right)^{-1}\right]} \\
&\qquad \sup_{\frac{2^K}{\xi^{k+1}} \le n_1 < \frac{2^K}{\xi^k}, (n_1,n_2) \in \mathbb{R}^d_{\kappa,\tau}} \left|\sigma^{\alpha}_{n_1,n_2} a(x_1,x_2)\right|^p \, dx_1 \, dx_2 \\
&\le C_p \sum_{k=0}^{\infty} 2^{-K} \kappa \left(\frac{2^K}{\xi^k}\right)^{-1} 2^K \kappa(2^K)^{1-p} \xi^{-kp} \kappa \left(\frac{2^K}{\xi^k}\right)^p \\
&\le C_p \sum_{k=0}^{\infty} \xi^{-kp} c_2^{k(1-p)}.
\end{aligned}
$$

Hence

$$
\int_{4I_1} \int_{\mathbb{T}\setminus 4I_2} \left|\sigma^{\alpha}_{\kappa} a(x_1,x_2)\right|^p \, dx_1 \, dx_2 \le C_p \qquad (p_1 < p \le 1).
$$

Integrating over $(\mathbb{T} \setminus 4I_1) \times (\mathbb{T} \setminus 4I_2)$, we decompose the integral as

$$
\begin{aligned}
&\int_{\mathbb{T}\setminus 4I_1} \int_{\mathbb{T}\setminus 4I_2} \left|\sigma^{\alpha}_{\kappa} a(x_1,x_2)\right|^p \, dx_1 \, dx_2 \\
&\le \int_{\mathbb{T}\setminus 4I_1} \int_{\mathbb{T}\setminus 4I_2} \sup_{n_1 \ge 2^K, (n_1,n_2) \in \mathbb{R}^d_{\kappa,\tau}} \left|\sigma^{\alpha}_{n_1,n_2} a(x_1,x_2)\right|^p \, dx_1 \, dx_2 \\
&\quad + \int_{\mathbb{T}\setminus 4I_1} \int_{\mathbb{T}\setminus 4I_2} \sup_{n_2 < 2^K, (n_1,n_2) \in \mathbb{R}^d_{\kappa,\tau}} \left|\sigma^{\alpha}_{n_1,n_2} a(x_1,x_2)\right|^p \, dx_1 \, dx_2 \\
&=: (C) + (D).
\end{aligned}
$$

Notice that

$$
(C) \le \sum_{i=1}^{2^K-1} \sum_{j=1}^{\kappa(2^K)/2-1} \int_{\pi i 2^{-K}}^{\pi(i+1)2^{-K}} \int_{\pi j \kappa(2^K)^{-1}}^{\pi(j+1)\kappa(2^K)^{-1}} \sup_{n_1 \ge 2^K} |\sigma^{\alpha}_{n_1,n_2} a(x_1,x_2)|^p \, dx_1 \, dx_2.
$$

For $x_1 \in [\pi i 2^{-K}, \pi(i+1)2^{-K})$ and $x_2 \in [\pi j \kappa(2^K)^{-1}, \pi(j+1)\kappa(2^K)^{-1})$, we have by (3.3.2) and (3.3.5) that

$$
\begin{aligned}
\left|\sigma^{\alpha}_{n_1,n_2} a(x_1,x_2)\right| &\le C_p 2^{K/p} \kappa(2^K)^{1/p} \int_{I_1} \frac{1}{n_1^{\alpha} |x_1 - t_1|^{\alpha+1}} \, dt_1 \\
&\qquad \int_{I_2} \frac{1}{n_2^{\alpha} |x_2 - t_2|^{\alpha+1}} \, dt_2 \\
&\le C_p \frac{2^{K/p + K\alpha} \kappa(2^K)^{1/p+\alpha}}{n_1^{\alpha} n_2^{\alpha} i^{\alpha+1} j^{\alpha+1}} \\
&\le C_p \frac{2^{K/p} \kappa(2^K)^{1/p}}{i^{\alpha+1} j^{\alpha+1}}.
\end{aligned} \tag{3.4.6}
$$

Then

$$(C) \le C_p \sum_{i=1}^{2^K-1} \sum_{j=1}^{\kappa(2^K)/2-1} \frac{1}{i^{(\alpha+1)p} j^{(\alpha+1)p}} < \infty$$

if $p > 1/(1+\alpha)$.

To consider (D) let us define $A_1(x_1, x_2)$, $A_2(x_1, x_2)$, $D^1_{n_1,n_2}(x_1, x_2)$, $D^2_{n_1,n_2}(x_1, x_2)$ and $D^3_{n_1,n_2}(x_1, x_2)$ as in (3.3.6) and (3.3.9), respectively, and let $I_1 = [-\mu, \mu]$, $I_2 = [-\nu, \nu]$. Then

$$|A_1(x_1, u)| \le 2^{K/p-K} \kappa(2^K)^{1/p}, \qquad |A_2(x_1, x_2)| \le 2^{K/p-K} \kappa(2^K)^{1/p-1}. \tag{3.4.7}$$

Obviously,

$$\begin{aligned}
&\int_{\mathbb{T}\setminus 4I_1} \int_{\mathbb{T}\setminus 4I_2} \sup_{n_1<2^K, (n_1,n_2)\in\mathbb{R}^d_{\kappa,\tau}} |D^1_{n_1,n_2}(x_1, x_2)|^p \, dx_1 \, dx_2 \\
&\le \sum_{k=0}^{\infty} \int_{\mathbb{T}\setminus 4I_1} \int_{\mathbb{T}\setminus 4I_2} \sup_{\frac{2^K}{\xi^{k+1}} \le n_1 < \frac{2^K}{\xi^k}, (n_1,n_2)\in\mathbb{R}^d_{\kappa,\tau}} |D^1_{n_1,n_2}(x_1, x_2)|^p \, dx_1 \, dx_2 \\
&\le \sum_{k=0}^{\infty} \sum_{j=1}^{\kappa(2^K)/2-1} \int_{\mathbb{T}\setminus\left[-\frac{\pi\xi^k}{2^K}, \frac{\pi\xi^k}{2^K}\right]} \int_{\pi j \kappa(2^K)^{-1}}^{\pi(j+1)\kappa(2^K)^{-1}} \\
&\qquad \sup_{n_1<2^K, (n_1,n_2)\in\mathbb{R}^d_{\kappa,\tau}} |D^1_{n_1,n_2}(x_1, x_2)|^p \, dx_1 \, dx_2 \\
&\le \sum_{k=0}^{\infty} \sum_{j=1}^{\kappa(2^K)/2-1} \int_{\left[-\frac{\pi\xi^k}{2^K}, \frac{\pi\xi^k}{2^K}\right]} \int_{\pi j \kappa(2^K)^{-1}}^{\pi(j+1)\kappa(2^K)^{-1}} \\
&\qquad \sup_{n_1<2^K, (n_1,n_2)\in\mathbb{R}^d_{\kappa,\tau}} |D^1_{n_1,n_2}(x_1, x_2)|^p \, dx_1 \, dx_2 \\
&=: (D_1) + (D_2).
\end{aligned}$$

It follows from (3.3.5), (3.3.10) and (3.4.7) that

$$\begin{aligned}
&|D^1_{n_1,n_2}(x_1, x_2)| \\
&\le C_p 2^{K/p-K} \kappa(2^K)^{1/p-2} \frac{1}{n_1^\alpha |x_1 - \mu|^{\alpha+1}} \frac{n_2^{\zeta+1+\alpha(\zeta-1)}}{|x_2 - \nu|^{(\alpha+1)(1-\zeta)}} \\
&\le C_p 2^{K/p-K} \kappa(2^K)^{1/p-2+(\alpha+1)(1-\zeta)} \frac{\left(\frac{2^K}{\xi^{k+1}}\right)^{-\alpha}}{|x_1 - \mu|^{\alpha+1}} \frac{\kappa\left(\frac{2^K}{\xi^k}\right)^{\zeta+1+\alpha(\zeta-1)}}{j^{(\alpha+1)(1-\zeta)}},
\end{aligned}$$

where $0 \le \zeta \le 1$. This leads to

$$(D_1) \le C_p \sum_{k=0}^{\infty} \sum_{j=1}^{\kappa(2^K)/2-1} \int_{\mathbb{T}\setminus\left[-\frac{\pi\xi^k}{2^K}, \frac{\pi\xi^k}{2^K}\right]} 2^{K(1-p-\alpha p)} \kappa(2^K)^{p(-2+(\alpha+1)(1-\zeta))} \xi^{k\alpha p}$$

$$|x_1-\mu|^{-(\alpha+1)p} \frac{\kappa\left(\frac{2^K}{\xi^k}\right)^{p(2+(\alpha+1)(\zeta-1))}}{j^{p(\alpha+1)(1-\zeta)}} \, dx_1$$

$$\le C_p \sum_{k=0}^{\infty} \sum_{j=1}^{\kappa(2^K)/2-1} 2^{K(1-p-\alpha p)} \xi^{k\alpha p} (\xi^k 2^{-K})^{-(\alpha+1)p+1} \frac{c_1^{-kp(2+(\alpha+1)(\zeta-1))}}{j^{p(\alpha+1)(1-\zeta)}}$$

$$\le C_p \sum_{k=0}^{\infty} \sum_{j=1}^{\kappa(2^K)/2-1} \frac{\xi^{k(1-p-\omega_1 p(2+(\alpha+1)(\zeta-1)))}}{j^{p(\alpha+1)(1-\zeta)}},$$

which is convergent if

$$p > \frac{1}{1+\omega_1(2+(\alpha+1)(\zeta-1))} \qquad \text{and} \qquad p > \frac{1}{(\alpha+1)(1-\zeta)}.$$

After some computation, we can see that the optimal bound is reached if

$$\zeta = \frac{\alpha - \omega_1 + \alpha\omega_1}{1+\alpha+\omega_1+\alpha\omega_1},$$

which means that

$$p > \frac{1+\omega_1}{1+2\omega_1}.$$

Considering (D_2), we estimate as follows:

$$|D^1_{n_1,n_2}(x_1,x_2)| \le C_p 2^{K/p-K} \kappa(2^K)^{1/p-2} n_1 \frac{n_2^{\zeta+1+\alpha(\zeta-1)}}{|x_2-\nu|^{(\alpha+1)(1-\zeta)}}$$

$$\le C_p 2^{K/p} \kappa(2^K)^{1/p-2+(\alpha+1)(1-\zeta)} \xi^{-k} \frac{\kappa\left(\frac{2^K}{\xi^k}\right)^{\zeta+1+\alpha(\zeta-1)}}{j^{(\alpha+1)(1-\zeta)}}$$

and

$$(D_2) \le C_p \sum_{k=0}^{\infty} \sum_{j=1}^{\kappa(2^K)/2-1}$$

$$\int_{\left[-\frac{\pi\xi^k}{2^K}, \frac{\pi\xi^k}{2^K}\right]} 2^K \kappa(2^K)^{p(-2+(\alpha+1)(1-\zeta))} \xi^{-kp} \frac{\kappa\left(\frac{2^K}{\xi^k}\right)^{p(2+(\alpha+1)(\zeta-1))}}{j^{p(\alpha+1)(1-\zeta)}} \, dx_1$$

$$\leq C_p \sum_{k=0}^{\infty} \sum_{j=1}^{\kappa(2^K)/2-1} \frac{\zeta^{k(1-p-\omega_1 p(2+(\alpha+1)(\zeta-1)))}}{j^{p(\alpha+1)(1-\zeta)}}$$
$$\leq C_p$$

as above.

The term $D^2_{n_1,n_2}$ can be handled similarly. We obtain

$$\int_{\mathbb{T}\setminus 4I_1} \int_{\mathbb{T}\setminus 4I_2} \sup_{n_1<2^K,(n_1,n_2)\in\mathbb{R}^d_{\kappa,\tau}} |D^2_{n_1,n_2}(x_1,x_2)|^p \, dx_1 \, dx_2 \leq C_p$$

if

$$p > \frac{1+\omega_2}{2+\omega_2}.$$

Using (3.3.3), we estimate $D^3_{n_1,n_2}$ in the same way as (C) in (3.4.6). Now the exponents of n_1 and n_2 are non-negative and so they can be estimated by 2^K and $\kappa(2^K)$ as in (3.4.6). This proves that

$$\int_{\mathbb{T}\setminus 4I_1} \int_{\mathbb{T}\setminus 4I_2} \left|\sigma^\alpha_\kappa a(x_1,x_2)\right|^p \, dx_1 \, dx_2 \leq C_p$$

which completes the proof. ■

Remark 3.4.8 In the d-dimensional case, the constant p_1 appears if we investigate the terms corresponding to $D^1_{n_1,n_2}$ and $D^2_{n_1,n_2}$. Indeed, let $\prod_{j=1}^d I_j$ be centered at 0 and the support of the atom a, A be the integral of a, $I_j =: [-\mu_j, \mu_j]$ and

$$\bar{t}_j := \begin{cases} \mu_j, & j \in H; \\ t_j, & j \in H^c, \end{cases}$$

$H \subset \{1,\ldots,d\}$, $H \neq \emptyset$, $H \neq \{1,\ldots,d\}$. If we integrate the term

$$\int_{\prod_{j\in H^c} I_j} A(\bar{t}_1,\ldots,\bar{t}_d) \prod_{j\in H} K^\alpha_{n_j}(x_j-\mu_j) \prod_{i\in H^c} (K^\alpha_{n_i})'(x_i-t_i)\, dt$$

over $\prod_{j=1}^d (\mathbb{T}\setminus 4I_j)$, then we get that

$$p > \frac{\sum_{j\in H}\omega_{j,2} + \sum_{j\in H^c}\omega_{j,1}}{\sum_{j\in H}\omega_{j,2} + 2\sum_{j\in H^c}\omega_{j,1}}.$$

Moreover, considering the integral

$$\int_{\prod_{j\in H}(\mathbb{T}\setminus 4I_j)} \int_{\prod_{j\in H^c} 4I_j} |\sigma_\kappa^\alpha a(x)|^p \, dx,$$

we obtain

$$p > \frac{\sum_{j\in H} \omega_{j,2}}{\sum_{j\in H} \omega_{j,2} + \sum_{j\in H^c} \omega_{j,1}}.$$

However, this bound is less than p_1.

Remark 3.4.9 If $\omega_{j,1} = \omega_{j,2} = 1$ for all $j = 1, \ldots, d$, then we obtain in Theorem 3.4.7 the bound

$$\max\left\{\frac{d}{d+1}, \frac{1}{\alpha+1}\right\}.$$

In particular, this holds if $\kappa_j = \mathcal{I}$ for all $j = 1, \ldots, d$, i.e., if we consider a cone. This bound was obtained for cones in Theorem 3.3.4.

Corollary 3.4.10 *If* $0 < \alpha \leq 1$ *and* $1 < p < \infty$, *then*

$$\left\|\sigma_\kappa^\alpha f\right\|_p \leq C_p \|f\|_p \qquad (f \in L_p(\mathbb{T}^d)).$$

We obtain similar results for the Riesz means (cf. Theorem 3.3.7). The details are left to the reader.

Theorem 3.4.11 *If* $0 < \alpha < \infty$, $\gamma \in \mathbb{P}$ *and*

$$\max\left\{p_1, \frac{1}{\alpha \wedge 1 + 1}\right\} < p \leq \infty,$$

then

$$\left\|\sigma_\kappa^{\alpha,\gamma} f\right\|_p \leq C_p \|f\|_{H_p^\kappa} \qquad (f \in H_p^\kappa(\mathbb{T}^d)).$$

Corollary 3.4.12 *Suppose that* $0 < \alpha < \infty$ *and* $\gamma \in \mathbb{P}$. *If* $1 < p < \infty$, *then*

$$\left\|\sigma_\kappa^{\alpha,\gamma} f\right\|_p \leq C_p \|f\|_p \qquad (f \in L_p(\mathbb{T}^d)).$$

Corollary 3.4.13 *If* $0 < \alpha \leq 1$, *then*

$$\sup_{\rho>0} \rho\lambda(\sigma_\kappa^\alpha f > \rho) \leq C\|f\|_1 \qquad (f \in L_1(\mathbb{T}^d).$$

If $0 < \alpha < \infty$ *and* $\gamma \in \mathbb{P}$, *then*

$$\sup_{\rho>0} \rho\lambda(\sigma_\kappa^{\alpha,\gamma} f > \rho) \leq C\|f\|_1 \qquad (f \in L_1(\mathbb{T}^d).$$

Corollary 3.4.14 *Suppose that $f \in L_1(\mathbb{T}^d)$. If $0 < \alpha \leq 1$, then*

$$\lim_{n\to\infty,\, n\in\mathbb{R}^d_{\kappa,\tau}} \sigma_n^\alpha f = f \qquad a.e.$$

If $0 < \alpha < \infty$ and $\gamma \in \mathbb{P}$, then

$$\lim_{n\to\infty,\, n\in\mathbb{R}^d_{\kappa,\tau}} \sigma_n^{\alpha,\gamma} f = f \qquad a.e.$$

In the two-dimensional case, Corollaries 3.4.13 and 3.4.14 were proved by Gát [119] for Fejér summability. In this case, he verified also that if the cone-like set $\mathbb{R}^d_{\kappa,\tau}$ is defined by $\tau_j(n_1)$ instead of τ_j and if $\tau_j(n_1)$ is not bounded, then Corollary 3.4.14 does not hold and the largest space for the elements of which we have almost everywhere convergence is $L\log L$. This means that under these conditions Theorem 3.4.7 cannot be true for any $p < 1$.

3.5 $H_p(\mathbb{T}^d)$ Hardy spaces

For the investigation of the unrestricted almost everywhere convergence of the rectangular summability means, we need a new type of Hardy spaces, the so-called product Hardy spaces.

Fix $\psi \in S(\mathbb{R})$ such that $\int_{\mathbb{R}} \psi(x)dx \neq 0$. We define the product radial maximal function, the product non-tangential maximal function and the hybrid maximal function of $f \in D(\mathbb{T}^d)$ by

$$\psi_+^*(f)(x) := \sup_{t_i\in(0,\infty),i=1,\ldots,d} \left|(f * (\psi_{t_1} \otimes \cdots \otimes \psi_{t_d}))(x)\right|,$$

$$\psi_\nabla^*(f)(x) := \sup_{t_i\in(0,\infty),|x_i-y_i|<t_i,i=1,\ldots,d} \left|(f * (\psi_{t_1} \otimes \cdots \otimes \psi_{t_d}))(y)\right|$$

and

$$\psi_{\sharp_i}^*(f)(x) := \sup_{t_k\in(0,\infty),k=1,\ldots,d;k\neq i} \left|(f * (\psi_{t_1} \otimes \cdots \otimes \psi_{t_{i-1}} \otimes \psi_{t_{i+1}} \otimes \cdots \otimes \psi_{t_d}))(x)\right|,$$

respectively, $(i = 1, \ldots, d)$.

Definition 3.5.1 For $0 < p < \infty$, the product Hardy spaces $H_p(\mathbb{T}^d)$, product weak Hardy spaces $H_{p,\infty}(\mathbb{T}^d)$ and the hybrid Hardy spaces $H_p^i(\mathbb{T}^d)$ $(i = 1, \ldots, d)$ consist of all distributions $f \in D(\mathbb{T}^d)$ for which

$$\|f\|_{H_p} := \left\|\psi_+^*(f)\right\|_p < \infty,$$

$$\|f\|_{H_{p,\infty}} := \left\|\psi_+^*(f)\right\|_{p,\infty} < \infty$$

and

$$\|f\|_{H_p^i} := \left\|\psi_{\sharp_i}^*(f)\right\|_p < \infty.$$

The Hardy spaces are independent of ψ_i, more exactly, different functions ψ_i give the same space with equivalent norms. For $f \in D(\mathbb{T}^d)$, let

$$P_+^*(f)(x) := \sup_{t_i \in (0,\infty), i=1,\dots,d} \left|(f * (P_{t_1} \otimes \cdots \otimes P_{t_d}))(x)\right|,$$

$$P_\nabla^*(f)(x) := \sup_{t_i \in (0,\infty), |x_i - y_i| < t_i, i=1,\dots,d} \left|(f * (P_{t_1} \otimes \cdots \otimes P_{t_d}))(x)\right|$$

and

$$P_{\sharp_i}^*(f)(x) := \sup_{t_k \in (0,\infty), k=1,\dots,d; k\neq i} \left|(f * (P_{t_1} \otimes \cdots \otimes P_{t_{i-1}} \otimes P_{t_{i+1}} \otimes \cdots \otimes P_{t_d}))(x)\right|,$$

respectively ($i = 1, \dots, d$), where the Poisson kernel P_{t_i} was defined before Theorem 2.4.14. The next theorems were proved in Chang and Fefferman [54, 55], Gundy and Stein [155] or Weisz [346], so we omit the proofs.

Theorem 3.5.2 *Let $0 < p < \infty$. Fix $\psi \in S(\mathbb{R})$ such that $\int_{\mathbb{R}} \psi(x)dx \neq 0$. Then $f \in H_p(\mathbb{T}^d)$ if and only if $\psi_\nabla^*(f) \in L_p(\mathbb{T}^d)$ or $P_+^*(f) \in L_p(\mathbb{T}^d)$ or $P_\nabla^*(f) \in L_p(\mathbb{T}^d)$. We have the following equivalences of norms:*

$$\|f\|_{H_p^\square} \sim \|\psi_\nabla^*(f)\|_p \sim \|P_+^*(f)\|_p \sim \|P_\nabla^*(f)\|_p.$$

The same holds for the weak Hardy spaces:

$$\|f\|_{H_{p,\infty}^\square} \sim \|\psi_\nabla^*(f)\|_{p,\infty} \sim \|P_+^*(f)\|_{p,\infty} \sim \|P_\nabla^*(f)\|_{p,\infty}$$

and for the hybrid Hardy spaces:

$$\|f\|_{H_p^i} \sim \|P_{\sharp_i}^*(f)\|_p \qquad (i = 1, \dots, d).$$

As we can see from the next theorem, in the theory of product Hardy spaces, the hybrid Hardy spaces $H_p^i(\mathbb{T}^d)$ will play the role of the $L_1(\mathbb{T}^d)$ spaces in some sense.

Theorem 3.5.3 *If $1 < p < \infty$ and $i = 1, \dots, d$, then $H_p(\mathbb{T}^d) \sim H_p^i(\mathbb{T}^d) \sim L_p(\mathbb{T}^d)$ and*

$$\|f\|_p \le \|f\|_{H_p^i} \le \|f\|_{H_p} \le C_p \|f\|_p.$$

For $p = 1$, $H_1(\mathbb{T}^d) \subset H_1^i(\mathbb{T}^d) \subset H_{1,\infty}^\square(\mathbb{T}^d) \cap L_1(\mathbb{T}^d)$ and

$$\|f\|_{H_1^i} \le \|f\|_{H_1} \qquad (f \in H_1(\mathbb{T}^d)),$$

$$\|f\|_{H_{1,\infty}} \le C\|f\|_{H_1^i} \qquad (f \in H_1^i(\mathbb{T}^d)).$$

Definition 3.5.4 The set $L(\log L)^{d-1}(\mathbb{T}^d)$ contains all measurable functions for which

$$\left\| |f|(\log^+ |f|)^{d-1} \right\|_1 < \infty.$$

Theorem 3.5.5 *$H_1^i(\mathbb{T}^d) \supset L(\log L)^{d-1}(\mathbb{T}^d)$ for all $i = 1, \dots, d$ and*

$$\|f\|_{H_1^i} \le C + C \left\| |f|(\log^+ |f|)^{d-1} \right\|_1 \qquad (f \in L(\log L)^{d-1}(\mathbb{T}^d)).$$

A straightforward generalization of the atoms would be the following:

(i) $\operatorname{supp} a \subset I$, $I \subset \mathbb{T}^d$ is a rectangle,
(ii) $\|a\|_\infty \le |I|^{-1/p}$,
(iii) $\int_{\mathbb{T}} a(x) x_i^k \, dx_i = 0$, for all $i = 1, \dots, d$.

However, the space $H_p(\mathbb{T}^d)$ do not have atomic decomposition with respect to these atoms (see Weisz [327]). The atomic decomposition for $H_p(\mathbb{T}^d)$ is much more complicated. One reason of this is that the support of an atom is not a rectangle but an open set. Moreover, here we have to choose the atoms from $L_2(\mathbb{T}^d)$ instead of $L_\infty(\mathbb{T}^d)$.

First of all, we introduce some notations. By a dyadic interval we mean one of the form $[k2^{-n}, (k+1)2^{-n})$ for some $k, n \in \mathbb{Z}$. A dyadic rectangle is the Cartesian product of d dyadic intervals. Suppose that $F \subset \mathbb{T}^d$ is an open set. Let $\mathcal{M}_1(F)$ denote those dyadic rectangles $R = I \times S \subset F$, $I \subset \mathbb{T}$ is a dyadic interval, $S \subset \mathbb{T}^{d-1}$ is a dyadic rectangle that are maximal in the first direction. In other words, if $I' \times S \supset R$ is a dyadic subrectangle of F (where $I' \subset \mathbb{T}$ is a dyadic interval) then $I = I'$. Define $\mathcal{M}_i(F)$ similarly. Denote by $\mathcal{M}(F)$ the maximal dyadic subrectangles of F in the above sense.

Recall that if $I \subset \mathbb{T}$ is an interval, then rI is the interval with the same center as I and with length $r|I|$ ($r \in \mathbb{N}$). For a rectangle $R = I_1 \times \ldots \times I_d \subset \mathbb{T}^d$ let $rR := rI_1 \times \ldots \times rI_d$. Instead of $2^r R$ we write R^r ($r \in \mathbb{N}$).

Definition 3.5.6 A function $a \in L_2(\mathbb{R}^d)$ is an H_p-atom ($0 < p \le 1$) if

(i) $\operatorname{supp} a \subset F$ for some open set $F \subset \mathbb{T}^d$ with finite measure,
(ii) $\|a\|_2 \le |F|^{1/2-1/p}$,
(iii) a can be decomposed further into the sum of "elementary particles" $a_R \in L_2(\mathbb{R}^d)$,

$$a = \sum_{R \in \mathcal{M}(F)} a_R,$$

satisfying

(a) $\operatorname{supp} a_R \subset 5R$,

(b) for all $R \in \mathcal{M}(F)$, $i = 1, \ldots, d$ and almost every fixed $x_1, \ldots, x_{i-1}, x_{i+1}, \ldots, x_d$,

$$\int_{\mathbb{T}} a_R(x) x_i^k \, dx_i = 0 \qquad (k = 0, \ldots, M(p) \geq \lfloor 2/p - 3/2 \rfloor),$$

(c) for every disjoint partition $\mathcal{P}_l$ $(l \in \mathbb{P})$ of $\mathcal{M}(F)$,

$$\left(\sum_{l \in \mathbb{P}} \left\| \sum_{R \in \mathcal{P}_l} a_R \right\|_2^2 \right)^{1/2} \leq |F|^{1/2-1/p}.$$

Theorem 3.5.7 *A distribution $f \in D(\mathbb{T}^d)$ is in $H_p(\mathbb{T}^d)$ $(0 < p \leq 1)$ if and only if there exist a sequence $(a^k, k \in \mathbb{N})$ of H_p-atoms and a sequence $(\mu_k, k \in \mathbb{N})$ of real numbers such that*

$$\sum_{k=0}^{\infty} |\mu_k|^p < \infty \qquad \text{and} \qquad \sum_{k=0}^{\infty} \mu_k a_k = f \quad \text{in } D(\mathbb{T}^d).$$

Moreover,

$$\|f\|_{H_p} \sim \inf \left(\sum_{k=0}^{\infty} |\mu_k|^p \right)^{1/p},$$

where the infimum is taken over all decompositions of f.

The result corresponding to Theorem 2.4.19 for the $H_p(\mathbb{T}^d)$ space is much more complicated. Since the definition of the H_p-atom is very complex, to obtain a usable condition about the boundedness of the operators, we have to introduce simpler atoms (see also the definition right after Theorem 3.5.5).

Definition 3.5.8 A function $a \in L_2(\mathbb{T}^d)$ is a simple H_p-atom or a rectangle H_p-atom if

(i) supp $a \subset R$ for a rectangle $R \subset \mathbb{T}^d$,
(ii) $\|a\|_2 \leq |R|^{1/2-1/p}$,
(iii) $\int_{\mathbb{T}} a(x) x_i^k \, dx_i = 0$ for $i = 1, \ldots, d, k = 0, \ldots, M(p) \geq \lfloor 2/p - 3/2 \rfloor$ and for almost every fixed x_j, $j = 1, \ldots, d$, $j \neq i$.

Note that $H_p(\mathbb{T}^d)$ cannot be decomposed into rectangle p-atoms, a counterexample can be found in Weisz [327]. However, the following result says that for an operator V to be bounded from $H_p(\mathbb{T}^2)$ to $L_p(\mathbb{T}^2)$ $(0 < p \leq 1)$, it is enough to check V on simple H_p-atoms and the boundedness of V on $L_2(\mathbb{T}^2)$. We omit the proof because it can be found for all dimensions in Weisz [332, 346] (see also Fefferman [98]).

Theorem 3.5.9 *Let $d = 2$, $0 < p_0 \le 1$, $K_n \in L_1(\mathbb{T}^2)$ and $V_n f := f * K_n$ ($n \in \mathbb{N}^2$). Suppose that there exists $\eta > 0$ such that for every simple H_{p_0}-atom a and for every $r \ge 1$*

$$\int_{\mathbb{T}^2 \setminus R^r} |V_* a|^{p_0} \, d\lambda \le C_p 2^{-\eta r},$$

where R is the support of a. If V_* is bounded from $L_2(\mathbb{T}^2)$ to $L_2(\mathbb{T}^2)$, then

$$\|V_* f\|_p \le C_{p_0} \|f\|_{H_p} \qquad (f \in H_p(\mathbb{T}^2))$$

for all $p_0 \le p \le 2$.

Note that Theorem 2.4.16 holds also for $H_p(\mathbb{T}^d)$ spaces with a very similar proof.

Theorem 3.5.10 *If $K \in L_1(\mathbb{T}^d)$, $0 < p < \infty$ and*

$$\lim_{k \to \infty} f_k = f \quad \text{in the } H_p(\mathbb{T}^d)\text{-norm},$$

then

$$\lim_{k \to \infty} f_k * K = f * K \quad \text{in } D(\mathbb{T}^d).$$

Corollary 3.5.11 *If $p_0 < 1$ in Theorem 3.5.9, then for all $f \in H_1^i(\mathbb{T}^2)$ and $i = 1, 2$,*

$$\sup_{\rho > 0} \rho \lambda(|V_* f| > \rho) \le C \|f\|_{H_1^i}.$$

Proof Using the preceding theorem and interpolation, we conclude that the operator

$$V_* \quad \text{is bounded from} \quad H_{p,\infty}(\mathbb{T}^2) \quad \text{to} \quad L_{p,\infty}(\mathbb{T}^2)$$

when $p_0 < p < 2$. Thus, it holds also for $p = 1$. By Theorem 3.5.3,

$$\sup_{\rho > 0} \rho \lambda(|V_* f| > \rho) = \|V_* f\|_{1,\infty} \le C \|f\|_{H_{1,\infty}} \le C \|f\|_{H_1^i}$$

for all $f \in H_1^i(\mathbb{T}^2)$, $i = 1, 2$. ■

Note that for higher dimensions, we have to modify slightly Theorem 3.5.9, Corollary 3.5.11 as well as the definition of simple H_p-atoms (see Weisz [332, 346]).

3.6 Almost Everywhere Unrestricted Summability

For the almost everywhere unrestricted summability, we introduce the next maximal operators.

Definition 3.6.1 We define the unrestricted maximal Cesàro and unrestricted maximal Riesz operator by

$$\sigma_*^\alpha f := \sup_{n\in\mathbb{N}^d} |\sigma_n^\alpha f|$$

and

$$\sigma_*^{\alpha,\gamma} f := \sup_{n\in\mathbb{N}^d} |\sigma_n^{\alpha,\gamma} f|,$$

respectively.

For $\alpha = \gamma = 1$, the operator is called unrestricted maximal Fejér operator and denoted by $\sigma_* f$.

We will first prove that the operator σ_*^α is bounded from $L_p(\mathbb{T}^d)$ to $L_p(\mathbb{T}^d)$ $(1 < p \le \infty)$ and then that it is bounded from $H_p(\mathbb{T}^d)$ to $L_p(\mathbb{T}^d)$ $(1/(\alpha+1) < p \le 1)$. To this end, we introduce the next one-dimensional operators.

Definition 3.6.2 Let

$$\tau_n^\alpha f(x) := f * \left|K_n^\alpha\right|(x),$$

$$\tau_n^{\alpha,\gamma} f(x) := f * \left|K_n^{\alpha,\gamma}\right|(x)$$

and

$$\tau_*^\alpha f := \sup_{n\in\mathbb{N}} \left|\tau_n^\alpha f\right|,$$

$$\tau_*^{\alpha,\gamma} f := \sup_{n\in\mathbb{N}} \left|\tau_n^{\alpha,\gamma} f\right|.$$

Obviously,

$$|\sigma_n^\alpha f| \le \tau_n^\alpha |f| \quad (n \in \mathbb{N}) \qquad \text{and} \qquad \sigma_*^\alpha f \le \tau_*^\alpha |f|.$$

The same holds for the operators $\sigma_*^{\alpha,\gamma}$ and $\tau_*^{\alpha,\gamma}$. The next result can be proved similar to Theorem 3.3.4.

Theorem 3.6.3 *If* $0 < \alpha \le 1$ *and* $1/(\alpha+1) < p \le \infty$, *then*

$$\left\|\tau_*^\alpha f\right\|_p \le C_p \|f\|_{H_p} \qquad (f \in H_p(\mathbb{T})).$$

Proof It is easy to see that

$$\left\|\tau_*^\alpha f\right\|_\infty \le C \|f\|_\infty \qquad (f \in L_\infty(\mathbb{T})).$$

Let a be an arbitrary H_p-atom with support $I \subset \mathbb{T}$ and

$$[-\pi 2^{-K-2}, \pi 2^{-K-2}] \subset I \subset [-\pi 2^{-K-1}, \pi 2^{-K-1}].$$

Then

$$\begin{aligned}\int_{\mathbb{T}\setminus 4I_1} |\tau_*^\alpha a(x)|^p \, dx &\le \sum_{|i|=1}^{2^K-1} \int_{\pi i 2^{-K}}^{\pi(i+1)2^{-K}} \sup_{n\ge 2^K} |\tau_n^\alpha a(x)|^p \, dx \\ &\quad + \sum_{|i|=1}^{2^K-1} \int_{\pi i 2^{-K}}^{\pi(i+1)2^{-K}} \sup_{n< 2^K} |\tau_n^\alpha a(x)|^p \, dx \\ &=: (A) + (B).\end{aligned}$$

Using (3.3.2) and (3.3.5), we can see that

$$\begin{aligned}\left|\tau_n^\alpha a(x)\right| &= \left|\int_I a(t)\left|K_n^\alpha(x-t)\right| \, dt\right| \\ &\le C_p 2^{K/p} \int_I \frac{1}{n^\alpha |x-t|^{\alpha+1}} \, dt \\ &\le C_p 2^{K/p} \frac{1}{i^{\alpha+1}}\end{aligned}$$

and

$$(A) \le C_p \sum_{i=1}^{2^K-1} 2^{-K} 2^K \frac{1}{i^{(\alpha+1)p}} \le C_p$$

as in Theorem 3.3.4.

To estimate (B), observe that by (iii) of the definition of the atom,

$$\tau_n^\alpha a(x) = \int_I a(t)\left|K_n^\alpha(x-t)\right| \, dt = \int_I a(t)\Big(\left|K_n^\alpha(x-t)\right| - \left|K_n^\alpha(x)\right|\Big) \, dt.$$

Thus,

$$\left|\tau_n^\alpha a(x)\right| \le \int_I |a(t)| \left|K_n^\alpha(x-t) - K_n^\alpha(x)\right| dt.$$

Using Lagrange's mean value theorem and (3.3.3), we conclude

$$\begin{aligned}\left|K_n^\alpha(x-t) - K_n^\alpha(x)\right| &= \left|(K_n^\alpha)'(x-\xi)\right| |t| \\ &\le \frac{C_p 2^{-K}}{n^{\alpha-1}|x-\xi|^{\alpha+1}} \le \frac{C_p 2^K}{i^{\alpha+1}},\end{aligned}$$

where $\xi \in I$ and $x \in [\pi i 2^{-K}, \pi(i+1)2^{-K})$. Consequently,

$$\left|\tau_n^\alpha a(x)\right| \le C_p 2^{K/p-K} \frac{2^K}{i^{\alpha+1}}$$

and

$$(B) \le C_p \sum_{i=1}^{2^K-1} 2^{-K} 2^K \frac{1}{i^{(\alpha+1)p}} \le C_p,$$

which proves the theorem. ■

We can verify in the same way

Theorem 3.6.4 *If* $0 < \alpha < \infty$, $\gamma \in \mathbb{P}$ *and* $1/(\alpha \wedge 1 + 1) < p \le \infty$, *then*

$$\left\|\tau_*^{\alpha,\gamma} f\right\|_p \le C_p \|f\|_{H_p} \qquad (f \in H_p(\mathbb{T})).$$

The next result can be obtained by interpolation.

Corollary 3.6.5 *Suppose that* $1 < p \le \infty$. *If* $0 < \alpha \le 1$, *then*

$$\sup_{\rho>0} \rho\, \lambda(\tau_*^{\alpha} f > \rho) \le C\|f\|_1 \qquad (f \in L_1(\mathbb{T}))$$

and

$$\left\|\tau_*^{\alpha} f\right\|_p \le C_p \|f\|_p \qquad (f \in L_p(\mathbb{T})).$$

If $0 < \alpha < \infty$ *and* $\gamma \in \mathbb{P}$, *then*

$$\sup_{\rho>0} \rho\, \lambda(\tau_*^{\alpha,\gamma} f > \rho) \le C\|f\|_1 \qquad (f \in L_1(\mathbb{T}))$$

and

$$\left\|\tau_*^{\alpha,\gamma} f\right\|_p \le C_p \|f\|_p \qquad (f \in L_p(\mathbb{T})).$$

Now, we turn to the higher dimensional case and verify the $L_p(\mathbb{T}^d)$ boundedness of σ_*^{α} and $\sigma_*^{\alpha,\gamma}$.

Theorem 3.6.6 *Suppose that* $1 < p \le \infty$. *If* $0 < \alpha < \infty$, *then*

$$\left\|\sigma_*^{\alpha} f\right\|_p \le C_p \|f\|_p \qquad (f \in L_p(\mathbb{T}^d)).$$

If $0 < \alpha < \infty$, $\gamma \in \mathbb{P}$ *and* $1 < p \le \infty$, *then*

$$\left\|\sigma_*^{\alpha,\gamma} f\right\|_p \le C_p \|f\|_p \qquad (f \in L_p(\mathbb{T}^d)).$$

Proof For $0 < \alpha \le 1$, let us apply Corollary 3.6.5 to obtain

$$
\begin{aligned}
&\int_{\mathbb{T}}\int_{\mathbb{T}} \sup_{n_1,n_2\in\mathbb{N}} \left| \int_{\mathbb{T}}\int_{\mathbb{T}} f(t_1,t_2) K_{n_1}^{\alpha}(x_1-t_1) K_{n_2}^{\alpha}(x_2-t_2)\, dt_1\, dt_2 \right|^p dx_1\, dx_2 \\
&\leq \int_{\mathbb{T}}\int_{\mathbb{T}} \sup_{n_2\in\mathbb{N}} \\
&\quad \left(\int_{\mathbb{T}} \left(\sup_{n_1\in\mathbb{N}} \left| \int_{\mathbb{T}} f(t_1,t_2) K_{n_1}^{\alpha}(x_1-t_1)\, dt_1 \right| \right) \left| K_{n_2}^{\alpha}(x_2-t_2) \right| dt_2 \right)^p dx_2\, dx_1 \\
&\leq C_p \int_{\mathbb{T}}\int_{\mathbb{T}} \sup_{n_1\in\mathbb{N}} \left| \int_{\mathbb{T}} f(t_1,x_2) K_{n_1}^{\alpha}(x_1-t_1)\, dt_1 \right|^p dx_1\, dx_2 \\
&\leq C_p \int_{\mathbb{T}}\int_{\mathbb{T}} |f(x_1,x_2)|^p\, dx_1\, dx_2.
\end{aligned}
$$

The inequality for $1 < \alpha < \infty$ follows from Lemma 3.1.9. The result for $\sigma_*^{\alpha,\gamma}$ can be proved in the same way. ■

The next result is due to the author [331, 332].

Theorem 3.6.7 *If* $0 < \alpha < \infty$ *and* $1/(\alpha+1) < p \leq \infty$, *then*

$$\left\| \sigma_*^{\alpha} f \right\|_p \leq C_p \|f\|_{H_p} \qquad (f \in H_p(\mathbb{T}^d)).$$

Proof By Theorem 3.1.8,

$$\left| \left(K_{n_j}^{\alpha} \right)^{(s)}(t) \right| \leq \frac{C}{n_j^{\alpha-s} |t|^{\alpha+1}} \tag{3.6.1}$$

for $0 < \alpha \leq s+1, n_j \in \mathbb{P}$ and $t \in \mathbb{T}, t \neq 0$. Choose a simple H_p-atom a with support $R = I_1 \times I_2$, where I_1 and I_2 are intervals with

$$2^{-K_i-1} < |I_i|/\pi \leq 2^{-K_i} \qquad (K_i \in \mathbb{N}, i = 1, 2)$$

and

$$[-\pi 2^{-K_i-2}, \pi 2^{-K_i-2}] \subset I_i \subset [-\pi 2^{-K_i-1}, \pi 2^{-K_i-1}].$$

We assume that $r \geq 2$ is an arbitrary integer. Theorem 3.6.6 implies that the operator σ_*^{α} is bounded from $L_2(\mathbb{T}^d)$ to $L_2(\mathbb{T}^d)$. By Theorem 3.5.9, we have to integrate $\left|\sigma_*^{\alpha} a\right|^p$ over

$$
\begin{aligned}
\mathbb{T}^2 \setminus R^r = (\mathbb{T} \setminus I_1^r) \times I_2 &\bigcup (\mathbb{T} \setminus I_1^r) \times (\mathbb{T} \setminus I_2) \\
&\bigcup I_1 \times (\mathbb{T} \setminus I_2^r) \bigcup (\mathbb{T} \setminus I_1) \times (\mathbb{T} \setminus I_2^r).
\end{aligned}
$$

First, we integrate over $(\mathbb{T} \setminus I_1^r) \times I_2$:

$$
\begin{aligned}
&\int_{\mathbb{T}\setminus 4I_1}\int_{I_2}\left|\sigma_*^\alpha a(x_1,x_2)\right|^p\,dx_1\,dx_2\\
&\quad\le \int_{\mathbb{T}\setminus 4I_1}\int_{I_2}\sup_{n_1\ge 2^{K_1},n_2\in\mathbb{N}}\left|\sigma_n^\alpha a(x_1,x_2)\right|^p\,dx_1\,dx_2\\
&\qquad+\int_{\mathbb{T}\setminus 4I_1}\int_{I_2}\sup_{n_1< 2^{K_1},n_2\in\mathbb{N}}\left|\sigma_n^\alpha a(x_1,x_2)\right|^p\,dx_1\,dx_2\\
&\quad\le \sum_{|i_1|=2^{r-2}}^{2^{K_1}-1}\int_{\pi i_1 2^{-K_1}}^{\pi(i_1+1)2^{-K_1}}\int_{I_2}\sup_{n_1\ge 2^{K_1},n_2\in\mathbb{N}}\left|\sigma_n^\alpha a(x_1,x_2)\right|^p\,dx_1\,dx_2\\
&\qquad+\sum_{|i_1|=2^{r-2}}^{2^{K_1}-1}\int_{\pi i_1 2^{-K_1}}^{\pi(i_1+1)2^{-K_1}}\int_{I_2}\sup_{n_1< 2^{K_1},n_2\in\mathbb{N}}\left|\sigma_n^\alpha a(x_1,x_2)\right|^p\,dx_1\,dx_2\\
&\quad=: (A)+(B).
\end{aligned}
$$

Here we may suppose that $i_1>0$. For $k,l\in\mathbb{N}$ let $A_{0,0}(x):=a(x)$,

$$
A_{1,0}(x_1,x_2):=\int_{-\pi}^{x_1}a(t,x_2)\,dt\qquad A_{0,1}(x_1,x_2):=\int_{-\pi}^{x_2}a(x_1,u)\,du
$$

and

$$
A_{k,l}(x_1,x_2):=\int_{-\pi}^{x_1}A_{k-1,l}(t,x_2)\,dt=\int_{-\pi}^{x_2}A_{k,l-1}(x_1,u)\,du.
$$

By (iii) of the definition of the simple H_p-atom, we can show that supp $A_{k,l}\subset R$ and $A_{k,l}(x_1,x_2)$ is zero if x_1 is at the boundary of I_1 or x_2 is at the boundary of I_2 for $k,l=0,\ldots,M(p)+1$ $(i=1,2)$, where $M(p)\ge\lfloor 2/p-3/2\rfloor$. Moreover, using (ii), we can compute that

$$
\left\|A_{k,l}\right\|_2\le |I_1|^{k+1/2-1/p}|I_2|^{l+1/2-1/p}\qquad (k,l=0,\ldots,M(p)+1). \tag{3.6.2}
$$

We may suppose that $M(p)\ge\alpha+1$ and choose $N\in\mathbb{N}$ such that $N<\alpha\le N+1$. For $x_1\in[\pi i_1 2^{-K_1},\pi(i_1+1)2^{-K_1})$, $t_1\in[-\pi 2^{-K_1-1},\pi 2^{-K_1-1})$, inequality (3.6.1) implies

$$
\left|(K_{n_1}^\alpha)^{(N)}(x_1-t_1)\right|\le\frac{Cn_1^{N-\alpha}2^{K_1(\alpha+1)}}{i_1^{\alpha+1}} \tag{3.6.3}
$$

and

$$
\left|(K_{n_1}^\alpha)^{(N+1)}(x_1-t_1)\right|\le\frac{Cn_1^{N+1-\alpha}2^{K_1(\alpha+1)}}{i_1^{\alpha+1}}. \tag{3.6.4}
$$

Integrating by parts, we can see that

$$\begin{aligned}|\sigma_n^\alpha a(x)| &= \left| \int_{I_1} \int_{I_2} A_{N,0}(t_1, t_2)(K_{n_1}^\alpha)^{(N)}(x_1 - t_1) K_{n_2}^\alpha(x_2 - t_2)\, dt_1\, dt_2 \right| \\ &\leq \frac{C n_1^{N-\alpha} 2^{K_1(\alpha+1)}}{i_1^{\alpha+1}} \int_{I_1} \left| \int_{I_2} A_{N,0}(t_1, t_2) K_{n_2}^\alpha(x_2 - t_2)\, dt_2 \right| dt_1\end{aligned}$$

whenever $x_1 \in [\pi i_1 2^{-K_1}, \pi(i_1 + 1)2^{-K_1})$. Hence, by Hölder's inequality and (3.6.3),

$$\begin{aligned}(A) &\leq C_p \sum_{i_1=2^r-2}^{2^{K_1}-1} 2^{-K_1} \frac{2^{K_1(N+1)p}}{i_1^{(\alpha+1)p}} \\ &\quad \int_{I_2} \left(\int_{I_1} \sup_{n_2 \in \mathbb{N}} \left| \int_{I_2} A_{N,0}(t_1, t_2) K_{n_2}^\alpha(x_2 - t_2)\, dt_2 \right| dt_1 \right)^p dx_2 \\ &\leq C_p |I_2|^{1-p} \sum_{i_1=2^r-2}^{2^{K_1}-1} \frac{2^{K_1((N+1)p-1)}}{i_1^{(\alpha+1)p}} \\ &\quad \left(\int_{I_2} \int_{I_1} \sup_{n_2 \in \mathbb{N}} \left| \int_{I_2} A_{N,0}(t_1, t_2) K_{n_2}^\alpha(x_2 - t_2)\, dt_2 \right| dt_1\, dx_2 \right)^p.\end{aligned}$$

Using again Hölder's inequality and the fact that σ_*^α is bounded on $L_2(\mathbb{T})$, we conclude

$$\begin{aligned}(A) &\leq C_p |I_2|^{1-p/2} \sum_{i_1=2^r-2}^{2^{K_1}-1} \frac{2^{K_1((N+1)p-1)}}{i_1^{(\alpha+1)p}} \\ &\quad \left(\int_{I_1} \left(\int_{I_2} \sup_{n_2 \in \mathbb{N}} \left| \int_{I_2} A_{N,0}(t_1, t_2) K_{n_2}^\alpha(x_2 - t_2)\, dt_2 \right|^2 dx_2 \right)^{1/2} dt_1 \right)^p \\ &\leq C_p |I_2|^{1-p/2} \sum_{i_1=2^r-2}^{2^{K_1}-1} \frac{2^{K_1((N+1)p-1)}}{i_1^{(\alpha+1)p}} \\ &\quad \left(\int_{I_1} \left(\int_{I_2} |A_{N,0}(t_1, x_2)|^2\, dx_2 \right)^{1/2} dt_1 \right)^p.\end{aligned}$$

Then (3.6.2) implies

$$(A) \le C_p |I_2|^{1-p/2} \sum_{i_1=2^{r-2}}^{2^{K_1}-1} 2^{-K_1 p/2} \frac{2^{K_1((N+1)p-1)}}{i_1^{(\alpha+1)p}}$$
$$\left(\int_{I_1} \int_{I_2} \left| A_{N,0}(t_1, x_2) \right|^2 dx_2 \, dt_1 \right)^{p/2}$$
$$\le C_p \sum_{i_1=2^{r-2}}^{2^{K_1}-1} \frac{1}{i_1^{(\alpha+1)p}} \le C_p 2^{-r((\alpha+1)p-1)}.$$

To estimate (B), we use (3.6.4):

$$(B) \le C_p \sum_{i_1=2^{r-2}}^{2^{K_1}-1} 2^{-K_1} \frac{2^{K_1(N+2)p}}{i_1^{(\alpha+1)p}}$$
$$\int_{I_2} \left(\int_{I_1} \sup_{n_2 \in \mathbb{N}} \left| \int_{I_2} A_{N+1,0}(t_1, t_2) K_{n_2}^{\alpha}(x_2 - t_2) \, dt_2 \right| dt_1 \right)^p dx_2$$
$$\le C_p |I_2|^{1-p/2} \sum_{i_1=2^{r-2}}^{2^{K_1}-1} \frac{2^{K_1((N+2)p-1)}}{i_1^{(\alpha+1)p}}$$
$$\left(\int_{I_1} \left(\int_{I_2} \sup_{n_2 \in \mathbb{N}} \left| \int_{I_2} A_{N+1,0}(t_1, t_2) K_{n_2}^{\alpha}(x_2 - t_2) \, dt_2 \right|^2 dx_2 \right)^{1/2} dt_1 \right)^p$$

and

$$(B) \le C_p |I_2|^{1-p/2} \sum_{i_1=2^{r-2}}^{2^{K_1}-1} \frac{2^{K_1((N+2)p-1)}}{i_1^{(\alpha+1)p}}$$
$$\left(\int_{I_1} \left(\int_{I_2} \left| A_{N+1,0}(t_1, x_2) \right|^2 dx_2 \right)^{1/2} dt_1 \right)^p$$
$$\le C_p |I_2|^{1-p/2} \sum_{i_1=2^{r-2}}^{2^{K_1}-1} 2^{-K_1 p/2} \frac{2^{K_1((N+2)p-1)}}{i_1^{(\alpha+1)p}}$$
$$\left(\int_{I_1} \int_{I_2} \left| A_{N+1,0}(t_1, x_2) \right|^2 dx_2 \, dt_1 \right)^{p/2}$$
$$\le C_p \sum_{i_1=2^{r-2}}^{2^{K_1}-1} \frac{1}{i_1^{(\alpha+1)p}} \le C_p 2^{-r((\alpha+1)p-1)}.$$

Next, we integrate over $(\mathbb{T} \setminus I_1^r) \times (\mathbb{T} \setminus I_2)$:

$$\begin{aligned}
&\int_{\mathbb{T}\setminus 4I_1}\int_{\mathbb{T}\setminus I_2} \left|\sigma_*^\alpha a(x_1,x_2)\right|^p \,dx_1\,dx_2 \\
&\quad\le \int_{\mathbb{T}\setminus 4I_1}\int_{\mathbb{T}\setminus I_2} \sup_{n_1\ge 2^{K_1},n_2\ge 2^{K_2}} \left|\sigma_n^\alpha a(x_1,x_2)\right|^p \,dx_1\,dx_2 \\
&\qquad+ \int_{\mathbb{T}\setminus 4I_1}\int_{\mathbb{T}\setminus I_2} \sup_{n_1\ge 2^{K_1},n_2< 2^{K_2}} \left|\sigma_n^\alpha a(x_1,x_2)\right|^p \,dx_1\,dx_2 \\
&\qquad+ \int_{\mathbb{T}\setminus 4I_1}\int_{\mathbb{T}\setminus I_2} \sup_{n_1< 2^{K_1},n_2\ge 2^{K_2}} \left|\sigma_n^\alpha a(x_1,x_2)\right|^p \,dx_1\,dx_2 \\
&\qquad+ \int_{\mathbb{T}\setminus 4I_1}\int_{\mathbb{T}\setminus I_2} \sup_{n_1< 2^{K_1},n_2< 2^{K_2}} \left|\sigma_n^\alpha a(x_1,x_2)\right|^p \,dx_1\,dx_2 \\
&\quad=: (C)+(D)+(E)+(F).
\end{aligned}$$

We will only consider the term (D):

$$\begin{aligned}
(D) \le \sum_{|i_1|=2^{r-2}}^{2^{K_1}-1}\sum_{|i_2|=1}^{2^{K_2}-1} &\int_{\pi i_1 2^{-K_1}}^{\pi(i_1+1)2^{-K_1}}\int_{\pi i_2 2^{-K_2}}^{\pi(i_2+1)2^{-K_2}} \\
&\sup_{n_1\ge 2^{K_1},n_2<2^{K_2}} \left|\sigma_n^\alpha a(x_1,x_2)\right|^p \,dx_1\,dx_2,
\end{aligned}$$

where we may suppose again that $i_1 > 0$ and $i_2 > 0$. Integrating by parts,

$$\begin{aligned}
&\left|\sigma_n^\alpha a(x)\right| \\
&= \left|\int_{I_1}\int_{I_2} A_{N,N+1}(t_1,t_2)(K_{n_1}^\alpha)^{(N)}(x_1-t_1)(K_{n_2}^\alpha)^{(N+1)}(x_2-t_2)\,dt_1\,dt_2\right| \\
&\le \frac{C2^{K_1(N+1)}2^{K_2(N+2)}}{i_1^{\alpha+1}i_2^{\alpha+1}} \int_{I_1}\int_{I_2} \left|A_{N,N+1}(t_1,t_2)\right| \,dt_1\,dt_2.
\end{aligned}$$

Thus

$$\begin{aligned}
(D) &\le C_p \sum_{i_1=2^{r-2}}^{2^{K_1}-1}\sum_{i_2=1}^{2^{K_2}-1} 2^{-K_1}2^{-K_2}\frac{2^{K_1(N+1)p}2^{K_2(N+2)p}}{i_1^{(\alpha+1)p}i_2^{(\alpha+1)p}} \\
&\quad \left(\int_{I_1}\int_{I_2} \left|A_{N,N+1}(t_1,t_2)\right| \,dt_1\,dt_2\right)^p \\
&\le C_p \sum_{i_1=2^{r-2}}^{2^{K_1}-1}\sum_{i_2=1}^{2^{K_2}-1} 2^{-K_1p/2}2^{-K_2p/2}\frac{2^{K_1((N+1)p-1)}2^{K_2((N+2)p-1)}}{i_1^{(\alpha+1)p}i_2^{(\alpha+1)p}} \\
&\quad \left(\int_{I_1}\int_{I_2} \left|A_{N,N+1}(t_1,t_2)\right|^2 \,dt_1\,dt_2\right)^{p/2}
\end{aligned}$$

$$\leq C_p \sum_{i_1=2^{r-2}}^{2^{K_1}-1} \sum_{i_2=1}^{2^{K_2}-1} \frac{1}{i_1^{(\alpha+1)p} i_2^{(\alpha+1)p}}$$
$$\leq C_p 2^{-r((\alpha+1)p-1)}.$$

All other integrals can be handled in the same way. Consequently,

$$\int_{\mathbb{T}^2 \setminus R^r} \left|\sigma_*^{\alpha} a(x_1, x_2)\right|^p \, dx_1 \, dx_2 \leq C_p 2^{-r((\alpha+1)p-1)},$$

which finishes the proof of the theorem. ∎

Theorem 3.6.8 *If* $0 < \alpha < \infty$, $\gamma \in \mathbb{P}$ *and* $1/(\alpha+1) < p \leq \infty$, *then*

$$\left\|\sigma_*^{\alpha,\gamma} f\right\|_p \leq C_p \|f\|_{H_p} \qquad (f \in H_p(\mathbb{T}^d)).$$

Proof Similar to (3.3.13), for $s \in \mathbb{N}$, $n_j \in \mathbb{P}$ and $t \in \mathbb{T}$, $t \neq 0$, we have

$$\left|\left(K_{n_j}^{\alpha,\gamma}\right)^{(s)}(t)\right| \leq \frac{C}{n_j^{\alpha-s} |t|^{\alpha+1}}.$$

The theorem can be proved as Theorem 3.6.7. ∎

Corollary 3.5.11 implies

Corollary 3.6.9 *Let* $f \in H_1^i(\mathbb{T}^d)$ *for some* $i = 1, \ldots, d$. *If* $0 < \alpha < \infty$, *then*

$$\sup_{\rho>0} \rho \lambda(\sigma_*^{\alpha} f > \rho) \leq C \|f\|_{H_1^i}.$$

If $0 < \alpha < \infty$ *and* $\gamma \in \mathbb{P}$, *then*

$$\sup_{\rho>0} \rho \lambda(\sigma_*^{\alpha,\gamma} f > \rho) \leq C \|f\|_{H_1^i}.$$

By the density argument, we get here almost everywhere convergence for functions from the spaces $H_1^i(\mathbb{T}^d)$ instead of $L_1(\mathbb{T}^d)$. In some sense, the Hardy space $H_1^i(\mathbb{T}^d)$ plays the role of $L_1(\mathbb{T}^d)$ in higher dimensions.

Corollary 3.6.10 *Let* $f \in H_1^i(\mathbb{T}^d)$ *for some* $i = 1, \ldots, d$. *If* $0 < \alpha < \infty$, *then*

$$\lim_{n \to \infty} \sigma_n^{\alpha} f = f \qquad a.e.$$

If $0 < \alpha < \infty$ *and* $\gamma \in \mathbb{P}$, *then*

$$\lim_{n \to \infty} \sigma_n^{\alpha,\gamma} f = f \qquad a.e.$$

The almost everywhere convergence is not true for all $f \in L_1(\mathbb{T}^d)$.

A counterexample, which shows that the almost everywhere convergence is not true for all integrable functions, is due to Gát [119]. Recall that

$$L_1(\mathbb{T}^d) \supset H_1^i(\mathbb{T}^d) \supset L(\log L)^{d-1}(\mathbb{T}^d) \supset L_p(\mathbb{T}^d) \qquad (1 < p \le \infty).$$

3.7 Rectangular θ-Summability

In this section, we introduce some new function spaces and then we generalize the rectangular Cesàro and Riesz means. As we will see in Definition 3.7.4, instead of condition (2.6.2), we have to suppose here that $\theta : \mathbb{R}^d \to \mathbb{R}$ is a d-dimensional function and

$$\sum_{k_1=-\infty}^{\infty} \cdots \sum_{k_d=-\infty}^{\infty} \left| \theta\left(\frac{k_1}{n_1}, \ldots, \frac{k_d}{n_d}\right)\right| < \infty \tag{3.7.1}$$

for all $n \in \mathbb{P}^d$. We will see that it is more convenient to suppose that θ is in the Wiener algebra $W(C, \ell_1)(\mathbb{R}^d)$. All summability methods considered in the literature satisfy the condition $\theta \in W(C, \ell_1)(\mathbb{R}^d)$.

Definition 3.7.1 A measurable function $f : \mathbb{R}^d \to \mathbb{R}$ belongs to the Wiener amalgam space $W(L_\infty, \ell_1)(\mathbb{R}^d)$ if

$$\|f\|_{W(L_\infty,\ell_1)} := \sum_{k\in\mathbb{Z}^d} \sup_{x\in[0,1)^d} |f(x+k)| < \infty.$$

The smallest closed subspace of $W(L_\infty, \ell_1)(\mathbb{R}^d)$ containing continuous functions is denoted by $W(C, \ell_1)(\mathbb{R}^d)$ and is called Wiener algebra.

Lemma 3.7.2 *If $1 \le p \le \infty$, then*

$$W(L_\infty, \ell_1)(\mathbb{R}^d) \subset L_p(\mathbb{R}^d) \qquad \text{and} \qquad \|f\|_p \le \|f\|_{W(L_\infty,\ell_1)}.$$

Moreover, $W(L_\infty, \ell_1)(\mathbb{R}^d)$ is dense in $L_p(\mathbb{R}^d)$ for $1 \le p < \infty$.

Proof For $p = \infty$, the statement is trivial. If $1 \le p < \infty$, then

$$\begin{aligned} \|f\|_p &= \left(\sum_{k\in\mathbb{Z}^d} \int_{k+[0,1)^d} |f(x)|^p dx\right)^{1/p} \\ &\le \left(\sum_{k\in\mathbb{Z}^d} \sup_{x\in[0,1)^d} |f(x+k)|^p\right)^{1/p} \end{aligned}$$

$$\leq \sum_{k \in \mathbb{Z}^d} \sup_{x \in [0,1)^d} |f(x+k)|$$
$$= \|f\|_{W(L_\infty, \ell_1)}.$$

Since $W(L_\infty, \ell_1)(\mathbb{R}^d)$ contains the space of continuous functions with compact support, $W(L_\infty, \ell_1)(\mathbb{R}^d)$ is dense in $L_p(\mathbb{R}^d)$ if $1 \leq p < \infty$. ■

The Wiener amalgam spaces and Wiener algebra are used quite often in Gabor analysis, because they provide convenient and general classes of windows (see, e.g., Walnut [323] and Gröchenig [152]).

Theorem 3.7.3 *(a) If $\theta \in W(C, \ell_1)(\mathbb{R}^d)$ then* (3.7.1) *holds.*

(b) If the one-dimensional function θ is continuous and $|\theta|$ can be estimated by an integrable function η which is non-decreasing on $(-\infty, c)$ and non-increasing on (c, ∞) then $\theta \in W(C, \ell_1)(\mathbb{R})$.

(c) There exists $\theta \notin W(C, \ell_1)(\mathbb{R})$ such that (3.7.1) *holds.*

Proof It is easy to see that

$$\sum_{k_1=-\infty}^{\infty} \cdots \sum_{k_d=-\infty}^{\infty} \left|\theta\left(\frac{k_1}{n_1}, \ldots, \frac{k_d}{n_d}\right)\right| \leq \sum_{l \in \mathbb{Z}^d} \left(\prod_{j=1}^{d} n_j\right) \sup_{x \in [0,1)^d} |\theta(x+l)|$$
$$= \left(\prod_{j=1}^{d} n_j\right) \|\theta\|_{W(C,\ell_1)} < \infty, \qquad (3.7.2)$$

which shows (a). Under the conditions of (b), $\|\theta\|_{W(C,\ell_1)} \leq \|\eta\|_1$.

To see (c), let $\theta \geq 0$ be continuous and even on $\mathbb{R}$, $\theta(0) := 0$,

$$\theta(x) := 0 \quad \text{if} \quad j + \frac{1}{j+1} \leq x \leq j+1 \quad (j \in \mathbb{N})$$

and

$$\sup_{[j,j+1]} \theta = \frac{1}{j+1} \quad (j \in \mathbb{N}).$$

Then $\theta \in L_1(\mathbb{R})$,

$$\|\theta\|_{W(C,\ell_1)} = 2 \sum_{k=0}^{\infty} \frac{1}{k+1} = \infty$$

and

$$\sum_{k=-\infty}^{\infty} \left|\theta\left(\frac{k}{n+1}\right)\right| \leq 2 \sum_{j=0}^{n} \frac{1}{j+1} \frac{n+1}{j+1} < \infty \qquad (n \in \mathbb{N}).$$

This finishes the proof of Theorem 3.7.3. ■

Definition 3.7.4 Suppose that $\theta \in W(C, \ell_1)(\mathbb{R}^d)$. For $f \in L_1(\mathbb{T}^d)$ and $n \in \mathbb{N}^d$, the nth rectangular θ-means $\sigma_n^\theta f$ of the Fourier series of f and the nth rectangular θ-kernel K_n^θ are introduced by

$$\sigma_n^\theta f(x) := \sum_{k_1 \in \mathbb{Z}} \cdots \sum_{k_d \in \mathbb{Z}} \theta\left(\frac{-k_1}{n_1}, \dots, \frac{-k_d}{n_d}\right) \widehat{f}(k) e^{\imath k \cdot x}$$

and

$$K_n^\theta(t) := \sum_{k_1 \in \mathbb{Z}} \cdots \sum_{k_d \in \mathbb{Z}} \theta\left(\frac{-k_1}{n_1}, \dots, \frac{-k_d}{n_d}\right) e^{\imath k \cdot t},$$

respectively.

By Theorem 3.7.3, the θ-kernels K_n^θ and the θ-means $\sigma_n^\theta f$ are well defined. We suppose often that

$$\theta = \theta_1 \otimes \cdots \otimes \theta_d,$$

where $\theta_i \in W(C, \ell_1)(\mathbb{R})$ for all $i = 1, \dots, d$. Then $\theta \in W(C, \ell_1)(\mathbb{R}^d)$ and

$$K_n^\theta = K_{n_1}^{\theta_1} \otimes \cdots \otimes K_{n_d}^{\theta_d}.$$

Lemma 3.7.5 *Suppose that $\theta \in W(C, \ell_1)(\mathbb{R}^d)$. For $f \in L_1(\mathbb{T}^d)$ and $n \in \mathbb{N}^d$, we have*

$$\sigma_n^\theta f(x) = \frac{1}{(2\pi)^d} \int_{\mathbb{T}^d} f(x-t) K_n^\theta(t)\, dt.$$

The θ-means can also be written as a convolution of f and the Fourier transform of θ in the following way.

Theorem 3.7.6 *If $\theta \in W(C, \ell_1)(\mathbb{R}^d)$ and $\widehat{\theta} \in L_1(\mathbb{R}^d)$, then*

$$\sigma_n^\theta f(x) = \left(\prod_{j=1}^d n_j\right) \int_{\mathbb{R}^d} f(x-t) \widehat{\theta}(n_1 t_1, \dots, n_d t_d)\, dt$$

for almost every $x \in \mathbb{T}^d$ and for all $n \in \mathbb{N}^d$ and $f \in L_1(\mathbb{T}^d)$.

Proof If $f(t) = e^{\imath k \cdot t}$ $(k \in \mathbb{Z}^d, t \in \mathbb{T}^d)$, then

$$\begin{aligned}
\sigma_n^\theta f(x) &= \theta\left(\frac{-k_1}{n_1}, \dots, \frac{-k_d}{n_d}\right) e^{\imath k \cdot x} \\
&= e^{\imath k \cdot x} \int_{\mathbb{R}^d} \left(\prod_{j=1}^d e^{-\imath k_j t_j / n_j}\right) \widehat{\theta}(t)\, dt
\end{aligned}$$

$$= \left(\prod_{j=1}^{d} n_j \right) \int_{\mathbb{R}^d} e^{\imath k \cdot (x-t)} \widehat{\theta}(n_1 t_1, \ldots, n_d t_d) \, dt.$$

Thus, the theorem holds also for trigonometric polynomials. The proof can be finished as in Theorem 2.2.30. ∎

We extend again the definition of the rectangular θ-means to distributions.

Definition 3.7.7 Suppose that $\theta \in W(C, \ell_1)(\mathbb{R}^d)$. For $f \in D(\mathbb{T}^d)$ and $n \in \mathbb{N}^d$, the nth rectangular θ-means $\sigma_n^\theta f$ of the Fourier series of f are given by

$$\sigma_n^\theta f := f * K_n^\theta.$$

3.7.1 Feichtinger's Algebra $S_0(\mathbb{R}^d)$

Theorem 3.7.6 is a fundamental result, so the condition $\widehat{\theta} \in L_1(\mathbb{R}^d)$ is of great importance. In this subsection, we give some sufficient conditions for a function θ to satisfy $\widehat{\theta} \in L_1(\mathbb{R}^d)$. In contrary to the other sections, we do not prove all results here. Some of them are presented without proof. Several such conditions are already known. The next one can be found in Bachman, Narici and Beckenstein [15, p. 323].

Theorem 3.7.8 *If $\theta \in L_1(\mathbb{R})$ is bounded on a neighborhood of 0 and $\widehat{\theta} \geq 0$, then $\widehat{\theta} \in L_1(\mathbb{R})$.*

Obviously, θ is bounded on a neighborhood of 0 if $\theta \in L_\infty(\mathbb{R})$ or θ is continuous at 0. Moreover, if $\theta \in L_1(\mathbb{R})$ has compact support and $\theta \in \text{Lip}(\alpha)$ for some $\alpha > 1/2$, then $\widehat{\theta} \in L_1(\mathbb{R})$ (see Natanson and Zuk [244, p. 176]).

Now we introduce a Banach space, called Feichtinger's algebra, the Fourier transforms of the elements of which are all integrable. This space was first considered in Feichtinger [100].

Definition 3.7.9 The short-time Fourier transform of $f \in L_2(\mathbb{R}^d)$ with respect to a window function $g \in L_2(\mathbb{R}^d)$ is defined by

$$S_g f(x, \omega) := \frac{1}{(2\pi)^d} \int_{\mathbb{R}^d} f(t) \overline{g(t-x)} e^{-\imath \omega \cdot t} \, dt \qquad (x, \omega \in \mathbb{R}^d).$$

Definition 3.7.10 Let $g_0(x) := e^{-\pi \|x\|_2^2}$ be the Gauss function. We define the Feichtinger's algebra $S_0(\mathbb{R}^d)$ by

$$S_0(\mathbb{R}^d) := \left\{ f \in L^2(\mathbb{R}^d) : \|f\|_{S_0} := \left\| S_{g_0} f \right\|_{L_1(\mathbb{R}^{2d})} < \infty \right\}.$$

Any other non-zero Schwartz function defines the same space and an equivalent norm. It is known that $S_0(\mathbb{R}^d)$ contains all Schwartz functions. Moreover, $S_0(\mathbb{R}^d)$ is isometrically invariant under translation, modulation and Fourier transform (see Feichtinger and Zimmermann [100, 106]). Actually, $S_0(\mathbb{R}^d)$ is the minimal Banach space having this property (see Feichtinger [100]). Furthermore, Feichtinger's algebra is a subspace of the Wiener algebra, the embedding $S_0(\mathbb{R}^d) \hookrightarrow W(C, \ell_1)(\mathbb{R}^d)$ is dense and continuous and

$$S_0(\mathbb{R}^d) \subsetneq W(C, \ell_1)(\mathbb{R}^d) \cap \mathcal{F}(W(C, \ell_1)(\mathbb{R}^d)),$$

where $\mathcal{F}$ denotes the Fourier transform and $\mathcal{F}(W(C, \ell_1)(\mathbb{R}^d))$ the set of Fourier transforms of the functions from $W(C, \ell_1)(\mathbb{R}^d)$ (see Feichtinger and Zimmermann [106], Losert [223] and Gröchenig [152]). Let us define the weight function

$$v_s(\omega) := \left(1 + \|\omega\|_2^2\right)^{d/2} \qquad (\omega \in \mathbb{R}^d, s \in \mathbb{R}).$$

Theorem 3.7.11 *(a) If $\theta \in S_0(\mathbb{R}^d)$, then $\widehat{\theta} \in S_0(\mathbb{R}^d) \subset L_1(\mathbb{R}^d)$.*
(b) If $\theta \in L_1(\mathbb{R}^d)$ and $\widehat{\theta}$ has compact support, then $\theta \in S_0(\mathbb{R}^d)$.
(c) If $\theta \in L_1(\mathbb{R}^d)$ has compact support and $\widehat{\theta} \in L_1(\mathbb{R}^d)$, then $\theta \in S_0(\mathbb{R}^d)$.
(d) If $\theta v_s, \widehat{\theta} v_s \in L_2(\mathbb{R}^d)$ for some $s > d$, then $\theta \in S_0(\mathbb{R}^d)$.
(e) If $\theta v_s, \widehat{\theta} v_s \in L_\infty(\mathbb{R}^d)$ for some $s > 3d/2$, then $\theta \in S_0(\mathbb{R}^d)$.

For more about Feichtinger's algebra see Feichtinger and Zimmermann [100, 106]).

Sufficient conditions can also be given with the help of Sobolev, fractional Sobolev and Besov spaces. We do not give a detailed description of these spaces. For the interested readers, we refer to Triebel [313], Runst and Sickel [267], Stein [289] and Grafakos [143]. The Sobolev space $W_p^k(\mathbb{R}^d)$ $(1 \le p \le \infty, k \in \mathbb{N})$ is defined by

$$W_p^k(\mathbb{R}^d) := \left\{\theta \in L_p(\mathbb{R}^d) : D^\alpha \theta \in L_p(\mathbb{R}^d), |\alpha| \le k\right\}$$

and endowed with the norm

$$\|\theta\|_{W_p^k} := \sum_{|\alpha| \le k} \|D^\alpha \theta\|_p,$$

where D denotes the distributional derivative.

This definition can be extended to every real s in the following way. The fractional Sobolev space $\mathcal{L}_p^s(\mathbb{R}^d)$ $(1 \le p \le \infty, s \in \mathbb{R})$ consists of all tempered distributions θ for which

$$\|\theta\|_{\mathcal{L}_p^s} := \left\|\mathcal{F}^{-1}\left((1 + |\cdot|^2)^{s/2} \widehat{\theta}\right)\right\|_p < \infty,$$

where $\mathcal{F}$ denotes the Fourier transform. It is known that

$$\mathcal{L}_p^s(\mathbb{R}^d) = W_p^k(\mathbb{R}^d) \quad \text{if} \quad s = k \in \mathbb{N} \quad \text{and} \quad 1 < p < \infty$$

with equivalent norms.

In order to define the Besov spaces, take a non-negative Schwartz function $\psi \in \mathcal{S}(\mathbb{R})$ with support $[1/2, 2]$ that satisfies

$$\sum_{k=-\infty}^{\infty} \psi(2^{-k}s) = 1 \quad \text{for all} \quad s \in \mathbb{R} \setminus \{0\}.$$

For $x \in \mathbb{R}^d$, let

$$\phi_k(x) := \psi(2^{-k}|x|) \quad \text{for} \quad k \geq 1 \quad \text{and} \quad \phi_0(x) = 1 - \sum_{k=1}^{\infty} \phi_k(x).$$

The Besov space $B_{p,r}^s(\mathbb{R}^d)$ $(0 < p, r \leq \infty, s \in \mathbb{R})$ is the space of all tempered distributions f for which

$$\|f\|_{B_{p,r}^s} := \left(\sum_{k=0}^{\infty} 2^{ksr} \left\| (\mathcal{F}^{-1}\phi_k) * f \right\|_p^r \right)^{1/r} < \infty.$$

The Sobolev, fractional Sobolev and Besov spaces are all quasi-Banach spaces, and if $1 \leq p, r \leq \infty$, then they are Banach spaces. All these spaces contain the Schwartz functions. The following facts are known: in the case $1 \leq p, r \leq \infty$, one has

$$W_p^m(\mathbb{R}^d),\ B_{p,r}^s(\mathbb{R}^d) \hookrightarrow L_p(\mathbb{R}^d) \qquad \text{if} \quad s > 0, m \in \mathbb{N},$$

$$W_p^{m+1}(\mathbb{R}^d) \hookrightarrow B_{p,r}^s(\mathbb{R}^d) \hookrightarrow W_p^m(\mathbb{R}^d) \qquad \text{if} \quad m < s < m+1, \tag{3.7.3}$$

$$B_{p,r}^s(\mathbb{R}^d) \hookrightarrow B_{p,r+\epsilon}^s(\mathbb{R}^d),\ B_{p,\infty}^{s+\epsilon}(\mathbb{R}^d) \hookrightarrow B_{p,r}^s(\mathbb{R}^d) \qquad \text{if} \quad \epsilon > 0, \tag{3.7.4}$$

$$B_{p_1,1}^{d/p_1}(\mathbb{R}^d) \hookrightarrow B_{p_2,1}^{d/p_2}(\mathbb{R}^d) \hookrightarrow C(\mathbb{R}^d) \qquad \text{if} \quad 1 \leq p_1 \leq p_2 < \infty. \tag{3.7.5}$$

For two quasi-Banach spaces $\mathbb{X}$ and $\mathbb{Y}$, the embedding $\mathbb{X} \hookrightarrow \mathbb{Y}$ means that $\mathbb{X} \subset \mathbb{Y}$ and $\|f\|_{\mathbb{Y}} \leq C\|f\|_{\mathbb{X}}$.

The connection between Besov spaces and Feichtinger's algebra is summarized in the next theorem.

Theorem 3.7.12 *We have*

(i) If $1 \leq p \leq 2$ and $\theta \in B_{p,1}^{d/p}(\mathbb{R}^d)$, then $\widehat{\theta} \in L_1(\mathbb{R}^d)$ and

$$\left\|\widehat{\theta}\right\|_1 \leq C \|\theta\|_{B_{p,1}^{d/p}}.$$

(ii) If $s > d$, then $\mathcal{L}_1^s(\mathbb{R}^d) \hookrightarrow S_0(\mathbb{R}^d)$.
(iii) If d' denotes the smallest even integer which is larger than d and $s > d'$, then

$$B_{1,\infty}^s(\mathbb{R}^d) \hookrightarrow W_1^{d'}(\mathbb{R}^d) \hookrightarrow S_0(\mathbb{R}^d).$$

Proof (i) was proved in Girardi and Weis [130] and (ii) in Okoudjou [250]. The first embedding of (iii) follows from (3.7.3) and (3.7.4). If k is even, then $W_1^k(\mathbb{R}^d) \hookrightarrow \mathcal{L}_1^k(\mathbb{R}^d)$ (see Stein [289, p. 160]). Then (ii) proves (iii). ∎

It follows from (i) and (3.7.3) that $\theta \in W_p^j(\mathbb{R}^d)$ ($j > d/p, j \in \mathbb{N}$) implies $\widehat{\theta} \in L_1(\mathbb{R}^d)$. If $j \geq d'$, then even $W_1^j(\mathbb{R}^d) \hookrightarrow S_0(\mathbb{R}^d)$ (see (iii)). Moreover, if $s > d'$ as in (iii), then

$$B_{1,\infty}^s(\mathbb{R}^d) \hookrightarrow B_{1,1}^d(\mathbb{R}^d) \hookrightarrow B_{p,1}^{d/p}(\mathbb{R}^d) \qquad (1 < p < \infty)$$

by (3.7.4) and (3.7.5). Theorem 3.7.12 says that $B_{1,\infty}^s(\mathbb{R}^d) \subset S_0(\mathbb{R}^d)$ ($s > d'$) and if we choose θ from the larger space $B_{p,1}^{d/p}(\mathbb{R}^d)$ ($1 \leq p \leq 2$), then $\widehat{\theta}$ is still integrable.

The embedding $W_1^2(\mathbb{R}) \hookrightarrow S_0(\mathbb{R})$ follows from (iii). With the help of the usual derivative, we give another useful sufficient condition for a function to be in $S_0(\mathbb{R}^d)$. As usual, we denote by $C^k(\mathbb{R}^d)$ the set of k times continuously differentiable functions.

Definition 3.7.13 A function θ is in $V_1^k(\mathbb{R})$ if there are numbers $-\infty = a_0 < a_1 < \cdots < a_n < a_{n+1} = \infty$, where $n = n(\theta)$ depends on θ and

$$\theta \in C^{k-2}(\mathbb{R}), \qquad \theta \in C^k(a_i, a_{i+1}), \qquad \theta^{(j)} \in L_1(\mathbb{R})$$

for all $i = 0, \ldots, n$ and $j = 0, \ldots, k$. The norm of this space is defined by

$$\|\theta\|_{V_1^k} := \sum_{j=0}^{k} \left\|\theta^{(j)}\right\|_1 + \sum_{i=1}^{n} \left|\theta^{(k-1)}(a_i + 0) - \theta^{(k-1)}(a_i - 0)\right|,$$

where $\theta^{(k-1)}(a_i \pm 0)$ denotes the right and left limits of $\theta^{(k-1)}$.

These limits do exist and are finite because $\theta^{(k)} \in C(a_i, a_{i+1}) \cap L_1(\mathbb{R})$ implies

$$\theta^{(k-1)}(x) = \theta^{(k-1)}(a) + \int_a^x \theta^{(k)}(t)\,dt$$

for some $a \in (a_i, a_{i+1})$. Since $\theta^{(k-1)} \in L_1(\mathbb{R})$, we establish that

$$\lim_{x\to-\infty} \theta^{(k-1)}(x) = \lim_{x\to\infty} \theta^{(k-1)}(x) = 0.$$

Similarly, $\theta^{(j)} \in C_0(\mathbb{R})$ for $j = 0, \ldots, k-2$.

Of course, $W_1^2(\mathbb{R})$ and $V_1^2(\mathbb{R})$ are not identical. For $\theta \in V_1^2(\mathbb{R})$, we have $\theta' = D\theta$; however, $\theta'' = D^2\theta$ only if $\lim_{x\to a_i+0}\theta'(x) = \lim_{x\to a_i-0}\theta'(x)$ $(i = 1, \dots, n)$.

Theorem 3.7.14 *We have* $V_1^2(\mathbb{R}) \hookrightarrow S_0(\mathbb{R})$.

Proof Integrating by parts, we have

$$
\begin{aligned}
&S_{g_0}\theta(x,\omega)\\
&= \frac{1}{2\pi}\int_{\mathbb{R}} \theta(t)\overline{g_0(t-x)}e^{-\imath\omega t}\,dt\\
&= \frac{1}{2\pi}\sum_{i=0}^{n}\int_{a_i}^{a_{i+1}} \theta(t)e^{-\pi(t-x)^2}e^{-\imath\omega t}\,dt\\
&= \frac{1}{2\pi}\sum_{i=0}^{n}\left[\theta(t)e^{-\pi(t-x)^2}\frac{e^{-\imath\omega t}}{-\imath\omega}\right]_{a_i}^{a_{i+1}}\\
&\quad - \frac{1}{2\pi}\sum_{i=0}^{n}\int_{a_i}^{a_{i+1}}\left(\theta'(t)e^{-\pi(t-x)^2} - 2\pi\theta(t)e^{-\pi(t-x)^2}(t-x)\right)\frac{e^{-\imath\omega t}}{-\imath\omega}\,dt.
\end{aligned}
$$

Observe that the first sum is 0. In the second sum, we integrate by parts again to obtain

$$
\begin{aligned}
S_{g_0}\theta(x,\omega) &= \frac{1}{2\pi}\sum_{i=0}^{n}\left[\left(\theta'(t)e^{-\pi(t-x)^2} - 2\pi\theta(t)e^{-\pi(t-x)^2}(t-x)\right)\frac{e^{-\imath\omega t}}{\omega^2}\right]_{a_i}^{a_{i+1}}\\
&\quad - \frac{1}{2\pi}\sum_{i=0}^{n}\int_{a_i}^{a_{i+1}}\left(\theta''(t)e^{-\pi(t-x)^2} - 4\pi\theta'(t)e^{-\pi(t-x)^2}(t-x)\right.\\
&\quad \left. - 2\pi\theta(t)\left(-2\pi e^{-\pi(t-x)^2}(t-x)^2 + e^{-\pi(t-x)^2}\right)\right)\frac{e^{-\imath\omega t}}{\omega^2}\,dt.
\end{aligned}
$$

The first sum is equal to

$$
\frac{1}{2\pi}\sum_{i=1}^{n}\Big(\theta'(a_i+0) - \theta'(a_i-0)\Big)e^{-\pi(a_i-x)^2}\frac{e^{-\imath\omega a_i}}{\omega^2}.
$$

Hence

$$
\int_{\mathbb{R}}\int_{\{|\omega|\ge 1\}} |S_{g_0}\theta(x,\omega)|\,dx\,d\omega \le C_s\|\theta\|_{V_1^2}.
$$

On the other hand,

$$
\begin{aligned}
\int_{\mathbb{R}}\int_{\{|\omega|<1\}} |S_{g_0}\theta(x,\omega)|\,dx\,d\omega &\le C_s\int_{\mathbb{R}}\int_{\{|\omega|<1\}}\int_{\mathbb{R}} |\theta(t)|g_0(t-x)\,dt\,dx\,d\omega\\
&\le C_s\|\theta\|_{V_1^2},
\end{aligned}
$$

which finishes the proof of Theorem 3.7.14. ■

The next Corollary follows from the definition of $S_0(\mathbb{R}^d)$ and from Theorem 3.7.14.

Corollary 3.7.15 *If each $\theta_j \in V_1^2(\mathbb{R})$ $(j = 1, \dots, d)$, then*

$$\theta = \theta_1 \otimes \cdots \otimes \theta_d \in S_0(\mathbb{R}^d).$$

3.7.2 Norm Convergence of the Rectangular θ-Means

First, we investigate the $L_2(\mathbb{T}^d)$-norm convergence of $\sigma_n^\theta f$ as $n \to \infty$ $(n \in \mathbb{N}^d)$ in Pringsheim's sense.

Theorem 3.7.16 *If $\theta \in W(C, \ell_1)(\mathbb{R}^d)$ and $\theta(0) = 1$, then*

$$\lim_{n\to\infty} \sigma_n^\theta f = f \quad \textit{in the } L_2(\mathbb{T}^d)\textit{-norm for all } f \in L_2(\mathbb{T}^d).$$

Proof It is easy to see that the norm of the operator

$$\sigma_n^\theta : L_2(\mathbb{T}^d) \to L_2(\mathbb{T}^d)$$

can be given by

$$\begin{aligned}
\sup_{f\in L_2(\mathbb{T}^d),\ \|f\|_2\le 1} \left\| f * K_n^\theta \right\|_2 &= \sup_{f\in L_2(\mathbb{T}^d),\ \|f\|_2\le 1} \left\| \widehat{f}\widehat{K}_n^\theta \right\|_2 \\
&= \sup_{\widehat{f}\in \ell_2(\mathbb{Z}^d),\ \|\widehat{f}\|_2\le 1} \left\| \widehat{f}\widehat{K}_n^\theta \right\|_2 \\
&= \left\| \widehat{K}_n^\theta \right\|_\infty \\
&= \sup_{k\in\mathbb{Z}^d} \left| \theta\left(\frac{-k_1}{n_1}, \dots, \frac{-k_d}{n_d}\right)\right| \\
&\le C.
\end{aligned}$$

Thus, the norms of σ_n^θ $(n \in \mathbb{N}^d)$ are uniformly bounded. Since θ is continuous, the convergence holds for all trigonometric polynomials. The set of the trigonometric polynomials are dense in $L_2(\mathbb{T}^d)$, so the usual density theorem proves Theorem 3.7.16. ■

Now, we give a sufficient and necessary condition for the uniform and $L_1(\mathbb{T}^d)$ convergence $\sigma_n^\theta f \to f$.

Theorem 3.7.17 *If $\theta \in W(C, \ell_1)(\mathbb{R}^d)$ and $\theta(0) = 1$, then the following conditions are equivalent:*

(i) $\widehat{\theta} \in L_1(\mathbb{R}^d)$,
(ii) $\sigma_n^\theta f \to f$ *uniformly for all* $f \in C(\mathbb{T}^d)$ *as* $n \to \infty$ *and* $n \in \mathbb{N}^d$,
(iii) $\sigma_n^\theta f(x) \to f(x)$ *for all* $x \in \mathbb{T}^d$ *and* $f \in C(\mathbb{T}^d)$ *as* $n \to \infty$ *and* $n \in \mathbb{N}^d$,
(iv) $\sigma_n^\theta f \to f$ *in the* $L_1(\mathbb{T}^d)$*-norm for all* $f \in L_1(\mathbb{T}^d)$ *as* $n \to \infty$ *and* $n \in \mathbb{N}^d$,
(v) $\sigma_n^\theta f \to f$ *uniformly for all* $f \in C(\mathbb{T}^d)$ *as* $n \to \infty$ *and* $n \in \mathbb{R}_\tau^d$,
(vi) $\sigma_n^\theta f(x) \to f(x)$ *for all* $x \in \mathbb{T}^d$ *and* $f \in C(\mathbb{T}^d)$ *as* $n \to \infty$ *and* $n \in \mathbb{R}_\tau^d$,
(vii) $\sigma_n^\theta f \to f$ *in the* $L_1(\mathbb{T}^d)$*-norm for all* $f \in L_1(\mathbb{T}^d)$ *as* $n \to \infty$ *and* $n \in \mathbb{R}_\tau^d$.

Recall the definition of R_τ^d from (3.3.1).

Proof We may suppose that $d = 1$, since the multi-dimensional case is similar. First, we verify the equivalence between (i), (ii), (iii) and (iv). If (i) holds, then by Theorem 3.7.6,

$$\left\|\sigma_n^\theta f\right\|_\infty \le \|f\|_\infty \left\|\widehat{\theta}\right\|_1 \qquad (f \in C(\mathbb{T}), n \in \mathbb{N})$$

and so the operators $\sigma_n : C(\mathbb{T}) \to C(\mathbb{T})$ are uniformly bounded. Since (ii) holds for all trigonometric polynomials and the set of the trigonometric polynomials are dense in $C(\mathbb{T})$, (ii) follows easily. (ii) implies (iii) trivially.

Suppose that (iii) is satisfied. We are going to prove (i). For a fixed $x \in \mathbb{T}$, the operators

$$U_n : C(\mathbb{T}) \to \mathbb{R}, \qquad U_n f := \sigma_n^\theta f(x) \qquad (n \in \mathbb{N})$$

are uniformly bounded by the Banach-Steinhaus theorem. We get by Lemma 3.7.5 that

$$\|U_n\| = \frac{1}{(2\pi)^d} \int_{\mathbb{T}} |K_n^\theta(x-t)|\,dt = \frac{1}{(2\pi)^d} \|K_n^\theta\|_1 \qquad (n \in \mathbb{N}).$$

Hence

$$\sup_{n\in\mathbb{N}} \|K_n^\theta\|_1 \le C.$$

Since K_n^θ is 2π-periodic, we have for $\alpha \le n/2$ that

$$\begin{aligned}\int_{-2\alpha\pi}^{2\alpha\pi} \frac{1}{n}\left|\sum_{k=-\infty}^{\infty} \theta\left(\frac{-k}{n}\right) e^{\imath tk/n}\right| dt &\le \int_{-n\pi}^{n\pi} \frac{1}{n}\left|\sum_{k=-\infty}^{\infty} \theta\left(\frac{-k}{n}\right) e^{\imath tk/n}\right| dt \\ &= \int_{-\pi}^{\pi} \left|\sum_{k=-\infty}^{\infty} \theta\left(\frac{-k}{n}\right) e^{\imath kx}\right| dx \\ &= \int_{\mathbb{T}} |K_n^\theta(x)|\,dx \le C. \end{aligned} \tag{3.7.6}$$

For a fixed $t \in \mathbb{R}$, let

$$h_n(t) := \frac{1}{n} \sum_{k=-\infty}^{\infty} \theta\left(\frac{-k}{n}\right) e^{\imath tk/n}$$

and

$$\varphi_n(t,u) := \sum_{k=-\infty}^{\infty} \theta\left(\frac{-k}{n}\right) e^{\imath tk/n} 1_{[\frac{k}{n},\frac{k+1}{n})}(u).$$

It is easy to see that

$$\lim_{n\to\infty} \varphi_n(t,u) = \theta(-u)e^{\imath tu}.$$

Moreover,

$$|\varphi_n(t,u)| \le \sum_{l=-\infty}^{\infty} \sup_{x\in[0,1)} |\theta(x-l-1)| \, 1_{[l,l+1)}(u)$$

and

$$\begin{aligned}\int_{-\infty}^{\infty} \sum_{l=-\infty}^{\infty} \sup_{x\in[0,1)} |\theta(x-l-1)| \, 1_{[l,l+1)}(u)\,du &= \sum_{l=-\infty}^{\infty} \sup_{x\in[0,1)} |\theta(x-l-1)| \\ &= \|\theta\|_{W(C,\ell_1)}.\end{aligned}$$

Lebesgue's dominated convergence theorem implies that

$$\lim_{n\to\infty} \int_{-\infty}^{\infty} \varphi_n(t,u)\,du = \int_{-\infty}^{\infty} \theta(-u)e^{\imath tu}\,du = (2\pi)^d \widehat{\theta}(t).$$

Obviously,

$$\int_{-\infty}^{\infty} \varphi_n(t,u)\,du = h_n(t)$$

and so

$$\lim_{n\to\infty} h_n(t) = (2\pi)^d \widehat{\theta}(t).$$

Of course, this holds for all $t \in \mathbb{R}$. We have by (3.7.2) that

$$|h_n(t)| \le \|\theta\|_{W(C,\ell_1)}.$$

Thus

$$\lim_{n\to\infty} \int_{-2\alpha\pi}^{2\alpha\pi} |h_n(t)|\,dt = (2\pi)^d \int_{-2\alpha\pi}^{2\alpha\pi} \left|\widehat{\theta}(t)\right|\,dt.$$

Inequality (3.7.6) yields that

$$\int_{-2\alpha\pi}^{2\alpha\pi} \left|\widehat{\theta}(t)\right|\,dt \le C \qquad \text{for all} \qquad \alpha > 0$$

and so

$$\int_{-\infty}^{\infty} \left|\widehat{\theta}(t)\right| dt \leq C,$$

which shows (i).

If $\widehat{\theta} \in L_1(\mathbb{R})$, then Theorem 3.7.6 implies

$$\left\|\sigma_n^\theta f\right\|_1 \leq \|f\|_1 \left\|\widehat{\theta}\right\|_1 \qquad (f \in L_1(\mathbb{T}), n \in \mathbb{N}).$$

Hence (iv) follows from (i) because the set of the trigonometric polynomials are dense in $L_1(\mathbb{T})$. The fact that (iv) implies (i) can be proved similarly as $(iii) \Rightarrow (i)$, since, by duality, the norm of the operator $\sigma_n^\theta : L_1(\mathbb{T}) \to L_1(\mathbb{T})$ is again

$$\left\|\sigma_n^\theta\right\| = \left\|K_n^\theta\right\|_1 .$$

It is easy to see that the equivalence between (i), (v), (vi) and (vii) can be proved in the same way. ∎

Note that the statement $(i) \Leftrightarrow (ii)$ was shown in the one-dimensional case by Natanson and Zuk [244] for θ having compact support. The situation in our general case is much more complicated and can be found in Feichtinger and Weisz [103]. One part of the preceding result can be generalized for $L_p(\mathbb{T}^d)$ spaces.

Theorem 3.7.18 *Assume that $\theta(0) = 1$, $\theta \in W(C, \ell_1)(\mathbb{R}^d)$ and $\widehat{\theta} \in L_1(\mathbb{R}^d)$. If $1 \leq p < \infty$ and $f \in L_p(\mathbb{T}^d)$, then*

$$\sup_{n\in\mathbb{N}} \left\|\sigma_n^\theta f\right\|_p \leq C\|f\|_p$$

and

$$\lim_{n\to\infty} \sigma_n^\theta f = f \qquad \text{in the } L_p(\mathbb{T}^d)\text{-norm.}$$

Proof For simplicity, we show the theorem for $d = 1$. Using Theorem 3.7.6, we conclude

$$\begin{aligned}\sigma_n^\theta f(x) - f(x) &= n \int_{\mathbb{R}} \Big(f(x-t) - f(x)\Big)\widehat{\theta}(nt)\, dt \\ &= \int_{\mathbb{R}} \left(f\left(x - \frac{t}{n}\right) - f(x)\right)\widehat{\theta}(t)\, dt\end{aligned}$$

and

$$\left\|\sigma_n^\theta f - f\right\|_p = \int_{\mathbb{R}} \left\|f\left(\cdot - \frac{t}{n}\right) - f(\cdot)\right\|_p \left|\widehat{\theta}(t)\right| dt.$$

The theorem follows from the Lebesgue dominated convergence theorem. ∎

Since $\theta \in S_0(\mathbb{R}^d)$ implies $\theta \in W(C, \ell_1)(\mathbb{R}^d)$ and $\widehat{\theta} \in S_0(\mathbb{R}^d) \subset L_1(\mathbb{R}^d)$, the next corollary follows from Theorems 3.7.17 and 3.7.18.

Corollary 3.7.19 *If $\theta \in S_0(\mathbb{R}^d)$ and $\theta(0) = 1$, then*

(i) $\sigma_n^\theta f \to f$ *uniformly for all* $f \in C(\mathbb{T}^d)$ *as* $n \to \infty$ *and* $n \in \mathbb{N}^d$,
(ii) $\sigma_n^\theta f \to f$ *in the* $L_1(\mathbb{T}^d)$*-norm for all* $f \in L_1(\mathbb{T}^d)$ *as* $n \to \infty$ *and* $n \in \mathbb{N}^d$,
(iii) $\sigma_n^\theta f \to f$ *in the* $L_p(\mathbb{T}^d)$*-norm for all* $f \in L_p(\mathbb{T}^d)$ $(1 < p < \infty)$ *as* $n \to \infty$ *and* $n \in \mathbb{N}^d$.

The next corollary follows from the fact that $\theta \in S_0(\mathbb{R}^d)$ is equivalent to $\widehat{\theta} \in L_1(\mathbb{R}^d)$, provided that θ has compact support (see, e.g., Feichtinger and Zimmermann [106]).

Corollary 3.7.20 *If $\theta \in C(\mathbb{R}^d)$ has compact support and $\theta(0) = 1$, then the following conditions are equivalent:*

(i) $\theta \in S_0(\mathbb{R}^d)$,
(ii) $\sigma_n^\theta f \to f$ *uniformly for all* $f \in C(\mathbb{T}^d)$ *as* $n \to \infty$ *and* $n \in \mathbb{N}^d$,
(iii) $\sigma_n^\theta f(x) \to f(x)$ *for all* $x \in \mathbb{T}^d$ *and* $f \in C(\mathbb{T}^d)$ *as* $n \to \infty$ *and* $n \in \mathbb{N}^d$,
(iv) $\sigma_n^\theta f \to f$ *in the* $L_1(\mathbb{T}^d)$*-norm for all* $f \in L_1(\mathbb{T}^d)$ *as* $n \to \infty$,
(v) $\sigma_n^\theta f \to f$ *uniformly for all* $f \in C(\mathbb{T}^d)$ *as* $n \to \infty$ *and* $n \in \mathbb{R}_\tau^d$,
(vi) $\sigma_n^\theta f(x) \to f(x)$ *for all* $x \in \mathbb{T}^d$ *and* $f \in C(\mathbb{T}^d)$ *as* $n \to \infty$ *and* $n \in \mathbb{R}_\tau^d$,
(vii) $\sigma_n^\theta f \to f$ *in the* $L_1(\mathbb{T}^d)$*-norm for all* $f \in L_1(\mathbb{T}^d)$ *as* $n \to \infty$ *and* $n \in \mathbb{R}_\tau^d$.

3.7.3 Almost Everywhere Convergence of the Rectangular θ-Means

Definition 3.7.21 For given κ, τ satisfying the conditions given in Sect. 3.4, we define the restricted maximal θ-operators by

$$\sigma_\square^\theta f := \sup_{n \in \mathbb{R}_\tau^d} \left|\sigma_n^\theta f\right|, \qquad \sigma_\kappa^\theta f := \sup_{n \in \mathbb{R}_{\kappa,\tau}^d} \left|\sigma_n^\theta f\right|.$$

The unrestricted maximal θ-operator is defined by

$$\sigma_*^\theta f := \sup_{n \in \mathbb{N}^d} \left|\sigma_n^\theta f\right|.$$

In this subsection, we suppose that

$$\theta(0) = 1, \qquad \theta = \theta_1 \otimes \cdots \otimes \theta_d, \qquad \theta_j \in W(C, \ell_1)(\mathbb{R}), \qquad j = 1, \ldots, d. \tag{3.7.7}$$

For the restricted convergence, we suppose in addition that

$$\mathcal{I}\theta_j \in W(C, \ell_1)(\mathbb{R}), \qquad j = 1, \ldots, d. \tag{3.7.8}$$

Here $\mathcal{I}$ denotes the identity function, so

$$\mathcal{I}(x) = x \quad \text{and} \quad (\mathcal{I}\theta_j)(x) = x\theta_j(x).$$

Similar to (2.6.6), assume that $\widehat{\theta_j}$ is $(N+1)$-times differentiable ($N \geq 0$) and there exists

$$N < \beta_j \leq N+1$$

such that

$$\left|\left(\widehat{\theta_j}\right)^{(i)}(x)\right| \leq C|x|^{-\beta_j-1} \qquad (x \neq 0) \tag{3.7.9}$$

for $i = N, N+1$ and all $j = 1, \ldots, d$.

Theorem 3.7.22 *Assume that* (3.7.7), (3.7.8) *and* (3.7.9) *are satisfied with* $N = 0$. *If*

$$\max\left\{\frac{d}{d+1}, \frac{1}{\beta_j+1}, j = 1, \ldots, d\right\} < p \leq \infty,$$

then

$$\left\|\sigma_{\square}^{\alpha} f\right\|_p \leq C_p \|f\|_{H_p^{\square}} \qquad (f \in H_p^{\square}(\mathbb{T}^d)).$$

Moreover,

$$\sup_{\rho>0} \rho\,\lambda(\sigma_{\square}^{\theta} f > \rho) \leq C\|f\|_1 \qquad (f \in L_1(\mathbb{T}^d)).$$

Proof Inequality (3.7.2) implies that

$$\left|K_{n_j}^{\theta_j}\right| \leq Cn_j \qquad (n_j \in \mathbb{N}).$$

Similarly,

$$\sum_{k=-\infty}^{\infty} \left|\frac{k}{n_j}\theta_j\left(\frac{k}{n_j}\right)\right| \leq n_j \left\|\mathcal{I}\theta_j\right\|_{W(C,\ell_1)} < \infty \qquad (n_j \in \mathbb{N}),$$

from which we get immediately that

$$\left|\left(K_{n_j}^{\theta_j}\right)'\right| \leq Cn_j^2 \qquad (n_j \in \mathbb{N}).$$

By Theorem 3.7.6,

$$K_{n_j}^{\theta_j}(x) = 2\pi n_j \sum_{k=-\infty}^{\infty} \widehat{\theta_j}\left(n_j(x+2k\pi)\right) \qquad (x \in \mathbb{T})$$

as in (2.2.34). From this, it follows that

$$\left|K_{n_j}^{\theta_j}(x)\right| \le \frac{C}{n_j^{\beta_j}|x|^{\beta_j+1}} \qquad (x \neq 0)$$

and

$$\left|\left(K_{n_j}^{\theta_j}\right)'(x)\right| \le \frac{C}{n_j^{\beta_j-1}|x|^{\beta_j+1}} \qquad (x \neq 0).$$

The proof can be finished as in Theorem 3.3.4. ■

Corollary 3.7.23 *Assume that* (3.7.7), (3.7.8) *and* (3.7.9) *are satisfied with* $N = 0$. *If* $f \in L_1(\mathbb{T}^d)$, *then*

$$\lim_{n\to\infty,\, n\in\mathbb{R}_\tau^d} \sigma_n^\theta f = f \qquad a.e.$$

Combining the proofs of Theorems 3.7.22 and 3.4.7, we obtain

Theorem 3.7.24 *Assume that* (3.7.7), (3.7.8) *and* (3.7.9) *are satisfied with* $N = 0$. *If*

$$\max\left\{p_1, \frac{1}{\beta_j+1}, j = 1, \ldots, d\right\} < p \le \infty,$$

then

$$\left\|\sigma_\kappa^\theta f\right\|_p \le C_p \|f\|_{H_p^\kappa} \qquad (f \in H_p^\kappa(\mathbb{T}^d)).$$

Moreover,

$$\sup_{\rho>0} \rho\,\lambda(\sigma_\kappa^\theta f > \rho) \le C\|f\|_1 \qquad (f \in L_1(\mathbb{T}^d)).$$

We recall that p_1 was defined in (3.4.4).

Corollary 3.7.25 *Assume that* (3.7.7), (3.7.8) *and* (3.7.9) *are satisfied with* $N = 0$. *If* $f \in L_1(\mathbb{T}^d)$, *then*

$$\lim_{n\to\infty,\, n\in\mathbb{R}_{\kappa,\tau}^d} \sigma_n^\theta f = f \qquad a.e.$$

For the unrestricted convergence, we can allow more general conditions for θ. The next theorem can be shown as Theorems 2.6.7 and 3.6.7.

Theorem 3.7.26 *If each* θ_j *satisfies* (2.6.2) *and* (2.6.3), *then*

$$\left\|\sigma_*^\theta f\right\|_p \le C_p \|f\|_{H_p} \qquad (f \in H_p(\mathbb{T}^d))$$

for $1/2 < p \le \infty$. *If* (3.7.7), (3.7.8) *and* (3.7.9) *are satisfied, then the preceding inequality holds for*

$$\max\left\{\frac{1}{\beta_j+1}, j = 1, \ldots, d\right\} < p \le \infty.$$

In both cases

$$\sup_{\rho>0} \rho\lambda(\sigma_*^\theta f > \rho) \le C\|f\|_{H_1^i} \qquad (f \in H_1^i(\mathbb{T}^d))$$

for all $i = 1, \dots, d$.

Corollary 3.7.27 *Under the conditions of Theorem 3.7.26,*

$$\lim_{n\to\infty} \sigma_n^\theta f = f \qquad a.e.$$

for all $f \in H_1^i(\mathbb{T}^d)$ *and* $i = 1, \dots, d$.

Note that these results are proved in Weisz [332, 333, 335].

3.7.4 Some Summability Methods

It is easy to see that $\theta \in V_1^2(\mathbb{R}) \subset S_0(\mathbb{R})$ for all examples 2.6.13–2.6.20 of Sect. 2.6.3 and Example 2.6.21 (the Riesz summation) with $1 \le \alpha < \infty$. Moreover, in Example 2.6.21, $\theta \in S_0(\mathbb{R})$ for all $0 < \alpha < \infty$. In the next examples, θ has d variables and $\theta \in S_0(\mathbb{R}^d)$.

Example 3.7.28 (Riesz summation]). Let

$$\theta(t) = \begin{cases} (1 - \|t\|_2^\gamma)^\alpha & \text{if } \|t\|_2 \le 1; \\ 0 & \text{if } \|t\|_2 > 1 \end{cases} \qquad (t \in \mathbb{R}^d)$$

for some $(d-1)/2 < \alpha < \infty$, $\gamma \in \mathbb{P}$ (see Fig. 3.4).

Example 3.7.29 (Weierstrass summation). Let

$$\theta(t) = e^{-\|t\|_2^2/2} \qquad \text{or} \qquad \theta(t) = e^{-\|t\|_2} \qquad (t \in \mathbb{R}^d)$$

(see Fig. 3.5). In the first case $\widehat{\theta}(x) = e^{-\|x\|_2^2/2}$ and in the second one, $\widehat{\theta}(x) = c_d/(1 + \|x\|_2^2)^{(d+1)/2}$ for some $c_d \in \mathbb{R}$ (see Stein and Weiss [293, p. 6.]).

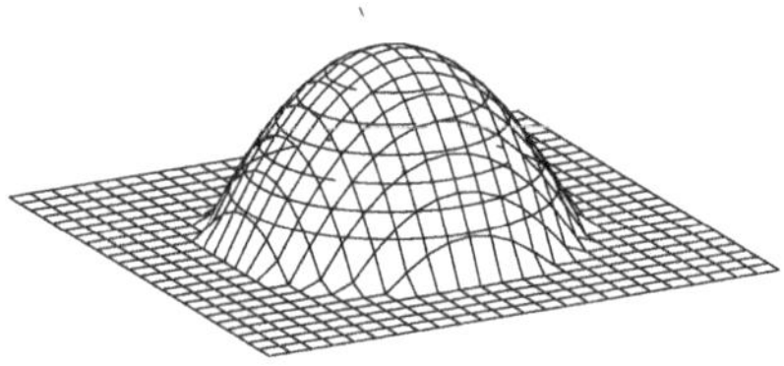

Fig. 3.4 Riesz summability function with $d = 2$, $\alpha = 1$, $\gamma = 2$

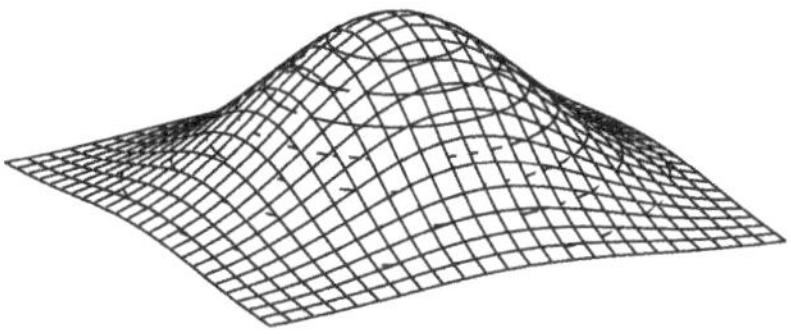

Fig. 3.5 Weierstrass summability function $\theta(t) = e^{-\|t\|_2^2/2}$

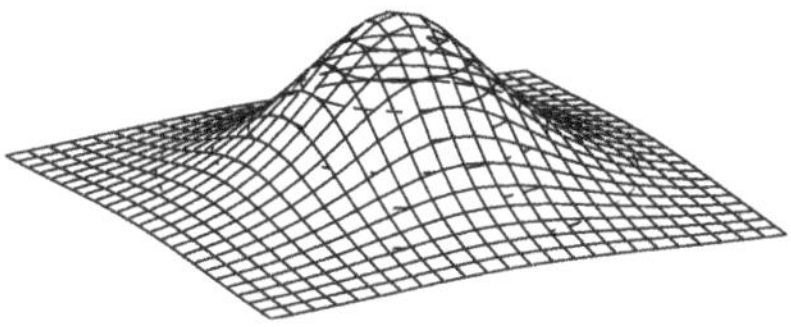

Fig. 3.6 Picard-Bessel summability function with $d = 2$

Example 3.7.30 (Picard and Bessel summations). Let

$$\theta_0(t) = \frac{1}{(1 + \|t\|_2^2)^{(d+1)/2}} \qquad (t \in \mathbb{R}^d)$$

(see Fig. 3.6). Here $\widehat{\theta_0}(x) = c_d e^{-\|x\|_2}$ for some $c_d \in \mathbb{R}^d$.

Lemma 3.7.31 *Let $\theta \in W(C, \ell_1)(\mathbb{R})$, $\mathcal{I}\theta \in W(C, \ell_1)(\mathbb{R})$ and θ be even and twice differentiable on the interval $(0, c)$, where $[-c, c]$ is the support of θ $(0 < c \le \infty)$. Suppose that*

$$\lim_{x \to c-0} x\theta(x) = 0, \quad \lim_{x \to +0} \theta' \in \mathbb{R}, \quad \lim_{x \to c-0} \theta' \in \mathbb{R} \quad \textit{and} \quad \lim_{x \to \infty} x\theta'(x) = 0.$$

If θ' and $\max(\mathcal{I}, 1)\theta''$ are integrable, then

$$\left|\widehat{\theta}(x)\right| \le \frac{C}{x^2}, \qquad \left|\left(\widehat{\theta}\right)'(x)\right| \le \frac{C}{x^2} \qquad (x \ne 0),$$

i.e., (3.7.9) *hold with $N = 0$ and $\beta_j = 0$.*

Proof By integrating by parts, we have

$$\begin{aligned}
\widehat{\theta}(x) &= \frac{2}{2\pi} \int_0^c \theta(t) \cos tx \, dt \\
&= \frac{1}{\pi x} \int_0^c \theta'(t) \sin tx \, dt \\
&= \frac{-1}{\pi x^2} [\theta'(t) \cos tx]_0^c + \frac{1}{\pi x^2} \int_0^c \theta''(t) \cos tx \, dt.
\end{aligned}$$

Similarly,

$$
\begin{aligned}
(\widehat{\theta})'(x) &= \frac{2}{2\pi} \int_0^c t\theta(t) \cos tx \, dt \\
&= \frac{1}{\pi x} \int_0^c (t\theta(t))' \sin tx \, dt \\
&= \frac{-1}{\pi x^2} [(t\theta(t))' \cos tx]_0^c + \frac{1}{\pi x^2} \int_0^c (t\theta(t))'' \cos tx \, dt,
\end{aligned}
$$

which proves the lemma. ■

Note that all examples 2.6.13–2.6.21 satisfy Lemma 3.7.31, (3.7.7), (3.7.8) and (3.7.9). Thus, all results of Sects. 3.7.2 and 3.7.3 hold.

Chapter 4
Lebesgue Points of Higher Dimensional Functions

In Theorem 1.5.4, we have proved the well known theorem of Lebesgue [197], i.e., for one-dimensional Fejér and Cesàro means and for all $f \in L_1(\mathbb{T})$,

$$\lim_{n\to\infty} \sigma_n^\alpha f(x) = f(x)$$

at each Lebesgue point of f. In this chapter, we generalize this result to higher dimensions and to all summability methods considered in Chaps. 2 and 3. We investigate a common generalization of the Cesàro, Riesz and θ-means and define

$$\sigma_n f(x) = \frac{1}{(2\pi)^d} \int_{\mathbb{T}^d} f(x-t) K_n(t)\, dt,$$

where $n \in \mathbb{N}$ or $n \in \mathbb{N}^d$ and $f \in L_1(\mathbb{T}^d)$, $K_n \in L_1(\mathbb{T}^d) \cap L_\infty(\mathbb{T}^d)$. We will give sufficient and/or necessary conditions for K_n such that $\sigma_n f$ is convergent at each Lebesgue point. We will study six versions of Lebesgue points, for different summability methods different Lebesgue points. We consider again the triangular, circular, cubic, the restricted (taken on a cone or cone-like set) and unrestricted rectangular summability as in the previous chapters. The proofs are very different for different summability methods, therefore each case needs new ideas. For each type of Lebesgue points, we introduce different and new type of Hardy-Littlewood maximal functions. We prove that these maximal operators are bounded from $L_p(\mathbb{T}^d)$ to $L_p(\mathbb{T}^d)$ with $1 < p \le \infty$ and we prove also a weak type inequality for $p = 1$. Using this, we obtain that almost every point is a Lebesgue point of an integrable function.

F. Weisz, *Lebesgue Points and Summability of Higher Dimensional Fourier Series*,
https://doi.org/10.1007/978-3-030-74636-0_4

4.1 ℓ_2-Summability

In this section, we use the notation

$$\sigma_n f(x) = \frac{1}{(2\pi)^d} \int_{\mathbb{T}^d} f(x-t) K_n(t)\, dt \qquad (n \in \mathbb{N}),$$

where $f \in L_1(\mathbb{T}^d)$ and $K_n \in L_1(\mathbb{T}^d) \cap L_\infty(\mathbb{T}^d)$ for all $n \in \mathbb{N}$. If $K_n = K_n^{2,\alpha,\gamma}$, $K_n^{2,\theta}$ or K_n is the one-dimensional Cesàro kernel K_n^α, then we obtain the ℓ_2-Riesz and θ-means $\sigma_n^{2,\alpha,\gamma} f$, $\sigma_n^{2,\theta} f$ or the one-dimensional Cesàro means $\sigma_n^\alpha f$, respectively. The higher dimensional ℓ_2-Riesz kernel $K_n^{2,\alpha,\gamma}$ and the one-dimensional Cesàro kernel K_n^α satisfy all conditions in this subsection. Under some conditions on θ, $K_n^{2,\theta}$ satisfies all conditions, too.

4.1.1 Hardy-Littlewood Maximal Functions

We generalize the Hardy-Littlewood maximal function for higher dimensions. As in the one-dimensional case, the Hardy-Littlewood maximal function is bounded on $L_p(\mathbb{T}^d)$ for $1 < p \leq \infty$ and it is of weak type $(1, 1)$. We denote by $B_r(c, h)$ $(c \in \mathbb{T}^d, h > 0)$ the r-ball

$$B_r(c, h) := \{x \in \mathbb{T}^d : \|x - c\|_r < h\} \qquad (1 \leq r \leq \infty)$$

with center c and radius h. For $r = 2$, we omit the index and write simply $B = B_2$. Similarly to the one-dimensional case, the Hardy-Littlewood maximal function can be given by

$$M_p^r f(x) = \sup_{x \in B_r} \left(\frac{1}{|B_r|} \int_{B_r} |f|^p \, d\lambda \right)^{1/p} \qquad (x \in \mathbb{T}^d),$$

where the supremum is taken over all r-balls B_r containing x and $1 \leq p, r \leq \infty$. In the special case when $r = \infty$, we have to take the supremum over all cubes I with sides parallel to the axes. Note that in the one-dimensional case this definition was given for $p = 1$, only. In this section, we will rather use the next equivalent centered version.

Definition 4.1.1 For $1 \leq p < \infty$ and $f \in L_p(\mathbb{T}^d)$, the Hardy-Littlewood maximal function is defined by

$$M_p f(x) := \sup_{h>0} \left(\frac{1}{(2h)^d} \int_{-h}^{h} \cdots \int_{-h}^{h} |f(x-t)|^p \, dt \right)^{1/p}.$$

If we take the supremum over all $0 < h < \pi$, then we get an equivalent definition. It is easy to see that

$$C_1 M_p f \leq M_p^r f \leq C_2 M_p f$$

for all $1 \leq p < \infty$ and $1 \leq r \leq \infty$. If $p = 1$, then we omit the notation p and write simply Mf. The next theorem can be proved exactly as Theorem 1.3.3 in the one-dimensional case.

Theorem 4.1.2 *If $1 \leq p < \infty$, then the maximal operator M_p is of weak type (p, p), i.e.,*

$$\sup_{\rho>0} \rho\lambda(M_p f > \rho)^{1/p} \leq C_p \|f\|_p \qquad (f \in L_p(\mathbb{T}^d)).$$

Moreover, if $p < r \leq \infty$, then

$$\left\|M_p f\right\|_r \leq C_r \|f\|_r \qquad (f \in L_r(\mathbb{T}^d)).$$

Using the density theorem of Marcinkiewicz and Zygmund (see Theorem 1.3.6), we can formulateLebesgue's differentiation theorem similarly to Corollary 1.3.8.

Corollary 4.1.3 *If $f \in L_1(\mathbb{T}^d)$, then*

$$\lim_{h\to 0} \frac{1}{(2h)^d} \int_{-h}^{h} \cdots \int_{-h}^{h} f(x-t)\,dt = f(x)$$

for almost every $x \in \mathbb{T}^d$.

This implies that the inequality

$$\|f\|_p \leq \|Mf\|_p \qquad (1 < p \leq \infty)$$

is trivial. Now we introduce the restricted Hardy-Littlewood maximal function by

$$M_{\square,p}^{\infty} f(x) := \left(\sup_{\substack{x\in I, \tau^{-1}\leq |I_i|/|I_j|\leq\tau \\ i,j=1,\dots,d}} \frac{1}{|I|} \int_I |f|^p \, d\lambda \right)^{1/p} \qquad (x \in \mathbb{T}^d)$$

for some fixed $\tau \geq 1$, where the supremum is taken over all appropriate rectangles

$$I = I_1 \times \cdots \times I_d$$

with sides parallel to the axes. The centered version is given in

Definition 4.1.4 For a fixed $\tau \geq 1$ and $f \in L_p(\mathbb{T}^d)$, the restricted Hardy-Littlewood maximal function is defined by

$$M_{\square,p} f(x) := \sup_{h \in \mathbb{R}_\tau^d} \left(\frac{1}{2^d \prod_{k=1}^d h_k} \int_{-h_1}^{h_1} \cdots \int_{-h_d}^{h_d} |f(x-t)|^p \, dt \right)^{1/p}.$$

Recall that

$$\mathbb{R}_\tau^d := \{x \in \mathbb{R}_+^d : \tau^{-1} \le x_i/x_j \le \tau, i, j = 1, \dots, d\}$$

was defined in (3.3.1). Taking the supremum over all $h \in \mathbb{R}_+^d$, we get a different maximal function, the so called strong Hardy-Littlewood maximal function. We will study this maximal operator in Sect. 4.2.1. Again, it is easy to see that

$$C_1 M_{\square,p} f \le M_{\square,p}^{\infty} f \le C_2 M_{\square,p} f$$

and

$$C_1 M_{\square,p} f \le M_p f \le C_2 M_{\square,p} f \tag{4.1.1}$$

for all $1 \le p \le \infty$. From this follows

Corollary 4.1.5 *If $\tau \ge 1$ is arbitrary and $1 \le p < \infty$, then*

$$\sup_{\rho>0} \rho \lambda (M_{\square,p} f > \rho)^{1/p} \le C \|f\|_p \qquad (f \in L_p(\mathbb{T}^d)).$$

Moreover, if $p < r \le \infty$, then

$$\left\| M_{\square,p} f \right\|_r \le C_r \|f\|_r \qquad (f \in L_r(\mathbb{T}^d)).$$

Corollary 4.1.6 *If $\tau \ge 1$ is arbitrary and $f \in L_1(\mathbb{T}^d)$, then*

$$\lim_{h \to 0, h \in \mathbb{R}_\tau^d} \frac{1}{2^d \prod_{k=1}^d h_k} \int_{-h_1}^{h_1} \cdots \int_{-h_d}^{h_d} f(x-t) \, dt = f(x)$$

for almost every $x \in \mathbb{T}$.

4.1.2 Lebesgue Points for the ℓ_2-Summability

First of all, we generalize the Herz spaces for higher dimensions.

Definition 4.1.7 For $1 \le q, r \le \infty$, the Herz space $E_q^r(\mathbb{R}^d)$ contains all functions f for which

$$\|f\|_{E_q^r} := \sum_{k=-\infty}^{\infty} 2^{kd(1-1/q)} \left\| f 1_{P_k^r} \right\|_q < \infty,$$

where

$$P_k^r := \{x \in \mathbb{R}^d : 2^{k-1}\pi \le \|x\|_r < 2^k\pi\} = B_r(0, 2^k\pi) \setminus B_r(0, 2^{k-1}\pi).$$

If we modify the definition of P_k^r,

$$P_k^r := \{x \in \mathbb{R}^d : 2^{k-1}\pi \le \|x\|_r < 2^k\pi\} \cap \mathbb{T}^d,$$

then we get the definition of the space $E_q^r(\mathbb{T}^d)$.

These spaces are special cases of the Herz spaces [166] (see also Garcia-Cuerva and Herrero [113]). We immediately obtain the next proposition.

Proposition 4.1.8 *For a fixed $1 \le q \le \infty$, the spaces $E_q^r(\mathbb{X}^d)$ are equivalent for all $1 \le r \le \infty$, where $\mathbb{X} = \mathbb{R}$ or $\mathbb{X} = \mathbb{T}$.*

For simplicity, we will use usually the sets P_k^∞ and the space $E_q^\infty(\mathbb{X}^d)$ ($\mathbb{X} = \mathbb{R}$ or $\mathbb{X} = \mathbb{T}$). These sets and spaces will be denoted by P_k and $E_q(\mathbb{X}^d)$. This means that we have to take the sum in the $E_q(\mathbb{T}^d)$-norm only for $k \le 0$, i.e.,

$$\|f\|_{E_q(\mathbb{T}^d)} = \sum_{k=-\infty}^{0} 2^{kd(1-1/q)} \left\| f 1_{P_k} \right\|_q < \infty.$$

It is easy to see that

$$L_1(\mathbb{X}^d) = E_1(\mathbb{X}^d) \hookleftarrow E_q(\mathbb{X}^d) \hookleftarrow E_{q'}(\mathbb{X}^d) \hookleftarrow E_\infty(\mathbb{X}^d)$$

for all $1 < q < q' < \infty$, where $\mathbb{X}$ denotes either $\mathbb{R}$ or $\mathbb{T}$. Moreover,

$$E_q(\mathbb{T}^d) \hookleftarrow L_q(\mathbb{T}^d) \qquad (1 \le q \le \infty). \tag{4.1.2}$$

Indeed,

$$\|f\|_{E_q(\mathbb{T}^d)} = \sum_{k=-\infty}^{0} 2^{kd(1-1/q)} \left\| f 1_{P_k} \right\|_q \le \sum_{k=-\infty}^{0} 2^{kd(1-1/q)} \|f\|_q \le \|f\|_q .$$

It is known in the one-dimensional case (see, e.g., Torchinsky [310]) that if there exists an even function η such that η is non-increasing on $\mathbb{R}_+$, $|\widehat{\theta}| \le \eta$, $\eta \in L_1(\mathbb{R})$, then σ_*^θ is of weak type $(1,1)$. Under similar conditions, we will generalize this result to the multi-dimensional setting.

Theorem 4.1.9 *For a measurable function f, let the non-increasing majorant be defined by*

$$\eta(x) := \sup_{\|t\|_r \ge \|x\|_r} |f(t)|$$

for some $1 \le r \le \infty$. *Then* $f \in E_\infty(\mathbb{X}^d)$ *if and only if* $\eta \in L_1(\mathbb{X}^d)$ *and*

$$C^{-1}\|\eta\|_1 \le \|f\|_{E_\infty(\mathbb{X}^d)} \le C\|\eta\|_1,$$

where $\mathbb{X} = \mathbb{R}$ *or* $\mathbb{X} = \mathbb{T}$.

Proof We prove the theorem for $\mathbb{X} = \mathbb{R}$. If $\eta \in L_1(\mathbb{R}^d)$, then

$$\|f\|_{E_\infty} \le \|\eta\|_{E_\infty} = \sum_{k=-\infty}^{\infty} 2^{kd}\left\|\eta 1_{P_k}\right\|_\infty = \sum_{k=-\infty}^{\infty} 2^{kd}\eta(2^{k-1}\pi) \le C\|\eta\|_1.$$

For the converse, denote by

$$a_k := \sup_{B_r(0,2^k\pi)\setminus B_r(0,2^{k-1}\pi)} |f| \quad \text{and} \quad \nu' := \sum_{k=-\infty}^{\infty} a_k 1_{B_r(0,2^k\pi)\setminus B_r(0,2^{k-1}\pi)}.$$

Let

$$\nu(x) := \sup_{\|t\|_r \ge \|x\|_r} \nu'(t) \qquad (x \in \mathbb{R}^d).$$

Since $f \in E_\infty(\mathbb{R}^d)$ implies $\lim_{k\to\infty} a_k = 0$, we conclude that there exists an increasing sequence $(n_k)_{k\in\mathbb{Z}}$ of integers such that $(a_{n_k})_{k\in\mathbb{Z}}$ is decreasing and ν can be written in the form

$$\nu = \sum_{k=-\infty}^{\infty} a_{n_k} 1_{B_r(0,2^{n_k}\pi)\setminus B_r(0,2^{n_{k-1}}\pi)}.$$

Thus

$$\begin{aligned}\|\eta\|_1 \le \|\nu\|_1 &= \sum_{k=-\infty}^{\infty} a_{n_k} \int_{B_r(0,2^{n_k}\pi)\setminus B_r(0,2^{n_{k-1}}\pi)} d\lambda \\ &= C\sum_{k=-\infty}^{\infty} \left(2^{dn_k} - 2^{dn_{k-1}}\right) a_{n_k}.\end{aligned}$$

By Abel rearrangement,

$$\|\eta\|_1 \le C\sum_{k=-\infty}^{\infty} 2^{dn_{k-1}}(a_{n_{k-1}} - a_{n_k}) \le C\|f\|_{E_\infty},$$

which proves the theorem. ∎

The maximal operator is introduced by

$$\sigma_* f := \sup_{n\in\mathbb{N}} |\sigma_n f|.$$

In the next theorem we show that, under some conditions, the maximal operator can be estimated by the Hardy-Littlewood maximal function pointwise.

Theorem 4.1.10 *If* $1 \le p < \infty$, $1/p + 1/q = 1$ *and*

$$\sup_{n\in\mathbb{N}} \|K_n\|_{E_q(\mathbb{T}^d)} \le C, \tag{4.1.3}$$

then

$$\sigma_* f(x) \le C \left(\sup_{n\in\mathbb{N}} \|K_n\|_{E_q(\mathbb{T}^d)} \right) M_p f(x)$$

for all $f \in L_p(\mathbb{T}^d)$ *and* $x \in \mathbb{T}^d$.

Proof By the definition of $\sigma_n f$,

$$\begin{aligned} |\sigma_n^\theta f(x)| &= \frac{1}{(2\pi)^d} \left| \int_{\mathbb{T}^d} f(x-t) K_n(t)\, dt \right| \\ &\le \frac{1}{(2\pi)^d} \sum_{k=-\infty}^{0} \int_{P_k} |f(x-t)||K_n(t)|\, dt. \end{aligned}$$

Recall that

$$P_k = P_k^\infty = \{x \in \mathbb{R}^d : 2^{k-1}\pi \le \|x\|_\infty < 2^k\pi\}.$$

By Hölder's inequality,

$$|\sigma_n f(x)| \le \frac{1}{(2\pi)^d} \sum_{k=-\infty}^{0} \left(\int_{P_k} |K_n(t)|^q\, dt \right)^{1/q} \left(\int_{P_k} |f(x-t)|^p\, dt \right)^{1/p}.$$

It is easy to see that if

$$G(u) := \left(\int_{-u}^{u} \cdots \int_{-u}^{u} |f(x-t)|^p\, dt \right)^{1/p} \qquad (u > 0),$$

then

$$\frac{G^p(u)}{(2u)^d} \le M_p^p f(x) \qquad (u > 0).$$

Therefore

$$\begin{aligned} |\sigma_n f(x)| &\le C \sum_{k=-\infty}^{0} \left(\int_{P_k} |K_n(t)|^q\, dt \right)^{1/q} G(2^k\pi) \\ &\le C \sum_{k=-\infty}^{0} 2^{kd/p} \left(\int_{P_k} |K_n(t)|^q\, dt \right)^{1/q} M_p f(x) \end{aligned}$$

$$= C \|K_n\|_{E_q(\mathbb{T}^d)} M_p f(x),$$

which shows the theorem. ∎

Note that $K_n \in L_\infty(\mathbb{T}^d) \subset E_\infty(\mathbb{T}^d) \subset E_q(\mathbb{T}^d)$ for all $n \in \mathbb{N}^d$, because of (4.1.2). Theorem 4.1.2 implies immediately

Theorem 4.1.11 *If $1 \le p < \infty$, $1/p + 1/q = 1$ and*

$$\sup_{n\in\mathbb{N}} \|K_n\|_{E_q(\mathbb{T}^d)} \le C,$$

then

$$\sup_{\rho>0} \rho\lambda(\sigma_* f > \rho)^{1/p} \le C_p \left(\sup_{n\in\mathbb{N}} \|K_n\|_{E_q(\mathbb{T}^d)}\right) \|f\|_p$$

for all $f \in L_p(\mathbb{T}^d)$. Moreover, for every $p < r \le \infty$,

$$\|\sigma_* f\|_r \le C \left(\sup_{n\in\mathbb{N}} \|K_n\|_{E_q(\mathbb{T}^d)}\right) \|f\|_r \qquad (f \in L_r(\mathbb{T}^d)).$$

Corollary 4.1.12 *Suppose that $1 \le p < \infty$, $1/p + 1/q = 1$ and*

$$\sup_{n\in\mathbb{N}} \|K_n\|_{E_q(\mathbb{T}^d)} \le C.$$

If

$$\lim_{n\to\infty} \widehat{K_n}(k) = 1$$

for all $k \in \mathbb{Z}^d$, then

$$\lim_{n\to\infty} \sigma_n f = f \quad a.e.$$

for all $f \in L_p(\mathbb{T}^d)$.

Proof For $f(x) = e^{\imath k\cdot x}$, we have

$$\lim_{n\to\infty} \sigma_n f(x) = \lim_{n\to\infty} \frac{1}{(2\pi)^d} \int_{\mathbb{T}^d} e^{\imath k\cdot(x-t)} K_n(t)\, dt = \lim_{n\to\infty} e^{\imath k\cdot x} \widehat{K_n}(k) = e^{\imath k\cdot x}.$$

This means that the convergence holds for all trigonometric polynomials. The corollary follows from Theorem 4.1.11 and from the density theorem. ∎

We consider the ℓ_2-θ-means given by

$$\sigma_n^{2,\theta} f(x) := \sum_{k\in\mathbb{Z}^d} \theta\left(\frac{\|k\|_2}{n}\right) \widehat{f}(k) e^{\imath k\cdot x},$$

where $\theta : \mathbb{R} \to \mathbb{R}$. As in Sect. 2.6, we suppose that

$$\sum_{k \in \mathbb{Z}^d} \left| \theta\left(\frac{\|k\|_2}{n} \right) \right| < \infty \tag{4.1.4}$$

and we use the notation

$$\theta_0(x) = \theta(\|x\|_2) \qquad (x \in \mathbb{R}^d).$$

Theorem 4.1.13 *Suppose that* $1 \le p < \infty$ *and* $1/p + 1/q = 1$. *If* (4.1.4) *is satisfied,* $\theta_0 \in L_1(\mathbb{R}^d)$ *and* $\widehat{\theta_0} \in E_q(\mathbb{R}^d)$, *then*

$$\sigma_*^{2,\theta} f(x) \le C \left\| \widehat{\theta_0} \right\|_{E_q(\mathbb{R}^d)} M_p f(x)$$

for all $f \in L_p(\mathbb{T}^d)$ *and* $x \in \mathbb{T}^d$.

Proof Similarly to Lemma 2.2.31,

$$\sigma_n^{2,\theta} f(x) = \frac{1}{(2\pi)^d} \int_{\mathbb{T}^d} f(x-t) K_n^{2,\theta}(t)\, dt = n^d \int_{\mathbb{R}^d} f(x-t) \widehat{\theta_0}(nt)\, dt$$

and

$$K_n^{2,\theta}(t) = (2\pi)^d n^d \sum_{j \in \mathbb{Z}^d} \widehat{\theta_0}(n(t_1 + 2j_1\pi), \dots, n(t_d + 2j_d\pi)). \tag{4.1.5}$$

We will prove that $\widehat{\theta_0} \in E_q(\mathbb{R}^d)$ implies

$$\left\| K_n^{2,\theta} \right\|_{E_q(\mathbb{T}^d)} \le C \left\| \widehat{\theta_0} \right\|_{E_q(\mathbb{R}^d)} \qquad (n \in \mathbb{N}). \tag{4.1.6}$$

First, we investigate the term $j = 0$ of the norm:

$$\begin{aligned}
&\left\| n^d \widehat{\theta_0}(nt_1, \dots, nt_d) \right\|_{E_q(\mathbb{T}^d)} \\
&\quad = \sum_{k=-\infty}^{0} 2^{kd(1-1/q)} n^d \left(\int_{P_k} |\widehat{\theta_0}(nt_1, \dots, nt_d)|^q \, dt \right)^{1/q} \\
&\quad \le C_q \sum_{k-\infty}^{0} 2^{kd(1-1/q)} n^{d(1-1/q)} \left(\int_{Q_k} |\widehat{\theta_0}(t_1, \dots, t_d)|^q \, dt \right)^{1/q},
\end{aligned}$$

where

$$Q_k := \prod_{j=1}^{d} \Big(-n2^k\pi, n2^k\pi \Big) \setminus \prod_{j=1}^{d} \Big(-n2^{k-1}\pi, n2^{k-1}\pi \Big).$$

Suppose that $2^{l-1} < n \leq 2^l$ for some $l \in \mathbb{N}$. If

$$Q_{k,l} := \prod_{j=1}^{d} \Big(-2^{k+l}\pi, 2^{k+l}\pi \Big) \setminus \prod_{j=1}^{d} \Big(-2^{k+l-2}\pi, 2^{k+l-2}\pi \Big),$$

then

$$\begin{aligned}
&\left\| n^d \widehat{\theta}_0(nt_1, \ldots, nt_d) \right\|_{E_q(\mathbb{T}^d)} \\
&\quad \leq C_q \sum_{k=-\infty}^{0} 2^{kd(1-1/q)} 2^{ld(1-1/q)} \left(\int_{Q_{k,l}} |\widehat{\theta}_0(t_1, \ldots, t_d)|^q \, dt \right)^{1/q} \\
&\quad \leq C_q \sum_{k=-\infty}^{0} 2^{(k+l)d(1-1/q)} \left(\sum_{i=k+l-1}^{k+l} \int_{P_i} |\widehat{\theta}_0(t_1, \ldots, t_d)|^q \, dt \right)^{1/q} \\
&\quad \leq C_q \sum_{k=-\infty}^{0} \sum_{i=k+l-1}^{k+l} 2^{id(1-1/q)} \left(\int_{P_i} |\widehat{\theta}_0(t_1, \ldots, t_d)|^q \, dt \right)^{1/q} \\
&\quad \leq C_q \sum_{i=-\infty}^{l} 2^{id(1-1/q)} \left(\int_{P_i} |\widehat{\theta}_0(t_1, \ldots, t_d)|^q \, dt \right)^{1/q} \\
&\quad \leq C_q \left\| \widehat{\theta}_0 \right\|_{E_q(\mathbb{R}^d)} .
\end{aligned} \tag{4.1.7}$$

Moreover,

$$\begin{aligned}
&\left\| n^d \sum_{j \in \mathbb{Z}^d, j \neq 0} \widehat{\theta}_0(n_1(t_1 + 2j_1\pi), \ldots, n_d(t_d + 2j_d\pi)) \right\|_{E_q(\mathbb{T}^d)} \\
&= \sum_{k=-\infty}^{0} 2^{kd(1-1/q)} n^d \\
&\qquad \left(\int_{P_k} \left| \sum_{j \in \mathbb{Z}^d, j \neq 0} \widehat{\theta}_0(n(t_1 + 2j_1\pi), \ldots, n(t_d + 2j_d\pi)) \right|^q \, dt \right)^{1/q} \\
&= \sum_{k=-\infty}^{0} 2^{kd(1-1/q)} n^d \\
&\qquad \left(\int_{\mathbb{T}^d} \left| \sum_{j \in \mathbb{Z}^d, j \neq 0} \widehat{\theta}_0(n(t_1 + 2j_1\pi), \ldots, n(t_d + 2j_d\pi)) \right|^q \, dt \right)^{1/q}
\end{aligned}$$

$$\leq C_q n^d \left(\int_{\mathbb{T}^d} \left| \sum_{j \in \mathbb{Z}^d, j \neq 0} \widehat{\theta}_0(n(t_1 + 2j_1\pi), \ldots, n(t_d + 2j_d\pi)) \right|^q dt \right)^{1/q}.$$

Let

$$R_i := \left\{ j \in \mathbb{Z}^d : j \neq 0, n(\mathbb{T} + 2j_1\pi) \times \ldots \times n(\mathbb{T} + 2j_d\pi) \cap P_i \neq 0 \right\}.$$

Since $|n(t_m + 2j_m\pi)| \geq 2^{l-1}\pi$ if $j_m \neq 0$, we conclude

$$\begin{aligned}
&\left\| n^d \sum_{j \in \mathbb{Z}^d, j \neq 0} \widehat{\theta}_0(n_1(t_1 + 2j_1\pi), \ldots, n_d(t_d + 2j_d\pi)) \right\|_{E_q(\mathbb{T}^d)} \\
&\quad \leq C_q n^d \left(\int_{\mathbb{T}^d} \left| \sum_{i=l}^{\infty} \sum_{j \in R_i} \widehat{\theta}_0(n(t_1 + 2j_1\pi), \ldots, n(t_d + 2j_d\pi)) \right|^q dt \right)^{1/q} \\
&\quad \leq C_q \sum_{i=l}^{\infty} n^d \left(\int_{\mathbb{T}^d} \left| \sum_{j \in R_i} \widehat{\theta}_0(n(t_1 + 2j_1\pi), \ldots, n(t_d + 2j_d\pi)) \right|^q dt \right)^{1/q}.
\end{aligned}$$

Since R_i has at most $C2^{id}/n$ members, we get that

$$\begin{aligned}
&\left\| n^d \sum_{j \in \mathbb{Z}^d, j \neq 0} \widehat{\theta}_0(n_1(t_1 + 2j_1\pi), \ldots, n_d(t_d + 2j_d\pi)) \right\|_{E_q(\mathbb{T}^d)} \\
&\leq C_q \sum_{i=l}^{\infty} n^d \left(\sum_{j \in R_i} \left(\frac{2^{id}}{n^d} \right)^{q-1} \int_{\mathbb{T}^d} \left| \widehat{\theta}_0(n(t_1 + 2j_1\pi), \ldots, n(t_d + 2j_d\pi)) \right|^q dt \right)^{1/q} \\
&\leq C_q \sum_{i=l}^{\infty} 2^{id(1-1/q)} \left(\sum_{j \in R_i} \int_{n_1(\mathbb{T}+2j_1\pi) \times \ldots \times n_d(\mathbb{T}+2j_d\pi)} \left| \widehat{\theta}_0(t_1, \ldots, t_d) \right|^q dt \right)^{1/q} \\
&\leq C_q \sum_{i=l}^{\infty} 2^{id(1-1/q)} \left(\int_{P_i} \left| \widehat{\theta}_0(t_1, \ldots, t_d) \right|^q dt \right)^{1/q} \\
&\leq C_q \left\| \widehat{\theta}_0 \right\|_{E_q(\mathbb{R}^d)},
\end{aligned} \tag{4.1.8}$$

which proves (4.1.6). The theorem follows from Theorem 4.1.10. ■

Note that

$$\widehat{\theta}_0 \in E_q(\mathbb{R}^d) \subset E_1(\mathbb{R}^d) \subset L_1(\mathbb{R}^d),$$

thus (2.6.5) is satisfied and θ_0 is continuous.

Theorem 4.1.14 *Suppose that $1 \le p < \infty$ and $1/p + 1/q = 1$. If (4.1.4) is satisfied, $\theta_0 \in L_1(\mathbb{R}^d)$ and $\widehat{\theta_0} \in E_q(\mathbb{R}^d)$, then*

$$\sup_{\rho>0} \rho \lambda(\sigma_*^{2,\theta} f > \rho)^{1/p} \le C_p \left\| \widehat{\theta_0} \right\|_{E_q(\mathbb{R}^d)} \|f\|_p$$

for all $f \in L_p(\mathbb{T}^d)$. Moreover, for every $p < r \le \infty$,

$$\left\| \sigma_*^{2,\theta} f \right\|_r \le C \left\| \widehat{\theta_0} \right\|_{E_q(\mathbb{R}^d)} \|f\|_r \qquad (f \in L_r(\mathbb{T}^d)).$$

Corollary 4.1.15 *Suppose that $1 \le p < \infty$ and $1/p + 1/q = 1$. If $\theta(0) = 1$, (4.1.4) is satisfied, $\theta_0 \in L_1(\mathbb{R}^d)$ and $\widehat{\theta_0} \in E_q(\mathbb{R}^d)$, then*

$$\lim_{n\to\infty} \sigma_n^{2,\theta} f = f \quad a.e.$$

for all $f \in L_p(\mathbb{T}^d)$.

Now we prove some converse type results. We know that the weak type inequality of Theorem 4.1.11 implies the almost everywhere convergence

$$\lim_{n\to\infty} \sigma_n f = f \quad \text{a.e.}$$

for all $f \in L_p(\mathbb{T}^d)$ (see Corollary 4.1.12). Conversely, if $1 \le p \le 2$ and the almost everywhere convergence holds for all $f \in L_p(\mathbb{T}^d)$, then σ_* is bounded from $L_p(\mathbb{T}^d)$ to $L_{p,\infty}(\mathbb{T}^d)$, as in Theorem 4.1.11 (see Stein [288]). The converse of Theorem 4.1.10 is given in the next result. More exactly, if $\sigma_* f$ can be estimated pointwise by $M_p f$, then (4.1.3) holds. Before proving this theorem, we need the following definition.

Definition 4.1.16 For $1 \le p < \infty$, we define the space $D_p(\mathbb{T}^d)$ with the norm

$$\|f\|_{D_p(\mathbb{T}^d)} := \sup_{0<r\le\pi} \left(\frac{1}{r^d} \int_{-r}^{r} \cdots \int_{-r}^{r} |f(t)|^p \, dt \right)^{1/p}.$$

Taking the supremum for all $0 < r < \infty$, we obtain the space $D_p(\mathbb{R}^d)$.

Lemma 4.1.17 *For $1 \le p < \infty$, the norm*

$$\|f\|_* = \sup_{k\le 0} 2^{-kd/p} \left\| f 1_{P_k} \right\|_p$$

is an equivalent norm on $D_p(\mathbb{T}^d)$.

Proof Choosing $r = 2^k\pi$ $(k \le 0)$, we conclude

$$2^{-kd/p} \left\| f 1_{P_k} \right\|_p \le C \left(\frac{1}{(2^k\pi)^d} \int_{-2^k\pi}^{2^k\pi} \cdots \int_{-2^k\pi}^{2^k\pi} |f(t)|^p \, dt \right)^{1/p} \le \|f\|_{D_p}.$$

On the other hand, suppose that $2^{N-1}\pi \le r < 2^N\pi$ for some $N \in \mathbb{N}$. Then

$$\begin{aligned}
\frac{1}{r^d}\int_{-r}^{r}\cdots\int_{-r}^{r}|f(t)|^p\,dt &\le C2^{-Nd}\int_{-2^N\pi}^{2^N\pi}\cdots\int_{-2^N\pi}^{2^N\pi}|f(t)|^p\,dt \\
&= C2^{-Nd}\sum_{k=-\infty}^{N}\int_{P_k}|f(t)|^p\,dt \\
&\le C2^{-Nd}\sum_{k=-\infty}^{N}2^{kd}\,\|f\|_*^p \le C\,\|f\|_*^p,
\end{aligned}$$

which shows the lemma. ■

We can see that $D_p(\mathbb{T}^d) \subset L_p(\mathbb{T}^d)$ and

$$\|f\|_p \le C\|f\|_{D_p(\mathbb{T}^d)} \qquad (f \in D_p(\mathbb{T}^d)).$$

Theorem 4.1.18 *If* $1 \le p < \infty$, $1/p + 1/q = 1$ *and*

$$\sigma_* f(0) \le CM_p f(0) \tag{4.1.9}$$

for all $f \in L_p(\mathbb{T}^d)$, *then*

$$\sup_{n\in\mathbb{N}}\|K_n\|_{E_q(\mathbb{T}^d)} \le C.$$

Proof It is easy to see by Lemma 4.1.17 that

$$\sup_{\|f\|_{D_p(\mathbb{T}^d)}\le 1}\left|\int_{\mathbb{T}^d} f(-t)K_n(t)\,dt\right| = \|K_n\|_{E_q(\mathbb{T}^d)}. \tag{4.1.10}$$

There exists a function $f \in D_p(\mathbb{T}^d)$ with $\|f\|_{D_p} < 1$ such that

$$\frac{\|K_n\|_{E_q(\mathbb{T}^d)}}{2} \le \left|\int_{\mathbb{T}^d} f(-t)K_n(t)\,dt\right|.$$

Since $f \in L_p(\mathbb{R}^d)$, by (4.1.9),

$$\left|\int_{\mathbb{T}^d} f(-t)K_n(t)\,dt\right| = |\sigma_n f(0)| \le CM_p f(0) \qquad (n \in N),$$

which implies

$$\|K_n\|_{E_q(\mathbb{T}^d)} \le CM_p f(0) \le C\|f\|_{D_p} \le C \qquad (n \in N).$$

This proves the result. ■

Note that the norm of $D_p(\mathbb{T}^d)$ is equivalent to

$$\|f\| = \sup_{r\in[0,\pi]^d\cap\mathbb{R}^d_\tau} \left(\frac{1}{\prod_{j=1}^d r_j}\int_{-r_1}^{r_1}\cdots\int_{-r_d}^{r_d}|f(t)|^p\,dt\right)^{1/p}.$$

Now, we introduce the first generalization of Lebesgue points for higher dimensions. Corollary 4.1.3 says that

$$\lim_{h\to 0}\frac{1}{(2h)^d}\int_{-h}^{h}\cdots\int_{-h}^{h}f(x-t)\,dt = f(x)$$

for almost every $x\in\mathbb{T}^d$, where $f\in L_1(\mathbb{T}^d)$. In other words,

$$\lim_{h\to 0}\frac{1}{(2h)^d}\int_{-h}^{h}\cdots\int_{-h}^{h}(f(x-t)-f(x))\,dt = 0,$$

which is equivalent to

$$\lim_{h\to 0}\frac{1}{(2h)^d}\left|\int_{-h}^{h}\cdots\int_{-h}^{h}(f(x-t)-f(x))\,dt\right| = 0.$$

In the next definition, we describe a stronger condition.

Definition 4.1.19 For $1\le p<\infty$, a point $x\in\mathbb{T}^d$ is called a p-Lebesgue point of $f\in L_p(\mathbb{T}^d)$ if

$$\lim_{h\to 0}\left(\frac{1}{(2h)^d}\int_{-h}^{h}\cdots\int_{-h}^{h}|f(x-t)-f(x)|^p\,dt\right)^{1/p} = 0.$$

For $p=1$, the points are said to be Lebesgue points. One can see that using the restricted maximal operator and Corollary 4.1.6, we get an equivalent definition:

$$\lim_{h\to 0, h\in\mathbb{R}^d_\tau}\left(\frac{1}{\prod_{j=1}^d(2h_j)}\int_{-h_1}^{h_1}\cdots\int_{-h_d}^{h_d}|f(x-t)-f(x)|^p\,dt\right)^{1/p} = 0.$$

If $p<r$ and x is an r-Lebesgue point of f, then it is also a p-Lebesgue point. Indeed, by Hölder's inequality,

$$\begin{aligned}&\left(\frac{1}{(2h)^d}\int_{-h}^{h}\cdots\int_{-h}^{h}|f(x-t)-f(x)|^p\,dt\right)^{1/p}\\ &\quad\le\left(\frac{1}{(2h)^d}\int_{-h}^{h}\cdots\int_{-h}^{h}|f(x-t)-f(x)|^r\,dt\right)^{1/r}.\end{aligned}$$

The following two results can be proved as in the one-dimensional case, see Theorem 1.3.11 and Lemma 1.3.12.

Theorem 4.1.20 *Almost every point $x \in \mathbb{T}^d$ is a p-Lebesgue point of $f \in L_p(\mathbb{T}^d)$* $(1 \le p < \infty)$.

Lemma 4.1.21 *If x is a p-Lebesgue point of $f \in L_p(\mathbb{T}^d)$, then $f(x)$ and $M_p f(x)$ are finite* $(1 \le p < \infty)$.

The next theorem generalizes Theorem 1.5.4.

Theorem 4.1.22 *Suppose that* $1 \le p < \infty$, $1/p + 1/q = 1$ *and*

$$\sup_{n\in\mathbb{N}} \|K_n\|_{E_q(\mathbb{T}^d)} \le C.$$

If for all $\delta > 0$

$$\lim_{n\to\infty} \|K_n\|_{L_q(\mathbb{T}^d \setminus (-\delta,\delta)^d)} = 0 \tag{4.1.11}$$

and

$$\lim_{n\to\infty} \widehat{K_n}(0) = 1, \tag{4.1.12}$$

then

$$\lim_{n\to\infty} \sigma_n f(x) = f(x)$$

for all p-Lebesgue points of $f \in L_p(\mathbb{T}^d)$.

Proof Now, set

$$G(u) := \left(\int_{-u}^{u} \cdots \int_{-u}^{u} |f(x-t) - f(x)|^p \, dt \right)^{1/p} \qquad (u > 0),$$

Since x is a p-Lebesgue point of f, for all $\epsilon > 0$, there exists $m \in \mathbb{Z}$, $m \le 0$ such that

$$\frac{G^p(u)}{(2u)^d} \le \epsilon \quad \text{if} \quad 0 < u \le 2^m \pi. \tag{4.1.13}$$

Observe that

$$\begin{aligned} \sigma_n f(x) - f(x) &= \frac{1}{(2\pi)^d} \int_{\mathbb{T}^d} (f(x-t) - f(x)) K_n(t) \, dt \\ &\quad + f(x) \left(\frac{1}{(2\pi)^d} \int_{\mathbb{T}^d} K_n(t) \, dt - 1 \right). \end{aligned}$$

Thus

$$
\begin{aligned}
|\sigma_n f(x) - f(x)| &\le C \int_{\mathbb{T}^d} |f(x-t) - f(x)||K_n(t)|\,dt + \left|f(x)\left(\widehat{K_n}(0) - 1\right)\right| \\
&= C \int_{-2^m\pi}^{2^m\pi} \cdots \int_{-2^m\pi}^{2^m\pi} |f(x-t) - f(x)||K_n(t)|\,dt \\
&\quad + C \int_{\mathbb{T}^d \setminus (-2^m\pi, 2^m\pi)^d} |f(x-t) - f(x)||K_n(t)|\,dt \\
&\quad + \left|f(x)\left(\widehat{K_n}(0) - 1\right)\right| \\
&=: A_1(x) + A_2(x) + A_3(x).
\end{aligned}
$$

We estimate $A_1(x)$ by

$$
\begin{aligned}
A_1(x) &= C \sum_{k=-\infty}^{m} \int_{P_k} |f(x-t) - f(x)||K_n(t)|\,dt \\
&\le C \sum_{k=-\infty}^{m} \left(\int_{P_k} |K_n(t)|^q\,dt \right)^{1/q} \left(\int_{P_k} |f(x-t) - f(x)|^p\,dt \right)^{1/p} \\
&\le C \sum_{k=-\infty}^{m} \left(\int_{P_k} |K_n(t)|^q\,dt \right)^{1/q} G(2^k\pi).
\end{aligned}
$$

Then, by (4.1.13),

$$
A_1(x) \le C_p \epsilon \sum_{k=-\infty}^{m} 2^{kd/p} \left(\int_{P_k} |K_n(t)|^q\,dt \right)^{1/q} \le C_p \epsilon \|K_n\|_{E_q(\mathbb{T}^d)}.
$$

For $0 < \delta < 2^m\pi$, we have

$$
\begin{aligned}
A_2(x) &\le C \int_{\mathbb{T}^d \setminus (-\delta,\delta)^d} |f(x-t) - f(x)||K_n(t)|\,dt \\
&\le C \left(\int_{\mathbb{T}^d \setminus (-\delta,\delta)^d} |K_n(t)|^q\,dt \right)^{1/q} \left(\|f\|_p + |f(x)| \right),
\end{aligned}
$$

which tends to 0 as $n \to \infty$. Moreover, $A_3(x) \to 0$ as $n \to \infty$, too. This completes the proof of the theorem. ■

Observe that (4.1.2) and $\delta' < 2^k\pi < \delta$ imply

$$
\begin{aligned}
\|K_n\|_{E_q(\mathbb{T}^d \setminus (-\delta,\delta)^d)} &\le \|K_n\|_{L_q(\mathbb{T}^d \setminus (-\delta,\delta)^d)} \\
&\le \|K_n\|_{L_q(\mathbb{T}^d \setminus (-2^k\pi, 2^k\pi)^d)} \\
&\le \left(\sum_{l=k+1}^{0} \int_{P_l} |K_n(t)|^q\,dt \right)^{1/q}
\end{aligned}
$$

$$\leq C_\delta \sum_{l=k+1}^{0} 2^{kd(1-1/q)} \left(\int_{P_l} |K_n(t)|^q \, dt \right)^{1/q}$$
$$\leq C_\delta \|K_n\|_{E_q(\mathbb{T}^d \setminus (-2^k\pi, 2^k\pi)^d)}$$
$$\leq C_\delta \|K_n\|_{E_q(\mathbb{T}^d \setminus (-\delta', \delta')^d)}. \qquad (4.1.14)$$

Then condition (4.1.11) is equivalent to

$$\lim_{n\to\infty} \|K_n\|_{E_q(\mathbb{T}^d \setminus (-\delta, \delta)^d)} = 0.$$

In the case $\widehat{\theta}_0 \in E_q(\mathbb{R}^d)$, we can formulate a somewhat simpler version of the preceding theorem.

Theorem 4.1.23 *Suppose that* $1 \leq p < \infty$ *and* $1/p + 1/q = 1$. *If* $\theta(0) = 1$, (4.1.4) *is satisfied,* $\theta_0 \in L_1(\mathbb{R}^d)$ *and* $\widehat{\theta}_0 \in E_q(\mathbb{R}^d)$, *then*

$$\lim_{n\to\infty} \sigma_n^{2,\theta} f(x) = f(x)$$

for all p-Lebesgue points of $f \in L_p(\mathbb{T}^d)$.

Proof We have seen in Theorem 4.1.13 that $\widehat{\theta}_0 \in E_q(\mathbb{R}^d)$ implies

$$\left\| K_n^{2,\theta} \right\|_{E_q(\mathbb{T}^d)} \leq C \left\| \widehat{\theta}_0 \right\|_{E_q(\mathbb{R}^d)} \qquad (n \in \mathbb{N}),$$

so the first condition of Theorem 4.1.22 is satisfied.

On the other hand, let $2^{k_0}\pi < \delta$ and $2^{l-1} \leq n < 2^l$ as in the proof of Theorem 4.1.13. We get similarly to (4.1.7) and (4.1.8) that

$$\left\| K_n^{2,\theta} \right\|_{E_q(\mathbb{T}^d \setminus (-\delta, \delta)^d)} \lesssim C_q \sum_{i=k_0+l-1}^{\infty} 2^{id(1-1/q)} \left(\int_{P_i} |\widehat{\theta}_0(t_1, \ldots, t_d)|^q \, dt \right)^{1/q}$$
$$+ C_q \sum_{i=l}^{\infty} 2^{id(1-1/q)} \left(\int_{P_i} |\hat{\theta}_0(t_1, \ldots, t_d)|^q \, dt \right)^{1/q},$$

which tends to 0 as $n \to \infty$, since $\widehat{\theta}_0 \in E_q(\mathbb{R}^d)$. Then (4.1.11) follows from (4.1.14). Finally, by (4.1.5),

$$\frac{1}{(2\pi)^d} \int_{\mathbb{T}^d} K_n(t) \, dt = n^d \sum_{j\in\mathbb{Z}^d} \int_{\mathbb{T}^d} \widehat{\theta}_0(n(t_1 + 2j_1\pi), \ldots, n(t_d + 2j_d\pi)) \, dt$$
$$= n^d \int_{\mathbb{R}^d} \widehat{\theta}_0(nt) \, dt = \theta_0(1) = 1,$$

which finishes the proof of our theorem. ■

Since each point of continuity is a Lebesgue point, we have

Corollary 4.1.24 *If the conditions of Theorem 4.1.22 or Theorem 4.1.23 are satisfied and if $f \in L_p(\mathbb{T}^d)$ is continuous at a point x, then*

$$\lim_{n\to\infty} \sigma_n f(x) = f(x).$$

The converse of Theorem 4.1.22 holds also.

Theorem 4.1.25 *Suppose that $1 \le p < \infty$ and $1/p + 1/q = 1$. If*

$$\lim_{n\to\infty} \sigma_n f(x) = f(x)$$

for all p-Lebesgue points of $f \in L_p(\mathbb{T}^d)$ then

$$\sup_{n\in\mathbb{N}} \|K_n\|_{E_q(\mathbb{T}^d)} \le C.$$

Proof The space $D_p^0(\mathbb{T}^d)$ consists of all functions $f \in D_p(\mathbb{T}^d)$ for which $f(0) = 0$ and 0 is a p-Lebesgue point of f, in other words

$$\lim_{h\to 0} \left(\frac{1}{(2h)^d} \int_{-h}^{h} \cdots \int_{-h}^{h} |f(t)|^p \, dt\right)^{1/p} = 0.$$

We will show that $D_p^0(\mathbb{T}^d)$ is a Banach space. Let (f_n) be a Cauchy sequence in $D_p^0(\mathbb{T}^d)$, i.e.,

$$\|f_n - f_m\|_{D_p^0(\mathbb{T}^d)} \to 0 \quad \text{as} \quad n, m \to \infty.$$

Then there exists a subsequence (f_{ν_n}) such that

$$\left\| f_{\nu_{n+1}} - f_{\nu_n} \right\|_{D_p^0(\mathbb{T}^d)} \le 2^{-n}.$$

Then

$$\left\| \sum_{n=0}^{\infty} |f_{\nu_{n+1}} - f_{\nu_n}| \right\|_{L_p(\mathbb{T}^d)} \le \left\| \sum_{n=0}^{\infty} |f_{\nu_{n+1}} - f_{\nu_n}| \right\|_{\mathbb{D}_p^0(\mathbb{T}^d)} \le 2,$$

thus the series

$$\sum_{n=0}^{\infty} |f_{\nu_{n+1}} - f_{\nu_n}|$$

is almost everywhere finite. That is to say the sequence (f_{ν_n}) is almost everywhere convergent. Let

$$f := \lim_{n\to\infty} f_{\nu_n} \quad \text{and} \quad f(0) = 0.$$

For all $\epsilon > 0$, there exists N such that

$$\|f - f_{\nu_N}\|_{D_p^0(\mathbb{T}^d)} \le \sum_{n=N}^{\infty} \|f_{\nu_{n+1}} - f_{\nu_n}\|_{D_p^0(\mathbb{T}^d)} \le \sum_{n=N}^{\infty} 2^{-n} < \epsilon.$$

If $h > 0$ is small enough, then

$$\left(\frac{1}{(2h)^d}\int_{-h}^{h}\cdots\int_{-h}^{h}|f_{\nu_N}(t)|^p\,dt\right)^{1/p} < \epsilon.$$

Hence

$$\begin{aligned}&\left(\frac{1}{(2h)^d}\int_{-h}^{h}\cdots\int_{-h}^{h}|f(t)|^p\,dt\right)^{1/p}\\&\le C\|f - f_{\nu_N}\|_{D_p^0(\mathbb{T}^d)} + \left(\frac{1}{(2h)^d}\int_{-h}^{h}\cdots\int_{-h}^{h}|f_{\nu_N}(t)|^p\,dt\right)^{1/p} < 2\epsilon,\end{aligned}$$

whenever h is small enough. From this it follows that $f \in D_p^0(\mathbb{T}^d)$ and 0 is a Lebesgue point of f. Thus $D_p^0(\mathbb{T}^d)$ is a Banach space, indeed.

We get from the conditions of the theorem that

$$\lim_{n\to\infty} \sigma_n f(0) = 0 \qquad \text{for all} \qquad f \in D_p^0(\mathbb{T}^d).$$

Thus the operators

$$U_n : D_p^0(\mathbb{T}^d) \to \mathbb{R}, \qquad U_n f := \sigma_n f(0) \qquad (n \in \mathbb{N})$$

are uniformly bounded by the Banach-Steinhaus theorem. Observe that in (4.1.10), we may suppose that f is 0 in a neighborhood of 0. Then

$$\begin{aligned}C &\ge \|U_n\|\\&= \sup_{\|f\|_{D_p^0(\mathbb{T}^d)}\le 1}\left|\int_{\mathbb{T}^d} f(-t)K_n(t)\,dt\right|\\&= \sup_{\|f\|_{D_p(\mathbb{T}^d)}\le 1}\left|\int_{\mathbb{T}^d} f(-t)K_n(t)\,dt\right|\\&= \|K_n\|_{E_q(\mathbb{T}^d)}\end{aligned}$$

for all $n \in \mathbb{N}$. ∎

Corollary 4.1.26 *Suppose that* $1 \le p < \infty$, $1/p + 1/q = 1$, (4.1.11) *and* (4.1.12) *hold. Then*

$$\lim_{n\to\infty} \sigma_n f(x) = f(x)$$

for all p-Lebesgue points of $f \in L_p(\mathbb{T}^d)$ *if and only if*

$$\sup_{n \in \mathbb{N}} \|K_n\|_{E_q(\mathbb{T}^d)} \leq C.$$

Note that these results can be found in Feichtinger and Weisz [104]. We know that our results can be applied to the one-dimensional Cesàro summability (see Sect. 1.5). Moreover, the Riesz, Weierstrass, Picard and Bessel summations given in Sect. 3.7.4 (Examples 3.7.28, 3.7.29, 3.7.30) satisfy all conditions of this section, too.

Corollary 4.1.27 *Suppose that* θ *is one of the Examples 3.7.28, 3.7.29 or 3.7.30. Then*

$$\lim_{n \to \infty} \sigma_n^{2,\theta} f(x) = f(x)$$

for all Lebesgue points of $f \in L_1(\mathbb{T}^d)$*. Moreover,*

$$\sup_{\rho > 0} \rho \lambda(\sigma_*^{2,\theta} > \rho) \leq C \left\|\widehat{\theta}_0\right\|_{E_\infty(\mathbb{R}^d)} \|f\|_1 \qquad (f \in L_1(\mathbb{T}^d))$$

and, for every $1 < p \leq \infty$,

$$\left\|\sigma_*^{2,\theta} f\right\|_p \leq C_p \left\|\widehat{\theta}_0\right\|_{E_\infty(\mathbb{R}^d)} \|f\|_p \qquad (f \in L_p(\mathbb{T}^d)).$$

4.2 Unrestricted Rectangular Summability

Here we study the operators

$$\sigma_n f(x) = \frac{1}{(2\pi)^d} \int_{\mathbb{T}^d} f(x-t) K_n(t)\, dt \qquad (n \in \mathbb{N}^d),$$

where $f \in L_1(\mathbb{T}^d)$ and $K_n \in L_1(\mathbb{T}^d) \cap L_\infty(\mathbb{T}^d)$ for all $n \in \mathbb{N}^d$. The higher dimensional rectangular Cesàro and Riesz kernels, K_n^α and $K_n^{\alpha,\gamma}$ satisfy the conditions of this section. The kernel K_n^θ is also investigated.

4.2.1 Strong Hardy-Littlewood Maximal Functions

A second generalization of the one-dimensional maximal function is the so-called strong Hardy-Littlewood maximal function given by

$$M'_{s,p} f(x) := \sup_{x \in I} \left(\frac{1}{|I|} \int_I |f|^p \, d\lambda \right)^{1/p} \qquad (x \in \mathbb{T}^d),$$

where $f \in L_p(\mathbb{T}^d)$ and the supremum is taken over all rectangles

$$I = I_1 \times \cdots \times I_d \subset \mathbb{T}^d$$

with sides parallel to the axes and containing x. This maximal function is different from M_p and from $M_{\square,p}$ defined in Sect. 4.1.1, it remains bounded on $L_p(\mathbb{T}^d)$ with $1 < p \le \infty$, but it is not of weak type $(1, 1)$. The reason for this is that in the definition the ratio of the sides of the rectangles can be large. We will use again the next centered version of the strong maximal function.

Definition 4.2.1 For $1 \le p < \infty$ and $f \in L_p(\mathbb{T}^d)$ the strong Hardy-Littlewood maximal function is defined by

$$M_{s,p} f(x) := \sup_{h\in\mathbb{R}_+^d} \left(\frac{1}{\prod_{j=1}^d (2h_j)} \int_{-h_1}^{h_1} \cdots \int_{-h_d}^{h_d} |f(x-t)|^p \, dt \right)^{1/p}.$$

Taking the supremum over all $h \in (0, \pi)^d$, we get an equivalent definition. It is easy to see that

$$C_1 M_{s,p} f \le M'_{s,p} f \le C_2 M_{s,p} f$$

for all $1 \le p < \infty$. If $p = 1$, then we omit the notation p and write simply $M_s f$. In the one-dimensional case M_s is the usual Hardy-Littlewood maximal function and so, it is of weak type $(1, 1)$. For higher dimensions it is known that there is a function $f \in L_1(\mathbb{T}^d)$ such that $M_s f = \infty$ almost everywhere (see Jessen, Marcinkiewicz and Zygmund [177] and Saks [268]). Thus M_s cannot be of weak type $(1, 1)$, however, with the help of the $L_p(\log L)^k(\mathbb{T}^d)$ spaces, we can show a weak type inequality. Set $\log^+ u := \max(0, \log u)$.

Definition 4.2.2 For $k \in \mathbb{N}$ and $1 \le p < \infty$, a measurable function f is in the set $L_p(\log L)^k(\mathbb{T}^d)$ if

$$\|f\|_{L_p(\log L)^k} := \left(\int_{\mathbb{T}^d} |f|^p (\log^+ |f|)^k \, d\lambda \right)^{1/p} < \infty.$$

If $p = \infty$, then set $L_\infty(\log L)^k(\mathbb{T}^d) = L_\infty(\mathbb{T}^d)$.

For $k = 0$, we get back the $L_p(\mathbb{T}^d)$ spaces. We have for all $k \in \mathbb{P}$ and $1 \le p < r \le \infty$ that

$$L_p(\mathbb{T}^d) \supset L_p(\log L)^{k-1}(\mathbb{T}^d) \supset L_p(\log L)^k(\mathbb{T}^d) \supset L_r(\mathbb{T}^d).$$

Theorem 4.2.3 *If $f \in L(\log L)^{d-1}(\mathbb{T}^d)$, then*

$$\sup_{\rho>0} \rho \lambda(M_s f > \rho) \le C + C \left\| |f| \left(\log^+ |f|\right)^{d-1} \right\|_1.$$

Moreover, for $1 < p \le \infty$, we have

$$\|M_s f\|_p \le C_p \|f\|_p \qquad (f \in L_p(\mathbb{T}^d)).$$

Proof Let us denote the one-dimensional Hardy-Littlewood maximal function in the ith dimension by $M^{(i)}$. Then

$$M_s f \le M^{(1)} \circ M^{(2)} \circ \cdots \circ M^{(d)} f.$$

By Theorems 1.3.3 and 1.3.5,

$$\begin{aligned}
\sup_{\rho>0} \rho\lambda(M_s f > \rho) &= \sup_{\rho>0} \rho\lambda(M^{(1)} \circ M^{(2)} \circ \cdots \circ M^{(d)} f > \rho) \\
&\le \left\| M^{(2)} \circ \cdots \circ M^{(d)} f \right\|_1 \\
&\le C + C \left\| M^{(3)} \circ \cdots \circ M^{(d)} f \right\|_{L_1(\log L)(\mathbb{T}^d)} \\
&\le \ldots \le C + C \|f\|_{L_1(\log L)^{d-1}(\mathbb{T}^d)} .
\end{aligned}$$

The second inequality of Theorem 4.2.3 follows similarly. ■

Similarly to Corollary 4.1.3, we obtain

Corollary 4.2.4 *If $f \in L_1(\log L)^{d-1}(\mathbb{T}^d)$, then*

$$\lim_{h\to 0} \frac{1}{\prod_{j=1}^d (2h_j)} \int_{-h_1}^{h_1} \cdots \int_{-h_d}^{h_d} f(x-t)\,dt = f(x)$$

for almost every $x \in \mathbb{T}^d$.

Note that this convergence result does not hold for al $f \in L_1(\mathbb{T}^d)$ (see Jessen, Marcinkiewicz and Zygmund [177] and Saks [268]). Since $M_{s,p}^p f = M_s(|f|^p)$ for $1 \le p < \infty$, we have

Corollary 4.2.5 *If $1 \le p < \infty$ and $f \in L_p(\log L)^{d-1}(\mathbb{T}^d)$, then*

$$\sup_{\rho>0} \rho\lambda(M_{s,p} f > \rho)^{1/p} \le C_p + C_p \|f\|_{L_p(\log L)^{d-1}} .$$

For $p < r \le \infty$,

$$\left\| M_{s,p} f \right\|_r \le C_r \|f\|_r \qquad (f \in L_r(\mathbb{T}^d)).$$

4.2.2 Lebesgue Points for the Unrestricted Rectangular Summability

To formulate the generalization of Lebesgue's theorem for the unrestricted rectangular summability, we have to modify slightly the definition of the space $E_q(\mathbb{R}^d)$.

Definition 4.2.6 For $1 \le q \le \infty$, the Herz space $E'_q(\mathbb{R}^d)$ resp. $E'_q(\mathbb{T}^d)$ contains all functions f for which

$$\|f\|_{E'_q(\mathbb{R}^d)} := \sum_{k_1=-\infty}^{\infty} \cdots \sum_{k_d=-\infty}^{\infty} \left(\prod_{j=1}^{d} 2^{k_j(1-1/q)} \right) \left\| f 1_{P_k} \right\|_q < \infty$$

resp.

$$\|f\|_{E'_q(\mathbb{T}^d)} := \sum_{k_1=-\infty}^{0} \cdots \sum_{k_d=-\infty}^{0} \left(\prod_{j=1}^{d} 2^{k_j(1-1/q)} \right) \left\| f 1_{P_k} \right\|_q < \infty,$$

where

$$P_k := P_{k_1} \times \cdots \times P_{k_d} \qquad (k \in \mathbb{Z}^d)$$

and

$$P_i = \{x \in \mathbb{R} : 2^{i-1}\pi \le |x| < 2^i \pi\} \qquad (i \in \mathbb{Z}).$$

Again,

$$L_1(\mathbb{X}^d) = E'_1(\mathbb{X}^d) \hookleftarrow E'_q(\mathbb{X}^d) \hookleftarrow E'_{q'}(\mathbb{X}^d) \hookleftarrow E'_\infty(\mathbb{X}^d), \qquad 1 < q < q' < \infty,$$

where $\mathbb{X} = \mathbb{R}$ or $\mathbb{T}$ and

$$E'_q(\mathbb{T}^d) \hookleftarrow L_q(\mathbb{T}^d) \qquad (1 \le q \le \infty).$$

It is easy to see that $E'_q(\mathbb{X}^d) \supset E_q(\mathbb{X}^d)$ and

$$\|f\|_{E'_q} \le C \|f\|_{E_q} \qquad (1 \le q \le \infty).$$

Here we will estimate pointwise the maximal operator

$$\sigma_* f := \sup_{n \in \mathbb{N}^d} |\sigma_n f|$$

by the strong Hardy-Littlewood maximal function. Since the condition (4.1.11) is not true for rectangular summability kernels (e.g., for the Cesàro or Riesz kernels, K_n^α, $K_n^{\alpha,\gamma}$), we use here other conditions and other ideas. We introduce the functions

$$
\begin{aligned}
\widetilde{K}_n(t) &:= \left(\prod_{j=1}^d n_j\right)^{-1} \left(1_{(-\pi,\pi)^d} K_n\right)\left(\frac{t_1}{n_1},\ldots,\frac{t_d}{n_d}\right) \\
&= \begin{cases} \left(\prod_{j=1}^d n_j\right)^{-1} K_n\left(\frac{t_1}{n_1},\ldots,\frac{t_d}{n_d}\right), & \text{if } |t_1| \le \pi n_1, \ldots, |t_d| \le \pi n_d; \\ 0, & \text{else.} \end{cases}
\end{aligned}
$$

Theorem 4.2.7 *For all $n \in \mathbb{N}^d$,*

$$
c \left\| K_n \right\|_{E_q'(\mathbb{T}^d)} \le \left\| \widetilde{K}_n \right\|_{E_q'(\mathbb{R}^d)} \le C \left\| K_n \right\|_{E_q'(\mathbb{T}^d)} .
$$

Proof We have

$$
\begin{aligned}
\left\| \widetilde{K}_n \right\|_{E_q'(\mathbb{R}^d)} &= \left(\prod_{j=1}^d n_j\right)^{-1} \sum_{k_1=-\infty}^{\infty} \cdots \sum_{k_d=-\infty}^{\infty} \left(\prod_{j=1}^d 2^{k_j(1-1/q)}\right) \\
&\qquad \left(\int_{P_{k_1}} \cdots \int_{P_{k_d}} \left|\left(1_{(-\pi,\pi)^d} K_n\right)\left(\frac{t_1}{n_1},\ldots,\frac{t_d}{n_d}\right)\right|^q \, dt\right)^{1/q} \\
&= \left(\prod_{j=1}^d n_j\right)^{-1+1/q} \sum_{k_1=-\infty}^{\infty} \cdots \sum_{k_d=-\infty}^{\infty} \left(\prod_{j=1}^d 2^{k_j(1-1/q)}\right) \\
&\qquad \left(\int_{P_{k_1}(n_1)} \cdots \int_{P_{k_d}(n_d)} \left|\left(1_{(-\pi,\pi)^d} K_n\right)(t)\right|^q \, dt\right)^{1/q},
\end{aligned}
$$

where

$$
P_{k_j}(n_j) := \{x \in \mathbb{R} : 2^{k_j-1}\pi/n_j \le |x| < 2^{k_j}\pi/n_j\} \qquad (j = 1,\ldots,d).
$$

Choosing $l_j \in \mathbb{N}$ such that $2^{l_j-1} < n_j \le 2^{l_j}$, we conclude that

$$
P_{k_j}(n_j) \subset \{x \in \mathbb{R} : 2^{k_j-l_j-1}\pi \le |x| < 2^{k_j-l_j+1}\pi\} =: R_{k_j,l_j} \qquad (j = 1,\ldots,d)
$$

and

$$
\begin{aligned}
\left\| \widetilde{K}_n \right\|_{E_q'(\mathbb{R}^d)} &\le C \sum_{k_1=-\infty}^{\infty} \cdots \sum_{k_d=-\infty}^{\infty} \left(\prod_{j=1}^d 2^{(k_j-l_j)(1-1/q)}\right) \\
&\qquad \left(\int_{R_{k_1,l_1}} \cdots \int_{R_{k_d,l_d}} \left|\left(1_{(-\pi,\pi)^d} K_n\right)(t)\right|^q \, dt\right)^{1/q}
\end{aligned}
$$

$$
\begin{aligned}
&\leq C \sum_{i_1=-\infty}^{\infty} \cdots \sum_{i_d=-\infty}^{\infty} \left(\prod_{j=1}^{d} 2^{i_j(1-1/q)} \right) \\
&\quad \left(\int_{P_{i_1}} \cdots \int_{P_{i_d}} \left| \left(1_{(-\pi,\pi)^d} K_n \right)(t) \right|^q \, dt \right)^{1/q} \\
&\leq C \left\| 1_{(-\pi,\pi)^d} K_n \right\|_{E'_q(\mathbb{R}^d)} \\
&= C \left\| K_n \right\|_{E'_q(\mathbb{T}^d)} .
\end{aligned}
$$

The other inequality can be shown in the same way. ■

Now we formulate the analogue of Theorem 4.1.10.

Theorem 4.2.8 *If* $1 \leq p < \infty$, $1/p + 1/q = 1$ *and*

$$
\sup_{n \in \mathbb{N}^d} \left\| K_n \right\|_{E'_q(\mathbb{T}^d)} \leq C, \tag{4.2.1}
$$

then

$$
\sigma_* f(x) \leq C \left(\sup_{n \in \mathbb{N}^d} \left\| K_n \right\|_{E'_q(\mathbb{T}^d)} \right) M_{s,p} f(x)
$$

for all $f \in L_p(\mathbb{T}^d)$ *and* $x \in \mathbb{T}^d$.

Proof Observe that

$$
\begin{aligned}
&\left| \sigma_n f(x) \right| \\
&= \frac{1}{(2\pi)^d} \left| \int_{\mathbb{R}^d} f(x-t) \left(1_{(-\pi,\pi)^d} K_n \right)(t) \, dt \right| \\
&= \frac{1}{(2\pi)^d} \sum_{k_1=-\infty}^{\infty} \cdots \sum_{k_d=-\infty}^{\infty} \int_{P_{k_1}(n_1)} \cdots \int_{P_{k_d}(n_d)} \left| f(x-t) \right| \left| \left(1_{(-\pi,\pi)^d} K_n \right)(t) \right| \, dt .
\end{aligned}
$$

By Hölder's inequality,

$$
\begin{aligned}
&\left| \sigma_n f(x) \right| \\
&\leq \frac{1}{(2\pi)^d} \sum_{k_1=-\infty}^{\infty} \cdots \sum_{k_d=-\infty}^{\infty} \left(\int_{P_{k_1}(n_1)} \cdots \int_{P_{k_d}(n_d)} \left| f(x-t) \right|^p \, dt \right)^{1/p} \\
&\quad \left(\int_{P_{k_1}(n_1)} \cdots \int_{P_{k_d}(n_d)} \left| \left(1_{(-\pi,\pi)^d} K_n \right)(t) \right|^q \, dt \right)^{1/q} \\
&= \frac{1}{(2\pi)^d} \sum_{k_1=-\infty}^{\infty} \cdots \sum_{k_d=-\infty}^{\infty} \left(\int_{P_{k_1}(n_1)} \cdots \int_{P_{k_d}(n_d)} \left| f(x-t) \right|^p \, dt \right)^{1/p}
\end{aligned}
$$

$$\left(\prod_{j=1}^{d} n_j\right)^{-1/q} \left(\int_{P_{k_1}} \cdots \int_{P_{k_d}} \left| \left(1_{(-\pi,\pi)^d} K_n\right)\left(\frac{t_1}{n_1}, \ldots, \frac{t_d}{n_d}\right)\right|^q dt\right)^{1/q}.$$

If we define

$$G(u) := \left(\int_{-u_1}^{u_1} \ldots \int_{-u_d}^{u_d} |f(x-t)|^p \, dt\right)^{1/p} \qquad (u \in \mathbb{R}_+^d),$$

then

$$\frac{G^p(u)}{\prod_{j=1}^d (2u_j)} \leq M_{s,p}^p f(x) \qquad (u \in \mathbb{R}_+^d).$$

Thus

$$\begin{aligned} |\sigma_n f(x)| &\leq \frac{1}{(2\pi)^d} \sum_{k_1=-\infty}^{\infty} \cdots \sum_{k_d=-\infty}^{\infty} G\left(\frac{2^{k_1}\pi}{n_1}, \ldots, \frac{2^{k_d}\pi}{n_d}\right) \left(\prod_{j=1}^{d} n_j\right)^{-1/q} \\ &\quad \left(\int_{P_{k_1}} \cdots \int_{P_{k_d}} \left| \left(1_{(-\pi,\pi)^d} K_n\right)\left(\frac{t_1}{n_1}, \ldots, \frac{t_d}{n_d}\right)\right|^q dt\right)^{1/q} \qquad (4.2.2) \\ &\leq \frac{1}{(2\pi)^d} \sum_{k_1=-\infty}^{\infty} \cdots \sum_{k_d=-\infty}^{\infty} \left(\prod_{j=1}^{d} 2^{k_j/p}\right) \left(\prod_{j=1}^{d} n_j\right)^{-1} M_{s,p} f(x) \\ &\quad \left(\int_{P_{k_1}} \cdots \int_{P_{k_d}} \left| \left(1_{(-\pi,\pi)^d} K_n\right)\left(\frac{t_1}{n_1}, \ldots, \frac{t_d}{n_d}\right)\right|^q dt\right)^{1/q} \\ &= C \left\| \widetilde{K}_n \right\|_{E_q'(\mathbb{R}^d)} M_{s,p} f(x). \end{aligned}$$

The result follows from Theorem 4.2.7. ■

The following result comes from Corollary 4.2.5.

Corollary 4.2.9 *If* $1 \leq p < \infty$, $1/p + 1/q = 1$ *and*

$$\sup_{n \in \mathbb{N}^d} \|K_n\|_{E_q'(\mathbb{T}^d)} \leq C,$$

then

$$\sup_{\rho > 0} \rho \lambda(\sigma_* f > \rho)^{1/p} \leq C_p \left(\sup_{n \in \mathbb{N}^d} \|K_n\|_{E_q'(\mathbb{T}^d)}\right) \left(1 + \|f\|_{L_p(\log L)^{d-1}}\right)$$

for all $f \in L_p(\log L)^{d-1}(\mathbb{T}^d)$. *Moreover, for every* $p < r \leq \infty$,

$$\|\sigma_* f\|_r \le C \left(\sup_{n\in\mathbb{N}^d} \|K_n\|_{E'_q(\mathbb{T}^d)} \right) \|f\|_r \qquad (f \in L_r(\mathbb{T}^d)).$$

Corollary 4.2.10 *Suppose that $1 \le p < \infty$, $1/p + 1/q = 1$ and*

$$\sup_{n\in\mathbb{N}^d} \|K_n\|_{E'_q(\mathbb{T}^d)} \le C,$$

If

$$\lim_{n\to\infty} \widehat{K_n}(k) = 1$$

for all $k \in \mathbb{Z}^d$, then

$$\lim_{n\to\infty} \sigma_n f = f \quad a.e.$$

for all $f \in L_p(\log L)^{d-1}(\mathbb{T}^d)$.

Recall that $L_p(\log L)^k(\mathbb{T}^d) \supset L_r(\mathbb{T}^d)$ with $1 \le p < r \le \infty$. In this section, we study the rectangular θ-means,

$$\sigma_n^\theta f(x) := \sum_{k_1\in\mathbb{Z}} \cdots \sum_{k_d\in\mathbb{Z}} \theta\left(\frac{-k_1}{n_1}, \ldots, \frac{-k_d}{n_d}\right) \widehat{f}(k) e^{\imath k\cdot x},$$

where $\theta \in W(C, \ell_1)(\mathbb{R}^d)$.

Theorem 4.2.11 *Suppose that $1 \le p < \infty$ and $1/p + 1/q = 1$. If $\theta \in W(C, \ell_1)(\mathbb{R}^d)$ and $\widehat{\theta} \in E'_q(\mathbb{R}^d)$, then*

$$\sigma_*^\theta f(x) \le C \left\|\widehat{\theta}\right\|_{E'_q(\mathbb{R}^d)} M_{s,p} f(x)$$

for all $f \in L_p(\mathbb{T}^d)$ and $x \in \mathbb{T}^d$.

Proof Since $\widehat{\theta} \in L_1(\mathbb{R}^d)$ and, by Theorem 3.7.6,

$$\sigma_n^\theta f(x) = \left(\prod_{j=1}^d n_j\right) \int_{\mathbb{R}^d} f(x-t) \widehat{\theta}(n_1 t_1, \ldots, n_d t_d)\, dt,$$

we can repeat the proof of Theorem 4.2.8 step by step. ■

Corollary 4.2.12 *Suppose that $1 \le p < \infty$ and $1/p + 1/q = 1$. If $\theta \in W(C, \ell_1)(\mathbb{R}^d)$ and $\widehat{\theta} \in E'_q(\mathbb{R}^d)$, then*

$$\sup_{\rho>0} \rho \lambda(\sigma_*^\theta f > \rho)^{1/p} \le C_p \left\|\widehat{\theta}\right\|_{E'_q(\mathbb{R}^d)} \left(1 + \|f\|_{L_p(\log L)^{d-1}}\right)$$

for all $f \in L_p(\log L)^{d-1}(\mathbb{T}^d)$. Moreover, for every $p < r \le \infty$,

$$\left\|\sigma_*^\theta f\right\|_r \le C \left\|\widehat{\theta}\right\|_{E_q'(\mathbb{R}^d)} \|f\|_r \qquad (f \in L_r(\mathbb{T}^d)).$$

Corollary 4.2.13 *Suppose that* $1 \le p < \infty$ *and* $1/p + 1/q = 1$. *If* $\theta \in W(C, \ell_1)$ $(\mathbb{R}^d)$, $\theta(0) = 1$ *and* $\widehat{\theta} \in E_q'(\mathbb{R}^d)$, *then*

$$\lim_{n\to\infty} \sigma_n^\theta f = f \quad a.e.$$

for all $f \in L_p(\log L)^{d-1}(\mathbb{T}^d)$.

For the converse theorems, we need

Definition 4.2.14 For $1 \le p < \infty$, we define the space $D_p'(\mathbb{T}^d)$ with the norm

$$\|f\|_{D_p'(\mathbb{T}^d)} := \sup_{r\in(0,\pi)^d} \left(\frac{1}{\prod_{j=1}^d r_j} \int_{-r_1}^{r_1} \cdots \int_{-r_d}^{r_d} |f(t)|^p \, dt \right)^{1/p}.$$

The next two results can be proved as Lemma 4.1.17 and Theorem 4.1.18.

Lemma 4.2.15 *For* $1 \le p < \infty$, *the norm*

$$\|f\|_* = \sup_{k_1\le 0,\dots,k_d\le 0} \left(\prod_{j=1}^d 2^{k_j/p} \right) \left\| f 1_{P_k} \right\|_p$$

is an equivalent norm on $D_p'(\mathbb{T}^d)$.

The converse of Theorem 4.2.8 reads as follows.

Theorem 4.2.16 *If* $1 \le p < \infty$, $1/p + 1/q = 1$ *and*

$$\sigma_* f(0) \le C M_{s,p} f(0)$$

for all $f \in L_p(\mathbb{T}^d)$, *then*

$$\sup_{n\in\mathbb{N}^d} \|K_n\|_{E_q'(\mathbb{T}^d)} \le C.$$

We are going to study the second generalization of Lebesgue points for higher dimensions. By Corollary 4.2.4,

$$\lim_{h\to 0} \frac{1}{\prod_{j=1}^d (2h_j)} \int_{-h_1}^{h_1} \cdots \int_{-h_d}^{h_d} f(x-t)\, dt = f(x)$$

for almost every $x \in \mathbb{T}^d$, where $f \in L(\log L)^{d-1}(\mathbb{T}^d)$. This is equivalent to

$$\lim_{h\to 0} \frac{1}{\prod_{j=1}^d (2h_j)} \left| \int_{-h_1}^{h_1} \cdots \int_{-h_d}^{h_d} (f(x-t) - f(x))\, dt \right| = 0.$$

Definition 4.2.17 For $1 \le p < \infty$, a point $x \in \mathbb{T}^d$ is called a strong p-Lebesgue point of $f \in L_p(\mathbb{T}^d)$ if

$$\lim_{h\to 0} \left(\frac{1}{\prod_{j=1}^d (2h_j)} \int_{-h_1}^{h_1} \cdots \int_{-h_d}^{h_d} |f(x-t) - f(x)|^p \, dt \right)^{1/p} = 0.$$

For $p = 1$, the points are called strong Lebesgue points. If $p < r$, then all strong r-Lebesgue points are strong p-Lebesgue points. The next result can be proved as Theorem 4.1.20

Theorem 4.2.18 *Almost every point $x \in \mathbb{T}^d$ is a strong p-Lebesgue point of $f \in L_p(\log L)^{d-1}(\mathbb{T}^d)$ $(1 \le p < \infty)$.*

This is not true for $f \in L_p(\mathbb{T}^d)$. The reason for this is again that in the definition of the strong Lebesgue points the ratio of the sides of the rectangles can be large. To be able to obtain convergence at strong Lebesgue points, we have to modify slightly condition (4.2.1).

Theorem 4.2.19 *Suppose that $1 \le p < \infty$, $1/p + 1/q = 1$ and*

$$\sum_{k_1=-\infty}^{\infty} \cdots \sum_{k_d=-\infty}^{\infty} \left(\prod_{j=1}^d 2^{k_j(1-1/q)} \right) \sup_{n\in\mathbb{N}^d} \left\| \widetilde{K}_n 1_{Q_k} \right\|_q \le C. \qquad (4.2.3)$$

If

$$\lim_{n\to\infty} \widehat{K_n}(0) = 1,$$

$M_{s,p} f(x)$ is finite and x is a strong p-Lebesgue point of $f \in L_p(\log L)^{d-1}(\mathbb{T}^d)$, then

$$\lim_{n\to\infty} \sigma_n f(x) = f(x).$$

Proof Similarly to Theorem 4.2.8, let

$$G(u) := \left(\int_{-u_1}^{u_1} \cdots \int_{-u_d}^{u_d} |f(x-t) - f(x)|^p \, dt \right)^{1/p} \qquad (u \in \mathbb{R}_+^d).$$

Since x is a strong p-Lebesgue point of f, for all $\epsilon > 0$, we can find an integer $m \le 0$ such that

$$\frac{G^p(u)}{\prod_{j=1}^d (2u_j)} \le \epsilon \quad \text{if} \quad 0 < u_j \le 2^m \pi, \; j = 1, \ldots, d. \qquad (4.2.4)$$

Let $\{\pi_1, \ldots, \pi_d\}$ be a permutation of $\{1, \ldots, d\}$ and $1 \le j \le d$. Then

$$\begin{aligned}|\sigma_n f(x) - f(x)| &\leq \frac{1}{(2\pi)^d} \int_{\mathbb{R}^d} |f(x-t) - f(x)| \left| \left(1_{(-\pi,\pi)^d} K_n\right)(t) \right| dt \\ &\quad + \left| f(x) \left(\frac{1}{(2\pi)^d} \int_{\mathbb{T}^d} K_n(t)\, dt - 1 \right) \right| \\ &= A_1(x) + A_2(x) + A_3(x),\end{aligned}$$

where

$$\begin{aligned}A_1(x) := \frac{1}{(2\pi)^d} &\sum_{k_1=-\infty}^{m+\lfloor \log_2 n_1 \rfloor} \cdots \sum_{k_d=-\infty}^{m+\lfloor \log_2 n_d \rfloor} \\ &\int_{P_{k_1}(n_1)} \cdots \int_{P_{k_d}(n_d)} |f(x-t) - f(x)| \left| \left(1_{(-\pi,\pi)^d} K_n\right)(t) \right| dt\end{aligned}$$

and

$$\begin{aligned}A_2(x) := \frac{1}{(2\pi)^d} &\sum_{\pi_1,\ldots,\pi_d} \sum_{k_{\pi_1}=m+\lfloor \log_2 n_{\pi_1} \rfloor+1}^{\infty} \cdots \sum_{k_{\pi_j}=m+\lfloor \log_2 n_{\pi_j} \rfloor+1}^{\infty} \sum_{k_{\pi_{j+1}}=-\infty}^{\infty} \cdots \sum_{k_{\pi_d}=-\infty}^{\infty} \\ &\int_{P_{k_1}(n_1)} \cdots \int_{P_{k_d}(n_d)} |f(x-t) - f(x)| \left| \left(1_{(-\pi,\pi)^d} K_n\right)(t) \right| dt,\end{aligned}$$

and

$$A_3(x) := \left| f(x) \left(\frac{1}{(2\pi)^d} \int_{\mathbb{T}^d} K_n(t)\, dt - 1 \right) \right| = \left| f(x) \left(\widehat{K_n}(0) - 1 \right) \right|.$$

It is clear that

$$\lim_{n\to\infty} A_3(x) = 0.$$

As in (4.2.2),

$$\begin{aligned}&A_1(x) \\ &\leq C \sum_{k_1=-\infty}^{m+\lfloor \log_2 n_1 \rfloor} \cdots \sum_{k_d=-\infty}^{m+\lfloor \log_2 n_d \rfloor} \left(\int_{P_{k_1}(n_1)} \cdots \int_{P_{k_d}(n_d)} |f(x-t) - f(x)|^p \, dt \right)^{1/p} \\ &\quad \left(\prod_{j=1}^d n_j \right)^{-1/q} \left(\int_{P_{k_1}} \cdots \int_{P_{k_d}} \left| \left(1_{(-\pi,\pi)^d} K_n\right) \left(\frac{t_1}{n_1}, \ldots, \frac{t_d}{n_d} \right) \right|^q dt \right)^{1/q} \\ &\leq C \sum_{k_1=-\infty}^{m+\lfloor \log_2 n_1 \rfloor} \cdots \sum_{k_d=-\infty}^{m+\lfloor \log_2 n_d \rfloor} G\left(\frac{2^{k_1}\pi}{n_1}, \ldots, \frac{2^{k_d}\pi}{n_d} \right) \left(\prod_{j=1}^d n_j \right)^{-1/q}\end{aligned}$$

$$\left(\int_{P_{k_1}} \cdots \int_{P_{k_d}} \left| \left(1_{(-\pi,\pi)^d} K_n \right) \left(\frac{t_1}{n_1}, \ldots, \frac{t_d}{n_d} \right) \right|^q dt \right)^{1/q}.$$

Inequality (4.2.4) and $2^{k_j}/n_j \le 2^m n_j/n_j = 2^m$ imply

$$A_1(x) \le C_p \epsilon \sum_{k_1=-\infty}^{m+\lfloor \log_2 n_1 \rfloor} \cdots \sum_{k_d=-\infty}^{m+\lfloor \log_2 n_d \rfloor} \left(\prod_{j=1}^d 2^{k_j/p} \right) \left(\prod_{j=1}^d n_j \right)^{-1}$$

$$\left(\int_{P_{k_1}} \cdots \int_{P_{k_d}} \left| \left(1_{(-\pi,\pi)^d} K_n \right) \left(\frac{t_1}{n_1}, \ldots, \frac{t_d}{n_d} \right) \right|^q dt \right)^{1/q}$$

$$\le C_p \epsilon \left\| \widetilde{K}_n \right\|_{E_q'(\mathbb{R}^d)}.$$

Similarly,

$$A_2(x)$$

$$\le C \sum_{\pi_1,\ldots,\pi_d} \sum_{k_{\pi_1}=m+\lfloor \log_2 n_{\pi_1} \rfloor+1}^{\infty} \cdots \sum_{k_{\pi_j}=m+\lfloor \log_2 n_{\pi_j} \rfloor+1}^{\infty} \sum_{k_{\pi_{j+1}}=-\infty}^{\infty} \cdots \sum_{k_{\pi_d}=-\infty}^{\infty}$$

$$\left(\int_{P_{k_1}(n_1)} \cdots \int_{P_{k_d}(n_d)} |f(x-t) - f(x)|^p \, dt \right)^{1/p}$$

$$\left(\prod_{j=1}^d n_j \right)^{-1/q} \left(\int_{P_{k_1}} \cdots \int_{P_{k_d}} \left| \left(1_{(-\pi,\pi)^d} K_n \right) \left(\frac{t_1}{n_1}, \ldots, \frac{t_d}{n_d} \right) \right|^q dt \right)^{1/q}.$$

We supposed that $M_{s,p} f(x)$ is finite and x is a strong p-Lebesgue point of f, so we have

$$\left(\int_{P_{k_1}(n_1)} \cdots \int_{P_{k_d}(n_d)} |f(x-t) - f(x)|^p \, dt \right)^{1/p}$$

$$\le C_p \left(\prod_{j=1}^d \frac{2^{k_j}}{n_j} \right)^{1/p} \left(M_{s,p} f(x) + |f(x)| \right).$$

Consequently,

$$A_2(x)$$

$$\le C_p \sum_{\pi_1,\ldots,\pi_d} \sum_{k_{\pi_1}=m+\lfloor \log_2 n_{\pi_1} \rfloor+1}^{\infty} \cdots \sum_{k_{\pi_j}=m+\lfloor \log_2 n_{\pi_j} \rfloor+1}^{\infty} \sum_{k_{\pi_{j+1}}=-\infty}^{\infty} \cdots \sum_{k_{\pi_d}=-\infty}^{\infty}$$

$$\left(\prod_{j=1}^{d} 2^{k_j/p}\right)\left(M_{s,p}f(x)+|f(x)|\right)$$

$$\left(\int_{P_{k_1}}\cdots\int_{P_{k_d}}\left|\left(\prod_{j=1}^{d} n_j\right)^{-1}\left(1_{(-\pi,\pi)^d}K_n\right)\left(\frac{t_1}{n_1},\ldots,\frac{t_d}{n_d}\right)\right|^q dt\right)^{1/q}$$

$$\leq C_p \sum_{\pi_1,\ldots,\pi_d} \sum_{k_{\pi_1}=m+\lfloor\log_2 n_{\pi_1}\rfloor+1}^{\infty} \cdots \sum_{k_{\pi_j}=m+\lfloor\log_2 n_{\pi_j}\rfloor+1}^{\infty} \sum_{k_{\pi_{j+1}}=-\infty}^{\infty} \cdots \sum_{k_{\pi_d}=-\infty}^{\infty}$$

$$\left(\prod_{j=1}^{d} 2^{k_j/p}\right) \sup_{n\in\mathbb{N}^d}\left(\int_{P_{k_1}}\cdots\int_{P_{k_d}}|\widetilde{K}_n(t)|^q\,dt\right)^{1/q}\left(M_{s,p}f(x)+|f(x)|\right).$$

Therefore (4.2.3) and the fact $\lfloor\log_2 n_{\pi_j}\rfloor\to\infty$ as $T\to\infty$ imply that $A_2(x)\to 0$ as $n\to\infty$. ∎

Obviously, (4.2.3) implies

$$\sup_{n\in\mathbb{N}^d}\|\widetilde{K}_n\|_{E_q'(\mathbb{R}^d)}\leq C,$$

which is equivalent to

$$\sup_{n\in\mathbb{N}^d}\|K_n\|_{E_q'(\mathbb{T}^d)}\leq C,$$

by Theorem 4.2.7. If $\widetilde{K}_n$ can be estimated by a function $g\in E_q'(\mathbb{R}^d)$ which is independent of n, then (4.2.3) holds clearly. This is true for the Cesàro kernel $K_n^\alpha=K_{n_1}^\alpha\otimes\cdots\otimes K_{n_d}^\alpha$ and for the Riesz kernel $K_n^{\alpha,\gamma}=K_{n_1}^{\alpha,\gamma}\otimes\cdots\otimes K_{n_d}^{\alpha,\gamma}$. Indeed, for the one-dimensional Cesàro kernel functions

$$\begin{aligned}\frac{1}{n_j}\left|K_{n_j}^\alpha\left(\frac{t}{n_j}\right)\right| &\leq \frac{C}{n_j}\min\left\{n_j,\frac{n_j}{|t|^{\alpha+1}}\right\}\\ &= C\min\left\{1,|t|^{-\alpha-1}\right\}\in E_\infty(\mathbb{R})\end{aligned} \tag{4.2.5}$$

by Theorem 1.4.16. Similarly, by (3.3.12),

$$\begin{aligned}\frac{1}{n_j}\left|K_{n_j}^{\alpha,\gamma}\left(\frac{t}{n_j}\right)\right| &\leq \frac{C}{n_j}\min\left\{n_j,\frac{n_j}{|t|^{\min(\alpha,1)+1}}\right\}\\ &= C\min\left\{1,|t|^{-\min(\alpha,1)-1}\right\}\in E_\infty(\mathbb{R}).\end{aligned} \tag{4.2.6}$$

Hence, by (4.2.7), $K_n^\alpha, K_n^{\alpha,\gamma}\in E_\infty'(\mathbb{R}^d)$.

Corollary 4.2.20 *If $0<\alpha\leq 1$, $M_{s,p}f(x)$ is finite and x is a strong Lebesgue point of $f\in L_1(\log L)^{d-1}(\mathbb{T}^d)$, then*

$$\lim_{n\to\infty} \sigma_n^\alpha f(x) = f(x).$$

Moreover,

$$\sup_{\rho>0} \rho\lambda(\sigma_*^\alpha > \rho) \le C\left(1 + \|f\|_{L_1(\log L)^{d-1}}\right)$$

for all $f \in L_1(\log L)^{d-1}(\mathbb{T}^d)$ *and, for every* $1 < p \le \infty$,

$$\left\|\sigma_*^\alpha f\right\|_p \le C_p \|f\|_p \qquad (f \in L_p(\mathbb{T}^d)).$$

The same hold for the Riesz summation $\sigma_n^{\alpha,\gamma}$ *if* $0 < \alpha < \infty$ *and* $\gamma \in \mathbb{P}$.

Considering different parameters α_j in the jth coordinate, we obtain the same results. The next result can be proved in the same way as Theorem 4.2.19.

Theorem 4.2.21 *Suppose that* $1 \le p < \infty$, $1/p + 1/q = 1$, $\theta \in W(C, \ell_1)(\mathbb{R}^d)$ *and* $\widehat{\theta} \in E_q'(\mathbb{R}^d)$. *If* $\theta(0) = 1$, $M_{s,p} f(x)$ *is finite and* x *is a strong* p*-Lebesgue point of* $f \in L_p(\log L)^{d-1}(\mathbb{T}^d)$, *then*

$$\lim_{n\to\infty} \sigma_n^\theta f(x) = f(x).$$

Corollary 4.2.22 *If the conditions of Theorem 4.2.19 or Theorem 4.2.21 are satisfied and if* $f \in L_1(\log L)^{d-1}(\mathbb{T}^d)$ *is continuous at a point* x, *then*

$$\lim_{n\to\infty} \sigma_n^\theta f(x) = f(x).$$

Now we show the partial converse of Theorem 4.2.19.

Theorem 4.2.23 *Suppose that* $1 \le p < \infty$ *and* $1/p + 1/q = 1$. *If*

$$\lim_{n\to\infty} \sigma_n f(x) = f(x)$$

for all strong p*-Lebesgue points of* $f \in L_p(\mathbb{T}^d)$ *then*

$$\sup_{n\in\mathbb{N}^d} \|K_n\|_{E_q'(\mathbb{T}^d)} \le C.$$

Proof We define $D_p^{'0}(\mathbb{T}^d)$ as the set of all functions $f \in D_p'(\mathbb{T}^d)$ for which $f(0) = 0$ and 0 is a strong p-Lebesgue point of f, i.e.,

$$\lim_{h\to 0}\left(\frac{1}{\prod_{j=1}^d (2h_j)} \int_{-h_1}^{h_1} \cdots \int_{-h_d}^{h_d} |f(t)|^p \, dt\right)^{1/p} = 0.$$

Then we can show that $D_p^{'0}(\mathbb{T}^d)$ is a Banach space and the proof can be finished as in Theorems 4.1.25 and 4.2.16. ■

In the next subsection, we give some further examples for the θ-summation satisfying the above conditions.

4.2.3 Some Applications

Now we suppose that

$$K_n = K_{n_1}^{(1)} \otimes \cdots \otimes K_{n_d}^{(d)} \qquad (n \in \mathbb{N}^d)$$

and

$$\theta = \theta_1 \otimes \cdots \otimes \theta_d.$$

For these functions, we have

$$\|K_n\|_{E_q'(\mathbb{T}^d)} = \prod_{j=1}^{d} \left\| K_{n_j}^{(j)} \right\|_{E_q(\mathbb{T})} \qquad (n \in \mathbb{N}^d) \tag{4.2.7}$$

and a similar formula holds for $\|\theta\|_{E_q'(\mathbb{R}^d)}$. Hence for these functions, it is enough to consider the one-dimensional Herz spaces $E_q(\mathbb{X})$ ($\mathbb{X} = \mathbb{T}, \mathbb{R}$). As we have seen in Corollary 4.2.20, the rectangular Cesàro and Riesz summation satisfy the conditions of the preceding subsection.

Now, we present some sufficient condition on θ such that $\widehat{\theta} \in E_\infty(\mathbb{R})$. The next theorem was proved in Herz [166], Peetre [254] and Girardi and Weis [130].

Lemma 4.2.24 *If $\theta \in B_{1,1}^1(\mathbb{R})$, then $\widehat{\theta} \in E_\infty(\mathbb{R})$ and*

$$\left\|\widehat{\theta}\right\|_{E_\infty} \le C_p \|\theta\|_{B_{1,1}^1}.$$

A function f belongs to the weighted Wiener amalgam space $W(L_\infty, \ell_1^{v_s})(\mathbb{R})$ if

$$\|f\|_{W(L_\infty, \ell_1^{v_s})} := \sum_{k=-\infty}^{\infty} \sup_{x \in [0,1)} |f(x+k)| v_s(k) < \infty,$$

where $v_s(x) := (1+|x|)^s$ $(x \in \mathbb{R})$.

Lemma 4.2.25 *If $\theta \in W(L_\infty, \ell_1^{v_1})(\mathbb{R})$, then $\theta \in E_\infty(\mathbb{R})$ and*

$$\|\theta\|_{E_\infty} \le C \|\theta\|_{W(L_\infty, \ell_1^{v_1})}.$$

Proof The inequalities

$$\begin{aligned}\|\theta\|_{E_\infty} &= \sum_{k=-\infty}^{\infty} 2^k \sup_{P_k} |\theta| \\ &\leq 2 \sup_{(-\pi,\pi)} |\theta| + C \sum_{k=0}^{\infty} 2^k \sum_{j:(-\pi,\pi)+2j\pi \cap P_k \neq \emptyset} \sup_{(-\pi,\pi)+2j\pi} |\theta| \\ &\leq C \sum_{j=-\infty}^{\infty} (1+|j|) \sup_{(-\pi,\pi)^d+2j\pi} |\theta| \\ &= C\|\theta\|_{W(L_\infty,\ell_1^{v_1})}\end{aligned}$$

prove the result. ∎

We generalize Feichtinger's algebra and introduce its weighted version.

Definition 4.2.26 Let $g_0(x) := e^{-\pi\|x\|_2^2}$ be the Gauss function. We define the weighted Feichtinger's algebra or modulation space $M_1^{v_s}(\mathbb{R}^d)$ $(s \geq 0)$ by

$$M_1^{v_s}(\mathbb{R}^d) := \left\{ f \in L^2(\mathbb{R}^d) : \|f\|_{M_1^{v_s}} := \left\| S_{g_0} f \cdot v_s \right\|_{L_1(\mathbb{R}^{2d})} < \infty \right\},$$

where $v_s(x,\omega) := v_s(\omega) = (1+|\omega|)^s$ $(x,\omega \in \mathbb{R}^d)$.

Any other non-zero Schwartz function defines the same space and an equivalent norm (see, e.g., Feichtinger [100] and Gröchenig [152]).

Lemma 4.2.27 *If $\theta \in M_1^{v_1}(\mathbb{R})$, then $\widehat{\theta} \in E_\infty(\mathbb{R})$ and*

$$\left\|\widehat{\theta}\right\|_{E_\infty} \leq \|\theta\|_{M_1^{v_1}}.$$

Proof By Lemma 4.2.25,

$$\left\|\widehat{\theta}\right\|_{E_\infty} \leq C \left\|\widehat{\theta}\right\|_{W(L_\infty,\ell_1^{v_1})} \leq C \|f\|_{M_1^{v_1}},$$

where the second inequality can be found in Gröchenig [152, p. 249]. ∎

Corollary 4.2.28 *Suppose that $\theta = \theta_1 \otimes \cdots \otimes \theta_d \in W(C,\ell_1)(\mathbb{R}^d)$ and $\theta(0) = 1$. If $\theta_j \in M_1^{v_1}(\mathbb{R})$ for all $j = 1, \ldots, d$, $M_s f(x)$ is finite and x is a strong Lebesgue point of $f \in L_1(\log L)^{d-1}(\mathbb{T}^d)$, then*

$$\lim_{n\to\infty} \sigma_n^\theta f(x) = f(x).$$

Moreover,

$$\sup_{\rho>0} \rho\lambda(\sigma_*^\theta > \rho) \leq C \left(\prod_{j=1}^{d} \|\theta_j\|_{M_1^{v_1}} \right) \left(1 + \|f\|_{L_1(\log L)^{d-1}}\right)$$

for all $f \in L_1(\log L)^{d-1}(\mathbb{T}^d)$. *Also, for every* $1 < p \le \infty$,

$$\left\|\sigma_*^\theta f\right\|_p \le C_p \left(\prod_{j=1}^d \|\theta_j\|_{M_1^{v_1}}\right) \|f\|_p \qquad (f \in L_p(\mathbb{R}^d)).$$

In the next theorem, we give a sufficient result for θ to be in $M_1^{v_s}(\mathbb{R})$.

Theorem 4.2.29 *If* $\theta \in V_1^k(\mathbb{R})$ *for some* $k \ge 2$, *then* $\theta \in M_1^{v_s}(\mathbb{R})$ *for all* $0 \le s < k-1$ *and*

$$\|\theta\|_{M_1^{v_s}} \le C_s \|\theta\|_{V_1^k}.$$

This theorem can be proved as was Theorem 3.7.14. Note that $V_1^k(\mathbb{R}^d)$ was defined in Definition 3.7.13.

The space $V_1^2(\mathbb{R})$ is not contained in $M_1^{v_1}(\mathbb{R})$. However, the same results hold as in Corollary 4.2.28.

Theorem 4.2.30 *If* $\theta \in V_1^2(\mathbb{R})$, *then* $\widehat{\theta} \in E_\infty(\mathbb{R})$.

Proof The inequality

$$\left|\widehat{\theta}(x)\right| \le \frac{C}{x^2} \qquad (x \ne 0) \tag{4.2.8}$$

can be shown similarly to Theorem 3.7.14. $\widehat{\theta} \in E_\infty(\mathbb{R})$ follows from Theorem 4.1.9. ∎

Corollary 4.2.31 *Suppose that* $\theta = \theta_1 \otimes \cdots \otimes \theta_d \in W(C, \ell_1)(\mathbb{R}^d)$ *and* $\theta(0) = 1$. *If* $\theta_j \in V_1^2(\mathbb{R})$ *for all* $j = 1, \ldots, d$, *then the results of Corollary 4.2.28 holds.*

Note that for all examples of Sect. 2.6.3, we have $\theta \in V_1^2(\mathbb{R})$ or (4.2.8). This means that all results of Sect. 4.2.2 hold if each θ_j denotes either the Cesàro summation or one of the examples of Sect. 2.6.3.

4.3 Restricted Rectangular Summability over a Cone

Let again

$$\sigma_n f(x) = \frac{1}{(2\pi)^d} \int_{\mathbb{T}^d} f(x-t) K_n(t)\, dt \qquad (n \in \mathbb{N}^d),$$

where $f \in L_1(\mathbb{T}^d)$ and $K_n \in L_1(\mathbb{T}^d) \cap L_\infty(\mathbb{T}^d)$. Here we suppose that $n \in \mathbb{N}^d$ is in the cone $\mathbb{R}_\tau^d$. Recall that $\mathbb{R}_\tau^d$ is defined by

$$\mathbb{R}_\tau^d := \{x \in \mathbb{R}_+^d : \tau^{-1} \le x_i/x_j \le \tau, i, j = 1, \ldots, d\},$$

where $\tau \ge 1$ is fixed. The higher dimensional rectangular Cesàro and Riesz kernels, K_n^α and $K_n^{\alpha,\gamma}$ satisfy the conditions of this section.

4.3.1 Hardy-Littlewood Maximal Functions

It would be a straightforward idea that for the restricted rectangular summability, we use the restricted Hardy-Littlewood maximal function $M_{\square,p}$ defined in Theorem 4.1.4. However, this would be not useful because the restricted maximal function is equivalent to the usual maximal function $M_p f$ (see (4.1.1)). So we have to introduce a third generalization of the maximal function.

Definition 4.3.1 For $\omega > 0$, $1 \leq p < \infty$ and $f \in L_p(\mathbb{T}^d)$, the Hardy-Littlewood maximal function $\mathcal{M}_p^{\omega,1} f$ is given by

$$\mathcal{M}_p^{\omega,1} f(x) := \sup_{i \in \mathbb{N}^d, h>0} 2^{-\omega\|i\|_1} \left(\frac{1}{(2h)^d 2^{\|i\|_1}} \int_{-2^{i_1}h}^{2^{i_1}h} \cdots \int_{-2^{i_d}h}^{2^{i_d}h} |f(x-t)|^p \, dt \right)^{1/p}.$$

For $p = 1$, we write simply $\mathcal{M}^{\omega,1} f$. If $\omega = 0$, we get back the definition of the strong Hardy-Littlewood maximal function $M_{s,p} f$. In contrary to the strong maximal function, due to the weight $2^{-\omega\|i\|_1}$, the weak type (p, p) inequality will be true for $\mathcal{M}_p^{\omega}$. It is clear that

$$\mathcal{M}_p^{\omega_1,1} f \leq \mathcal{M}_p^{\omega_2,1} f \qquad \text{for } \omega_1 > \omega_2 > 0 \text{ and } 1 \leq p < \infty.$$

Let us point out the definition in the two-dimensional case. We have

$$\mathcal{M}_p^{\omega,1} f(x_1, x_2) = \sup_{i_1, i_2 \in \mathbb{N}, h>0} 2^{-\omega(i_1+i_2)} \Bigg(\frac{1}{4 \cdot 2^{i_1+i_2} h^2} \int_{-2^{i_1}h}^{2^{i_1}h} \int_{-2^{i_2}h}^{2^{i_2}h} |f(x_1 - t_1, x_2 - t_2)|^p \, dt \Bigg)^{1/p}.$$

To prove inequalities for $\mathcal{M}_p^{\omega,1} f$, we need another generalization of the maximal function $M_p f$. Let $\mu(h)$ and $\nu(h)$ be two continuous functions of $h \geq 0$, strictly increasing to ∞ and 0 at $h = 0$. Let

$$M_p^{1,\mu,\nu} f(x_1, x_2) := \sup_{h>0} \left(\frac{1}{4\mu(h)\nu(h)} \int_{-\mu(h)}^{\mu(h)} \int_{-\nu(h)}^{\nu(h)} |f(x_1 - t_1, x_2 - t_2)|^p \, dt \right)^{1/p},$$

where $f \in L_p(\mathbb{T}^2)$. If $\mu(h) = \nu(h) = h$, then we get back the usual Hardy-Littlewood maximal function $M_p f$ investigated in Sect. 4.1.1. The next result can be proved in the same way as Theorem 4.1.2.

Theorem 4.3.2 *If* $1 \leq p < \infty$, *then*

$$\sup_{\rho>0} \rho \lambda(M_p^{1,\mu,\nu} f > \rho)^{1/p} \leq C_p \|f\|_p \qquad (f \in L_p(\mathbb{T}^2)).$$

Moreover, if $p < r \leq \infty$, then

$$\left\| M_p^{1,\mu,\nu} f \right\|_r \leq C_r \, \|f\|_r \qquad (f \in L_r(\mathbb{T}^2)),$$

where the constants C_p and C_r are independent of μ and ν.

Using this theorem, we can prove the inequalities for $\mathcal{M}_p^{\omega,1}$.

Theorem 4.3.3 *If $\omega > 0$ and $1 \leq p < \infty$, then*

$$\sup_{\rho>0} \rho\lambda(\mathcal{M}_p^{\omega,1} f > \rho)^{1/p} \leq C\|f\|_p \qquad (f \in L_p(\mathbb{T}^d)).$$

Moreover, if $p < r \leq \infty$, then

$$\left\| \mathcal{M}_p^{\omega,1} f \right\|_r \leq C_r \|f\|_r \qquad (f \in L_r(\mathbb{T}^d)).$$

Proof Applying Theorem 4.3.2 to $\mu(h) = 2^{i_1} h$ and $\nu(h) = 2^{i_2} h$, we obtain

$$\begin{aligned}
\rho^p \lambda(\mathcal{M}_p^{\omega,1} f > \rho) &\leq \rho\lambda\Bigg(\bigcup_{i_1,i_2=0}^{\infty} 2^{-\omega(i_1+i_2)} \sup_{h>0} \Bigg(\frac{1}{4 \cdot 2^{i_1+i_2} h^2} \\
&\qquad \int_{-2^{i_1} h}^{2^{i_1} h} \int_{-2^{i_2} h}^{2^{i_2} h} |f(x_1 - t_1, x_2 - t_2)|^p \, dt \Bigg)^{1/p} > \rho \Bigg) \\
&\leq \rho^p \sum_{i_1=0}^{\infty} \sum_{i_2=0}^{\infty} \lambda\left(M_p^{1,\mu,\nu} f > 2^{\omega(i_1+i_2)} \rho \right) \\
&\leq C_p \sum_{i_1=0}^{\infty} \sum_{i_2=0}^{\infty} 2^{-\omega p(i_1+i_2)} \|f\|_p^p \\
&\leq C_p \|f\|_p^p
\end{aligned}$$

for all $f \in L_1(\mathbb{T}^2)$ and $\rho > 0$. The inequality

$$\left\| \mathcal{M}^{\omega,1} f \right\|_p \leq C_p \|f\|_p \qquad (f \in L_p(\mathbb{T}^2), 1 < p \leq \infty)$$

can be shown similarly. ■

4.3.2 Lebesgue Points for the Summability over a Cone

We briefly write $L_p^{\omega}(\mathbb{R}^d)$ ($\omega \geq 0$) instead of the weighted Lebesgue space $L_p^{\omega}(\mathbb{R}^d, \lambda)$ equipped with the norm

$$\|f\|_{L_p^\omega} := \left(\int_{\mathbb{R}^d} |f(x)(1+|x|)^\omega|^p \, dx \right)^{1/p} \qquad (1 \le p < \infty),$$

with the usual modification for $p = \infty$. If $\omega = 0$, then we get back the $L_p(\mathbb{R}^d)$ spaces. Clearly, $L_p(\mathbb{R}^d) \supset L_p^\omega(\mathbb{R}^d)$.

In this subsection, we introduce a new type of Herz spaces, the so-called weighted inhomogeneous Herz spaces.

Definition 4.3.4 For $\omega \ge 0$ and $1 \le q \le \infty$, the weighted Herz space $E_q^\omega(\mathbb{R}^d)$ contains all functions f for which

$$\|f\|_{E_q^\omega} := \sum_{k_1=0}^{\infty} \cdots \sum_{k_d=0}^{\infty} \left(\prod_{j=1}^{d} 2^{k_j(\omega+1-1/q)} \right) \left\| f 1_{Q_k} \right\|_q < \infty,$$

where

$$Q_k := Q_{k_1} \times \cdots \times Q_{k_d} \qquad (k \in \mathbb{N}^d)$$

and

$$Q_i = \{x \in \mathbb{R} : 2^{i-1}\pi \le |x| < 2^i \pi\} \qquad (i \in \mathbb{N}_+), \qquad Q_0 := (-\pi, \pi).$$

It is clear that

$$E_q^\omega(\mathbb{R}^d) \supset E_q^{\omega'}(\mathbb{R}^d) \qquad 0 \le \omega < \omega' < \infty$$

and

$$L_1(\mathbb{R}^d) \supset L_1^\omega(\mathbb{R}^d) = E_1^\omega(\mathbb{R}^d) \supset E_q^\omega(\mathbb{R}^d) \supset E_{q'}^\omega(\mathbb{R}^d) \supset E_\infty^\omega(\mathbb{R}^d)$$

for any $1 < q < q' < \infty$ with continuous embeddings. Moreover,

$$E_q^\omega(\mathbb{R}^d) \subset L_q^\omega(\mathbb{R}^d) \qquad \text{and} \qquad \|f\|_{L_q^\omega(\mathbb{R}^d)} \le C_q \, \|f\|_{E_q^\omega(\mathbb{R}^d)} .$$

Indeed,

$$\int_{\mathbb{R}^d} |f(t)|^q \, dt \le \sum_{k_1=0}^{\infty} \cdots \sum_{k_d=0}^{\infty} \left(\prod_{j=1}^{d} 2^{k_j(\omega+1-1/q)q} \right) \int_{Q_k} |f(t)|^q \, dt,$$

which implies the inequality. The connection between $E_q^0(\mathbb{R}^d)$ and $E_q'(\mathbb{R}^d)$ is the following. First of all,

$$E_q^0(\mathbb{R}^d) \subset E_q'(\mathbb{R}^d) \qquad \text{and} \qquad \|f\|_{E_q'(\mathbb{R}^d)} \le C_q \, \|f\|_{E_q^0(\mathbb{R}^d)} .$$

We prove this for one dimension, only:

$$\sum_{k=-\infty}^{0} 2^{k(1-1/q)} \left\| f 1_{P_k} \right\|_q \le \sum_{k=-\infty}^{0} 2^{k(1-1/q)} \left\| f 1_{(-\pi,\pi)} \right\|_q \le C_q \| f \|_{E_q^0(\mathbb{R})} .$$

We get the inequality for higher dimensions similarly. Summarizing these results, we can see that

$$E_q^0(\mathbb{R}^d) = L_q(\mathbb{R}^d) \cap E_q'(\mathbb{R}^d) \quad \text{and} \quad \| f \|_{E_q^0(\mathbb{R}^d)} \sim \| f \|_q + \| f \|_{E_q'(\mathbb{R}^d)}$$

with equivalent norms. Usually, $\widehat{\theta} \in L_1(\mathbb{R}^d) \cap C_0(\mathbb{R}^d)$, thus $\widehat{\theta} \in E_q'(\mathbb{R}^d)$ if and only if $\widehat{\theta} \in E_q^0(\mathbb{R}^d)$, but $\|\widehat{\theta}\|_{E_q'(\mathbb{R}^d)} \le C_q \|\widehat{\theta}\|_{E_q^0(\mathbb{R}^d)}$, $(1 \le q \le \infty)$.

We show that $f \in E_\infty^\omega(\mathbb{R})$ if and only if f has a decreasing majorant function belonging to $L_1^\omega(\mathbb{R})$.

Theorem 4.3.5 *Let $\omega \ge 0$ and $\eta(x) := \sup_{|t|\ge|x|} |f(t)|$. Then $f \in E_\infty^\omega(\mathbb{R})$ if and only if $\eta \in L_1^\omega(\mathbb{R})$ and*

$$C^{-1} \|\eta\|_{L_1^\omega} \le \| f \|_{E_\infty^\omega} \le C \|\eta\|_{L_1^\omega} + \eta(0).$$

Proof If $\eta \in L_1^\omega(\mathbb{R})$, then

$$\begin{aligned} \| f \|_{E_\infty^\omega} &\le \sum_{k=0}^{\infty} 2^{k(\omega+1)} \left\| \eta 1_{Q_k} \right\|_\infty \\ &= \sum_{k=1}^{\infty} 2^{k(\omega+1)} \eta(2^{k-1}\pi) + \eta(0) \le C \|\eta\|_{L_1^\omega} + \eta(0). \end{aligned}$$

For the converse, we use the function ν introduced in the proof of Theorem 4.1.9 to obtain

$$\begin{aligned} \|\eta\|_{L_1^\omega} \le \|\nu\|_{L_1^\omega} &= \sum_{k=0}^{\infty} a_{n_k} \int_{B(0,2^{n_k})\setminus B(0,2^{n_{k-1}})} (1+x)^\omega \, dx \\ &= C \sum_{k=0}^{\infty} 2^{n_k \omega} \left(2^{n_k} - 2^{n_{k-1}}\right) a_{n_k} \le C \| f \|_{E_\infty^\omega} , \end{aligned}$$

which proves the theorem. ∎

In this section, we investigate the restricted maximal operator

$$\sigma_\square f := \sup_{n \in \mathbb{R}_\tau^d} |\sigma_n f| ,$$

where $\tau \ge 1$ is fixed. Recall that

$$\widetilde{K}_n(t) := \left(\prod_{j=1}^d n_j\right)^{-1} \left(1_{(-\pi,\pi)^d} K_n\right)\left(\frac{t_1}{n_1}, \ldots, \frac{t_d}{n_d}\right).$$

Theorem 4.3.6 *If $\omega \geq 0$, $1 \leq p < \infty$, $1/p + 1/q = 1$ and*

$$\sup_{n \in \mathbb{R}_\tau^d} \|\widetilde{K}_n\|_{E_q^\omega(\mathbb{R}^d)} \leq C,$$

then

$$\sigma_\Box f(x) \leq C \left(\sup_{n \in \mathbb{R}_\tau^d} \|\widetilde{K}_n\|_{E_q^\omega(\mathbb{R}^d)}\right) \mathcal{M}_p^{\omega,1} f(x)$$

for all $f \in L_p(\mathbb{T}^d)$ and $x \in \mathbb{T}^d$.

Proof We have

$$\begin{aligned}
&|\sigma_n f(x)| \\
&= \frac{1}{(2\pi)^d} \left| \int_{\mathbb{R}^d} f(x-t) \left(1_{(-\pi,\pi)^d} K_n\right)(t)\, dt \right| \\
&\leq \frac{1}{(2\pi)^d} \sum_{k_1=0}^\infty \cdots \sum_{k_d=0}^\infty \int_{Q_{k_1}(n_1)} \cdots \int_{Q_{k_d}(n_d)} |f(x-t)| \left|\left(1_{(-\pi,\pi)^d} K_n\right)(t)\right| dt,
\end{aligned}$$

where

$$Q_i(n_j) := \{x \in \mathbb{R} : 2^{i-1}\pi/n_j \leq |x| < 2^i \pi/n_j\} \qquad (i \in \mathbb{N}_+)$$

and

$$Q_0(n_j) := (-\pi/n_j, \pi/n_j).$$

By Hölder's inequality,

$$\begin{aligned}
&|\sigma_n f(x)| \\
&\leq \frac{1}{(2\pi)^d} \sum_{k_1=0}^\infty \cdots \sum_{k_d=0}^\infty \left(\int_{Q_{k_1}(n_1)} \cdots \int_{Q_{k_d}(n_d)} |f(x-t)|^p\, dt\right)^{1/p} \\
&\quad \left(\int_{Q_{k_1}(n_1)} \cdots \int_{Q_{k_d}(n_d)} \left|\left(1_{(-\pi,\pi)^d} K_n\right)(t)\right|^q\, dt\right)^{1/q} \\
&= \frac{1}{(2\pi)^d} \sum_{k_1=0}^\infty \cdots \sum_{k_d=0}^\infty \left(\int_{Q_{k_1}(n_1)} \cdots \int_{Q_{k_d}(n_d)} |f(x-t)|^p\, dt\right)^{1/p}
\end{aligned}$$

$$\left(\prod_{j=1}^{d} n_j\right)^{-1/q} \left(\int_{Q_{k_1}} \cdots \int_{Q_{k_d}} \left| \left(1_{(-\pi,\pi)^d} K_n\right) \left(\frac{t_1}{n_1}, \ldots, \frac{t_d}{n_d}\right)\right|^q dt\right)^{1/q}. \quad (4.3.1)$$

Choose $s \in \mathbb{N}$ such that $2^{s-1} < \tau \leq 2^s$. Since $n \in \mathbb{R}^d_\tau$, we conclude

$$\begin{aligned}
&|\sigma_n f(x)| \\
&\leq \frac{1}{(2\pi)^d} \sum_{k_1=0}^{\infty} \cdots \sum_{k_d=0}^{\infty} \left(\int_{-2^{k_1+s}\pi/n_1}^{2^{k_1+s}\pi/n_1} \cdots \int_{-2^{k_d+s}\pi/n_1}^{2^{k_d+s}\pi/n_1} |f(x-t)|^p \, dt\right)^{1/p} \\
&\quad \left(\prod_{j=1}^{d} n_j\right)^{-1/q} \left(\int_{Q_{k_1}} \cdots \int_{Q_{k_d}} \left| \left(1_{(-\pi,\pi)^d} K_n\right) \left(\frac{t_1}{n_1}, \ldots, \frac{t_d}{n_d}\right)\right|^q dt\right)^{1/q}. \quad (4.3.2)
\end{aligned}$$

Let again

$$G(u) := \left(\int_{-u_1}^{u_1} \cdots \int_{-u_d}^{u_d} |f(x-t)|^p \, dt\right)^{1/p} \qquad (u \in \mathbb{R}^d_+).$$

Then

$$2^{-\omega(k_1+\ldots+k_d)p} n_1^d \frac{G^p(2^{k_1+s}\pi/n_1, \ldots, 2^{k_d+s}\pi/n_1)}{(2\pi)^d 2^{sd} 2^{k_1+\ldots+k_d}} \leq (\mathcal{M}_p^{\omega,1})^p f(x)$$

and so

$$\begin{aligned}
&|\sigma_n f(x)| \\
&\leq C \sum_{k_1=0}^{\infty} \cdots \sum_{k_d=0}^{\infty} G\left(\frac{2^{k_1+s}\pi}{n_1}, \ldots, \frac{2^{k_d+s}\pi}{n_1}\right) \\
&\quad \left(\prod_{j=1}^{d} n_j\right)^{-1/q} \left(\int_{Q_{k_1}} \cdots \int_{Q_{k_d}} \left| \left(1_{(-\pi,\pi)^d} K_n\right) \left(\frac{t_1}{n_1}, \ldots, \frac{t_d}{n_d}\right)\right|^q dt\right)^{1/q} \\
&\leq C \sum_{k_1=0}^{\infty} \cdots \sum_{k_d=0}^{\infty} \left(\prod_{j=1}^{d} 2^{k_j(\omega+1/p)}\right) n_1^{-d/p} \mathcal{M}_p^{\omega,1} f(x) \\
&\quad \left(\prod_{j=1}^{d} n_j\right)^{-1/q} \left(\int_{Q_{k_1}} \cdots \int_{Q_{k_d}} \left| \left(1_{(-\pi,\pi)^d} K_n\right) \left(\frac{t_1}{n_1}, \ldots, \frac{t_d}{n_d}\right)\right|^q dt\right)^{1/q}.
\end{aligned}$$

The fact $n \in \mathbb{R}^d_\tau$ implies

$$|\sigma_n f(x)| \leq C \sum_{k_1=0}^{\infty} \cdots \sum_{k_d=0}^{\infty} \left(\prod_{j=1}^{d} 2^{k_j(\omega+1/p)} \right) \mathcal{M}_p^{\omega,1} f(x) \left(\prod_{j=1}^{d} n_j \right)^{-1}$$

$$\left(\int_{Q_{k_1}} \cdots \int_{Q_{k_d}} \left| \left(1_{(-\pi,\pi)^d} K_n \right) \left(\frac{t_1}{n_1}, \ldots, \frac{t_d}{n_d} \right) \right|^q dt \right)^{1/q}$$

$$= C \left\| \widetilde{K}_n \right\|_{E_q^{\omega}(\mathbb{R}^d)} \mathcal{M}_p^{\omega,1} f(x),$$

which shows the theorem. ∎

Taking into account Theorem 4.3.3, we have

Theorem 4.3.7 *If $\omega > 0$, $1 \leq p < \infty$, $1/p + 1/q = 1$ and*

$$\sup_{n \in \mathbb{R}_\tau^d} \left\| \widetilde{K}_n \right\|_{E_q^{\omega}(\mathbb{R}^d)} \leq C,$$

then

$$\sup_{\rho>0} \rho \lambda(\sigma_\square f > \rho)^{1/p} \leq C_p \left(\sup_{n \in \mathbb{R}_\tau^d} \left\| \widetilde{K}_n \right\|_{E_q^{\omega}(\mathbb{R}^d)} \right) \|f\|_p$$

for all $f \in L_p(\mathbb{T}^d)$. Moreover, for every $p < r \leq \infty$,

$$\|\sigma_\square f\|_r \leq C \left(\sup_{n \in \mathbb{R}_\tau^d} \left\| \widetilde{K}_n \right\|_{E_q^{\omega}(\mathbb{R}^d)} \right) \|f\|_r \qquad (f \in L_r(\mathbb{T}^d)).$$

Corollary 4.3.8 *Suppose that $\omega > 0$, $1 \leq p < \infty$, $1/p + 1/q = 1$ and*

$$\sup_{n \in \mathbb{R}_\tau^d} \left\| \widetilde{K}_n \right\|_{E_q^{\omega}(\mathbb{R}^d)} \leq C.$$

If

$$\lim_{n \to \infty,\, n \in \mathbb{R}_\tau^d} \widehat{K_n}(k) = 1$$

for all $k \in \mathbb{Z}^d$, then

$$\lim_{n \to \infty,\, n \in \mathbb{R}_\tau^d} \sigma_n f = f \quad a.e.$$

for all $f \in L_p(\mathbb{T}^d)$.

For the rectangular θ-means

$$\sigma_n^{\theta} f(x) := \sum_{k_1 \in \mathbb{Z}} \cdots \sum_{k_d \in \mathbb{Z}} \theta \left(\frac{-k_1}{n_1}, \ldots, \frac{-k_d}{n_d} \right) \widehat{f}(k) e^{\imath k \cdot x},$$

we obtain

Theorem 4.3.9 *Suppose that $\omega \geq 0$, $1 \leq p < \infty$ and $1/p + 1/q = 1$. If $\theta \in W(C, \ell_1)(\mathbb{R}^d)$ and $\widehat{\theta} \in E_q^{\omega}(\mathbb{R}^d)$, then*

$$\sigma_{\square}^{\theta} f(x) \leq C \left\|\widehat{\theta}\right\|_{E_q^{\omega}(\mathbb{R}^d)} \mathcal{M}_p^{\omega,1} f(x)$$

for all $f \in L_p(\mathbb{T}^d)$ and $x \in \mathbb{T}^d$.

This inequality can be proved as Theorem 4.3.6 (see also Theorem 4.2.11).

Theorem 4.3.10 *Suppose that $\omega > 0$, $1 \leq p < \infty$ and $1/p + 1/q = 1$. If $\theta \in W(C, \ell_1)(\mathbb{R}^d)$ and $\widehat{\theta} \in E_q^{\omega}(\mathbb{R}^d)$, then*

$$\sup_{\rho > 0} \rho \lambda(\sigma_{\square}^{\theta} f > \rho)^{1/p} \leq C_p \left\|\widehat{\theta}\right\|_{E_q^{\omega}(\mathbb{R}^d)} \|f\|_p$$

for all $f \in L_p(\mathbb{T}^d)$. Moreover, for every $p < r \leq \infty$,

$$\left\|\sigma_{\square}^{\theta} f\right\|_r \leq C \left\|\widehat{\theta}\right\|_{E_q^{\omega}(\mathbb{R}^d)} \|f\|_r \qquad (f \in L_r(\mathbb{T}^d)).$$

Corollary 4.3.11 *Suppose that $\omega > 0$, $1 \leq p < \infty$ and $1/p + 1/q = 1$. If $\theta(0) = 1$, $\theta \in W(C, \ell_1)(\mathbb{R}^d)$ and $\widehat{\theta} \in E_q^{\omega}(\mathbb{R}^d)$, then*

$$\lim_{n \to \infty,\, n \in \mathbb{R}_\tau^d} \sigma_n^{\theta} f = f \quad a.e.$$

for all $f \in L_p(\mathbb{T}^d)$.

We introduce the third generalization of Lebesgue points as follows. Starting from the maximal function $\mathcal{M}_p^{\omega,1} f$, we introduce

$$U_{r,p}^{\omega,1} f(x) := \sup_{i \in \mathbb{N}^d, h > 0, 2^{i_k} h < r, k=1,\ldots d} 2^{-\omega \|i\|_1} \left(\frac{1}{(2h)^d 2^{\|i\|_1}} \int_{-2^{i_1} h}^{2^{i_1} h} \cdots \int_{-2^{i_d} h}^{2^{i_d} h} |f(x-t) - f(x)|^p \, dt \right)^{1/p}. \tag{4.3.3}$$

In case $p = 1$, we omit the notation p and write simply $U_r^{\omega,1} f$. In the two-dimensional case this definition reads as

$$U_{r,p}^{\omega,1} f(x_1, x_2) = \sup_{i_1, i_2 \in \mathbb{N}, h > 0, 2^{i_k} h < r, k=1,2} 2^{-\omega(i_1+i_2)} \left(\frac{1}{4 \cdot 2^{i_1+i_2} h^2} \int_{-2^{i_1} h}^{2^{i_1} h} \int_{-2^{i_2} h}^{2^{i_2} h} |f(x_1 - t_1, x_2 - t_2) - f(x_1, x_2)|^p \, dt \right)^{1/p}.$$

Note that the definitions of the p-Lebesgue points and strong p-Lebesgue points (see Definitions 4.1.19 and 4.2.17) can be rewritten as

$$\lim_{r\to 0}\sup_{0<h<r}\left(\frac{1}{(2h)^d}\int_{-h}^{h}\cdots\int_{-h}^{h}|f(x-t)-f(x)|^p\,dt\right)^{1/p}=0$$

and

$$\lim_{r\to 0}\sup_{0<h_j<r,j=1,\ldots,d}\left(\frac{1}{\prod_{j=1}^{d}(2h_j)}\int_{-h_1}^{h_1}\cdots\int_{-h_d}^{h_d}|f(x-t)-f(x)|^p\,dt\right)^{1/p}=0,$$

respectively. Similarly to this definition, we introduce a new type of Lebesgue points.

Definition 4.3.12 For $1\le p<\infty$ and $\omega>0$, a point $x\in\mathbb{T}^d$ is called a (p,ω)-Lebesgue point of $f\in L_p(\mathbb{T}^d)$ if

$$\lim_{r\to 0}U_{r,p}^{\omega,1}f(x)=0.$$

If $\omega=0$, then the $(p,0)$-Lebesgue points are the same as the strong p-Lebesgue points. It is easy to see that every (p,ω_2)-Lebesgue point is a (p,ω_1)-Lebesgue point if $\omega_1>\omega_2>0$. Moreover, if $p<r$, then every (r,ω)-Lebesgue point is a (p,ω)-Lebesgue point. If f is continuous at x, then x is a (p,ω)-Lebesgue point of f for all $1\le p<\infty$ and $\omega>0$.

Theorem 4.3.13 *For $1\le p<\infty$ and $\omega>0$, almost every point $x\in\mathbb{T}^d$ is a (p,ω)-Lebesgue point of $f\in L_p(\mathbb{T}^d)$.*

Proof If f is a continuous function, then x is obviously a (p,ω)-Lebesgue point. By Theorem 4.3.3,

$$\begin{aligned}\rho^p\lambda\left(\sup_{r>0}U_{r,p}^{\omega,1}f>\rho\right)&\le\rho^p\lambda(\mathcal{M}_p^{\omega,1}f>\rho/2)+2\rho^p\lambda(|f|>\rho/2)\\&\le C\left\|f\right\|_p^p.\end{aligned}$$

Since the result holds for continuous functions and the continuous functions are dense in $L_p(\mathbb{T}^d)$, the theorem follows from the usual density argument of Theorem 1.3.7. ■

Theorem 4.3.14 *Suppose that $\omega>0$, $1\le p<\infty$, $1/p+1/q=1$ and*

$$\sum_{k_1=0}^{\infty}\cdots\sum_{k_d=0}^{\infty}\left(\prod_{j=1}^{d}2^{k_j(\omega+1-1/q)}\right)\sup_{n\in\mathbb{R}^d_{\kappa,\tau}}\left\|\widetilde{K}_n 1_{Q_k}\right\|_q\le C.\tag{4.3.4}$$

If

$$\lim_{n\to\infty,\, n\in\mathbb{R}_\tau^d} \widehat{K_n}(0) = 1,$$

$\mathcal{M}_p^{\omega,1} f(x)$ is finite and x is a (p,ω)-Lebesgue points of $f \in L_p(\mathbb{T}^d)$, then

$$\lim_{n\to\infty,\, n\in\mathbb{R}_\tau^d} \sigma_n f(x) = f(x).$$

Proof Choose again $s \in \mathbb{N}$ such that $2^{s-1} < \tau \le 2^s$. Since x is a (p,ω)-Lebesgue point of f, we can fix a number $r < 1$ such that

$$U_{r2^s\pi,p}^{\omega,1} f(x) < \epsilon,$$

where $U_{r,p}^{\omega,1} f$ was introduced in (4.3.3). Let us denote by r_0 the largest number i, for which $r/2 \le 2^i/n_1 < r$. We have

$$\begin{aligned} |\sigma_n f(x) - f(x)| &\le \frac{1}{(2\pi)^d} \int_{\mathbb{R}^d} |f(x-t) - f(x)| \left|\left(1_{(-\pi,\pi)^d} K_n\right)(t)\right| dt \\ &\quad + \left| f(x) \left(\frac{1}{(2\pi)^d} \int_{\mathbb{T}^d} K_n(t)\, dt - 1 \right) \right| \\ &= A_1(x) + A_2(x) + A_3(x), \end{aligned}$$

where

$$\begin{aligned} A_1(x) := \frac{1}{(2\pi)^d} &\sum_{k_1=0}^{r_0} \cdots \sum_{k_d=0}^{r_0} \\ &\int_{Q_{k_1}(n_1)} \cdots \int_{Q_{k_d}(n_d)} |f(x-t) - f(x)| \left|\left(1_{(-\pi,\pi)^d} K_n\right)(t)\right| dt, \end{aligned}$$

$$\begin{aligned} A_2(x) := \frac{1}{(2\pi)^d} &\sum_{\pi_1,\ldots,\pi_d} \sum_{k_{\pi_1}=r_0+1}^{\infty} \cdots \sum_{k_{\pi_j}=r_0+1}^{\infty} \sum_{k_{\pi_{j+1}}=0}^{\infty} \cdots \sum_{k_{\pi_d}=0}^{\infty} \\ &\int_{Q_{k_1}(n_1)} \cdots \int_{Q_{k_d}(n_d)} |f(x-t) - f(x)| \left|\left(1_{(-\pi,\pi)^d} K_n\right)(t)\right| dt, \end{aligned}$$

and

$$A_3(x) := \left| f(x) \left(\frac{1}{(2\pi)^d} \int_{\mathbb{T}^d} K_n(t)\, dt - 1 \right) \right| = \left| f(x) \left(\widehat{K_n}(0) - 1 \right) \right|.$$

Here $\{\pi_1, \ldots, \pi_d\}$ denotes a permutation of $\{1, \ldots, d\}$ and $1 \le j \le d$. Obviously,

$$\lim_{n\to\infty,\, n\in\mathbb{R}_\tau^d} A_3(x) = 0.$$

Similarly to (4.3.1) and (4.3.2), we deduce

$$\begin{aligned}
&A_1(x) \\
&\le \frac{1}{(2\pi)^d} \sum_{k_1=0}^{r_0} \cdots \sum_{k_d=0}^{r_0} \left(\int_{Q_{k_1}(n_1)} \cdots \int_{Q_{k_d}(n_d)} |f(x-t) - f(x)|^p \, dt \right)^{1/p} \\
&\quad \left(\prod_{j=1}^d n_j \right)^{-1/q} \left(\int_{Q_{k_1}} \cdots \int_{Q_{k_d}} \left| \left(1_{(-\pi,\pi)^d} K_n\right) \left(\frac{t_1}{n_1}, \ldots, \frac{t_d}{n_d} \right) \right|^q \, dt \right)^{1/q} \\
&\le \frac{1}{(2\pi)^d} \sum_{k_1=0}^{r_0} \cdots \sum_{k_d=0}^{r_0} \left(\int_{-2^{k_1+s}\pi/n_1}^{2^{k_1+s}\pi/n_1} \cdots \int_{-2^{k_d+s}\pi/n_1}^{2^{k_d+s}\pi/n_1} |f(x-t) - f(x)|^p \, dt \right)^{1/p} \\
&\quad \left(\prod_{j=1}^d n_j \right)^{-1/q} \left(\int_{Q_{k_1}} \cdots \int_{Q_{k_d}} \left| \left(1_{(-\pi,\pi)^d} K_n\right) \left(\frac{t_1}{n_1}, \ldots, \frac{t_d}{n_d} \right) \right|^q \, dt \right)^{1/q}. \qquad (4.3.5)
\end{aligned}$$

Setting

$$G(u) := \left(\int_{-u_1}^{u_1} \cdots \int_{-u_d}^{u_d} |f(x-t) - f(x)|^p \, dt \right)^{1/p} \qquad (u \in \mathbb{R}_+^d),$$

we conclude

$$2^{-\omega(k_1+\ldots+k_d)} n_1^{d/p} \frac{G(2^{k_1+s}\pi/n_1, \ldots, 2^{k_d+s}\pi/n_1)}{\left((2\pi)^d 2^{sd} 2^{k_1+\ldots+k_d}\right)^{1/p}} \le U_{r2^s\pi,p}^{\omega,1} f(x) \qquad (4.3.6)$$

because $n \in \mathbb{R}_\tau^d$ and so

$$\frac{2^{k_j+s}}{n_1} \le \frac{2^{r_0+s}}{n_1} < r2^s \qquad (j = 1, \ldots, d).$$

Hence

$$\begin{aligned}
&A_1(x) \\
&\le \frac{1}{(2\pi)^d} \sum_{k_1=0}^{r_0} \cdots \sum_{k_d=0}^{r_0} G(2^{k_1+s}\pi/n_1, \ldots, 2^{k_d+s}\pi/n_1) \\
&\quad \left(\prod_{j=1}^d n_j \right)^{-1/q} \left(\int_{Q_{k_1}} \cdots \int_{Q_{k_d}} \left| \left(1_{(-\pi,\pi)^d} K_n\right) \left(\frac{t_1}{n_1}, \ldots, \frac{t_d}{n_d} \right) \right|^q \, dt \right)^{1/q}
\end{aligned}$$

$$\leq C \sum_{k_1=0}^{r_0} \cdots \sum_{k_d=0}^{r_0} \left(\prod_{j=1}^{d} 2^{k_j(\omega+1/p)} \right) n_1^{-d/p} U_{r2^s\pi,p}^{\omega,1} f(x)$$
$$\left(\prod_{j=1}^{d} n_j \right)^{-1/q} \left(\int_{Q_{k_1}} \cdots \int_{Q_{k_d}} \left| \left(1_{(-\pi,\pi)^d} K_n \right) \left(\frac{t_1}{n_1}, \ldots, \frac{t_d}{n_d} \right) \right|^q dt \right)^{1/q}.$$

From this it follows immediately that

$$A_1(x) \leq C \sum_{k_1=0}^{r_0} \cdots \sum_{k_d=0}^{r_0} \left(\prod_{j=1}^{d} 2^{k_j(\omega+1/p)} \right) U_{r2^s\pi,p}^{\omega,1} f(x)$$
$$\left(\prod_{j=1}^{d} n_j \right)^{-1} \left(\int_{Q_{k_1}} \cdots \int_{Q_{k_d}} \left| \left(1_{(-\pi,\pi)^d} K_n \right) \left(\frac{t_1}{n_1}, \ldots, \frac{t_d}{n_d} \right) \right|^q dt \right)^{1/q}$$
$$\leq C \left\| \widetilde{K}_n \right\|_{E_q^\omega(\mathbb{R}^d)} \epsilon.$$

Similarly to (4.3.5) and (4.3.6), we can see that

$$A_2(x)$$
$$\leq \frac{1}{(2\pi)^d} \sum_{\pi_1,\ldots,\pi_d} \sum_{k_{\pi_1}=r_0+1}^{\infty} \cdots \sum_{k_{\pi_j}=r_0+1}^{\infty} \sum_{k_{\pi_{j+1}}=0}^{\infty} \cdots \sum_{k_{\pi_d}=0}^{\infty}$$
$$\left(\int_{-2^{k_1+s}\pi/n_1}^{2^{k_1+s}\pi/n_1} \cdots \int_{-2^{k_d+s}\pi/n_1}^{2^{k_d+s}\pi/n_1} |f(x-t) - f(x)|^p \, dt \right)^{1/p}$$
$$\left(\prod_{j=1}^{d} n_j \right)^{-1/q} \left(\int_{Q_{k_1}} \cdots \int_{Q_{k_d}} \left| \left(1_{(-\pi,\pi)^d} K_n \right) \left(\frac{t_1}{n_1}, \ldots, \frac{t_d}{n_d} \right) \right|^q dt \right)^{1/q}$$

and

$$2^{-\omega(k_1+\ldots+k_d)} n_1^{d/p} \frac{G(2^{k_1+s}\pi/n_1, \ldots, 2^{k_d+s}\pi/n_1)}{\left((2\pi)^d 2^{sd} 2^{k_1+\ldots+k_d}\right)^{1/p}} \leq \mathcal{M}_p^{\omega,1} f(x) + |f(x)|.$$

Since $\mathcal{M}_p^{\omega,1} f(x)$ is finite, we have

$$A_2(x)$$
$$\leq C \sum_{\pi_1,\ldots,\pi_d} \sum_{k_{\pi_1}=r_0+1}^{\infty} \cdots \sum_{k_{\pi_j}=r_0+1}^{\infty} \sum_{k_{\pi_{j+1}}=0}^{\infty} \cdots \sum_{k_{\pi_d}=0}^{\infty}$$

$$
\left(\prod_{j=1}^{d} 2^{k_j(\omega+1/p)}\right) \left(\mathcal{M}_p^{\omega,1} f(x) + |f(x)|\right)
$$

$$
\left(\prod_{j=1}^{d} n_j\right)^{-1} \left(\int_{Q_{k_1}} \cdots \int_{Q_{k_d}} \left|\left(1_{(-\pi,\pi)^d} K_n\right)\left(\frac{t_1}{n_1}, \ldots, \frac{t_d}{n_d}\right)\right|^q dt\right)^{1/q}
$$

$$
\leq C \sum_{\pi_1,\ldots,\pi_d} \sum_{k_{\pi_1}=r_0+1}^{\infty} \cdots \sum_{k_{\pi_j}=r_0+1}^{\infty} \sum_{k_{\pi_{j+1}}=0}^{\infty} \cdots \sum_{k_{\pi_d}=0}^{\infty}
$$

$$
\left(\prod_{j=1}^{d} 2^{k_j(\omega+1/p)}\right) \left(\int_{Q_{k_1}} \cdots \int_{Q_{k_d}} |\widetilde{K}_n(t)|^q \, dt\right)^{1/q} \left(\mathcal{M}_p^{\omega,1} f(x) + |f(x)|\right).
$$

Since $r_0 \to \infty$ as $n_1 \to \infty$ and (4.3.4) holds, we conclude that

$$
\lim_{n\to\infty} A_2(x) = 0,
$$

which finishes the proof. ■

Note that (4.3.4) implies

$$
\sup_{n\in\mathbb{R}_\tau^d} \left\| \widetilde{K}_n \right\|_{E_q^\omega(\mathbb{R}^d)} \leq C.
$$

If $\widetilde{K}_n$ can be estimated by a function $g \in E_q^\omega(\mathbb{R}^d)$ which is independent of n, then (4.3.4) holds. This holds again for the Cesàro kernel $K_n^\alpha = K_{n_1}^\alpha \otimes \cdots \otimes K_{n_d}^\alpha$ and for the Riesz kernel $K_n^{\alpha,\gamma} = K_{n_1}^{\alpha,\gamma} \otimes \cdots \otimes K_{n_d}^{\alpha,\gamma}$. Indeed, taking into account (4.2.5) and (4.2.6), we can see that

$$
\min\left\{1, |t|^{-\alpha-1}\right\} \in E_\infty^\omega(\mathbb{R}^d)
$$

if $0 < \omega < \alpha \leq 1$ and

$$
\min\left\{1, |t|^{-\min(\alpha,1)-1}\right\} \in E_\infty^\omega(\mathbb{R}^d)
$$

if $0 < \omega < \min(\alpha, 1)$.

Corollary 4.3.15 *If $0 < \omega < \alpha \leq 1$, $\mathcal{M}^{\omega,1} f(x)$ is finite and x is a $(1, \omega)$-Lebesgue point of $f \in L_1(\mathbb{T}^d)$, then*

$$
\lim_{n\to\infty,\, n\in\mathbb{R}_\tau^d} \sigma_n^\alpha f(x) = f(x).
$$

Moreover,

$$
\sup_{\rho>0} \rho\lambda(\sigma_\square^\alpha > \rho) \leq C\|f\|_1 \qquad (f \in L_1(\mathbb{T}^d))
$$

and, for every $1 < p \le \infty$,

$$\left\|\sigma_{\square}^{\alpha} f\right\|_p \le C_p \|f\|_p \qquad (f \in L_p(\mathbb{T}^d)).$$

The same hold for the Riesz summation $\sigma_n^{\alpha,\gamma}$ *if* $0 < \omega < \min(\alpha, 1)$, $0 < \alpha < \infty$ *and* $\gamma \in \mathbb{P}$.

The proof of Theorem 4.3.14 shows also

Theorem 4.3.16 *Suppose that* $\omega > 0$, $1 \le p < \infty$, $1/p + 1/q = 1$, $\theta \in W(C, \ell_1)(\mathbb{R}^d)$ *and* $\widehat{\theta} \in E_q^{\omega}(\mathbb{R}^d)$. *If* $\theta(0) = 1$, $\mathcal{M}_p^{\omega,1} f(x)$ *is finite and* x *is a* (p, ω)*-Lebesgue point of* $f \in L_p(\mathbb{T}^d)$, *then*

$$\lim_{n\to\infty,\, n\in\mathbb{R}_\tau^d} \sigma_n^\theta f(x) = f(x).$$

Corollary 4.3.17 *If the conditions of Theorem 4.3.14 or Theorem 4.3.16 are satisfied and if* $f \in L_1(\mathbb{T}^d)$ *is continuous at a point* x, *then*

$$\lim_{n\to\infty,\, n\in\mathbb{R}_\tau^d} \sigma_n^\theta f(x) = f(x).$$

Taking into account Theorem 4.2.30, we obtain

Corollary 4.3.18 *Suppose that* $\theta = \theta_1 \otimes \cdots \otimes \theta_d \in W(C, \ell_1)(\mathbb{R}^d)$, $\theta(0) = 1$ *and* $\theta_j \in V_1^2(\mathbb{R})$ *for all* $j = 1, \ldots, d$. *If* $0 < \omega < 1$, $\mathcal{M}^{\omega,1} f(x)$ *is finite and* x *is a* $(1, \omega)$*-Lebesgue point of* $f \in L_1(\mathbb{T}^d)$, *then*

$$\lim_{n\to\infty,\, n\in\mathbb{R}_\tau^d} \sigma_n^\theta f(x) = f(x).$$

Moreover,

$$\sup_{\rho>0} \rho\lambda(\sigma_{\square}^{\theta} > \rho) \le C \left(\prod_{j=1}^{d} \left\|\theta_j\right\|_{E_\infty^\omega(\mathbb{R})}\right) \|f\|_1 \qquad (f \in L_1(\mathbb{T}^d))$$

and, for every $1 < p \le \infty$,

$$\left\|\sigma_{\square}^{\theta} f\right\|_p \le C_p \left(\prod_{j=1}^{d} \left\|\theta_j\right\|_{E_\infty^\omega(\mathbb{R})}\right) \|f\|_p \qquad (f \in L_p(\mathbb{T}^d)).$$

All examples of Sect. 2.6.3 satisfy the condition $\theta \in V_1^2(\mathbb{R})$. This means that all results of Sect. 4.3.2, especially Corollary 4.3.18 hold if each θ_j denotes either the Cesàro summation or one of the examples of Sect. 2.6.3.

4.4 Restricted Rectangular Summability over a Cone-Like Set

In this section, we investigate the operators

$$\sigma_n f(x) = \frac{1}{(2\pi)^d} \int_{\mathbb{T}^d} f(x-t) K_n(t)\, dt$$

over a cone-like set, i.e., we assume that $n \in \mathbb{R}^d_{\kappa,\tau}$. Recall that $\mathbb{R}^d_{\kappa,\tau}$ was defined in Sect. 3.4 by

$$\mathbb{R}^d_{\kappa,\tau} := \{x \in \mathbb{R}^d_+ : \tau_j^{-1} \kappa_j(n_1) \le n_j \le \tau_j \kappa_j(n_1),\ j = 2, \ldots, d\},$$

where κ_1 is the identity function and, for all $j = 2, \ldots, d$, $\kappa_j : \mathbb{R}_+ \to \mathbb{R}_+$ are strictly increasing and continuous functions such that

$$\lim_{j\to\infty} \kappa_j = \infty \qquad \text{and} \qquad \lim_{j\to+0} \kappa_j = 0.$$

Instead of (3.4.1), we will suppose that there exist $c_j, \xi > 1$ such that

$$\kappa_j(\xi x) = c_j \kappa_j(x) \qquad (x > 0). \tag{4.4.1}$$

The higher dimensional rectangular Cesàro and Riesz kernels, K_n^α and $K_n^{\alpha,\gamma}$ will satisfy again the conditions of this section.

4.4.1 *Hardy-Littlewood Maximal Functions*

We generalize the definition of $\mathcal{M}_p^{\omega,1} f$ as follows.

Definition 4.4.1 For $\omega > 0$, $1 \le p < \infty$ and $f \in L_p(\mathbb{T}^d)$, the Hardy-Littlewood maximal function $\mathcal{M}_p^{\kappa,\omega} f$ is given by

$$\mathcal{M}_p^{\kappa,\omega} f(x) := \sup_{i\in\mathbb{N}^d, h>0} \left(\prod_{j=1}^d \kappa_j(\xi^{i_j})^{-\omega}\right) \left(\frac{1}{\prod_{j=1}^d (2\kappa_j(\xi^{i_j} h)\pi)} \int_{-\kappa_1(\xi^{i_1} h)\pi}^{\kappa_1(\xi^{i_1} h)\pi} \cdots \int_{-\kappa_d(\xi^{i_d} h)\pi}^{\kappa_d(\xi^{i_d} h)\pi} |f(x-t)|^p\, dt\right)^{1/p}.$$

We show that the operator $\mathcal{M}_p^{\kappa,\omega}$ is of weak type (p, p) as was $\mathcal{M}_p^{\omega,1}$.

Theorem 4.4.2 *If* $1 \le p < \infty$, *then*

$$\sup_{\rho>0} \rho\lambda(\mathcal{M}_p^{\kappa,\omega} f > \rho)^{1/p} \leq C_p \|f\|_p \qquad (f \in L_p(\mathbb{T}^d)).$$

Moreover, if $p < r \leq \infty$, *then*

$$\left\|\mathcal{M}_p^{\kappa,\omega} f\right\|_r \leq C_r \|f\|_r \qquad (f \in L_r(\mathbb{R}^d)).$$

Proof Choosing

$$\mu(h) := \kappa_1(\xi^{i_1} h), \qquad \nu(h) := \kappa_2(\xi^{i_2} h)$$

in the definition of $M_p^{1,\mu,\nu} f$ (see Theorem 4.3.2), we get that

$$\begin{aligned}
&\rho^p \lambda(\mathcal{M}_p^{\kappa,\omega} f > \rho) \\
&\leq \rho^p \lambda\Bigg(\bigcup_{i_1,i_2=0}^{\infty} \left(\prod_{j=1}^{2} \kappa_j(\xi^{i_j})^{-\omega}\right) \sup_{h>0} \Bigg(\frac{1}{\prod_{j=1}^2 (2\kappa_j(\xi^{i_j} h))} \\
&\quad \int_{-\kappa_1(\xi^{i_1} h)}^{\kappa_1(\xi^{i_1} h)} \int_{-\kappa_2(\xi^{i_2} h)}^{\kappa_2(\xi^{i_2} h)} |f(x_1 - t_1, x_2 - t_2)|^p \, dt \Bigg)^{1/p} > \rho \Bigg) \\
&\leq \rho^p \sum_{i_1=0}^{\infty} \sum_{i_2=0}^{\infty} \lambda\left(M_p^{1,\mu,\nu} f > \left(\prod_{j=1}^{2} \kappa_j(\xi^{i_j})^{\omega}\right) \rho \right) \\
&\leq C_p \sum_{i\in\mathbb{N}^d} \left(\prod_{j=1}^{2} \kappa_j(\xi^{i_j})^{-\omega p}\right) \|f\|_p^p \\
&\leq C_p \sum_{i\in\mathbb{N}^d} \left(\prod_{j=1}^{2} c_j^{-\omega p i_j} \kappa_j(1)^{-\omega p}\right) \|f\|_p^p \leq C_p \|f\|_p^p
\end{aligned}$$

because $c_j > 1$ $(j = 1, 2)$. ■

4.4.2 Lebesgue Points for the Summability over a Cone-Like Set

Using the functions κ_j, we modify slightly the norm of the Herz spaces $E_q^{\omega}(\mathbb{R}^d)$.

Definition 4.4.3 For $\omega \geq 0$ and $1 \leq q \leq \infty$, the weighted Herz space $E_q^{\kappa,\omega}(\mathbb{R}^d)$ contains all functions f for which

$$\|f\|_{E_q^{\kappa,\omega}} := \sum_{k_1=0}^{\infty} \cdots \sum_{k_d=0}^{\infty} \left(\prod_{j=1}^{d} \kappa_j(\xi^{k_j})^{\omega+1-1/q}\right) \left\|f 1_{Q_k}\right\|_q < \infty,$$

where

$$Q_k := Q_{1,k_1} \times \cdots \times Q_{d,k_d} \qquad (k \in \mathbb{N}^d),$$

$$Q_{j,0} := Q_{j,0}^{\kappa_j} := \left(-\kappa_j(1)\pi, \kappa_j(1)\pi\right) \qquad (j = 1, \ldots, d)$$

and

$$Q_{j,i} := Q_{j,i}^{\kappa_j} := \{x \in \mathbb{R} : \kappa_j(\xi^{i-1})\pi \leq |x| < \kappa_j(\xi^i)\pi\} \qquad (i \in \mathbb{N}_+).$$

However, these spaces are equivalent to the Herz spaces $E_q^\omega(\mathbb{R}^d)$ studied in Sect. 4.3.2.

Theorem 4.4.4 *The spaces $E_q^\omega(\mathbb{R}^d)$ and $E_q^{\kappa,\omega}(\mathbb{R}^d)$ are equivalent for all $1 \leq q \leq \infty, \omega \geq 0$.*

Proof It is enough to show the result for one dimension. Then we denote the function κ_j and the corresponding constant c_j simply by κ and c, and the sets $Q_{j,k}^{\kappa_j}$ by Q_k^κ. For a fixed k, let ν be the smallest natural number l for which $\kappa(\xi^l) = c^l\kappa(1) \geq 2^k$ and μ be the largest natural number l for which $\kappa(\xi^l) \leq 2^{k-1}$. Then

$$2^{k(\omega+1-1/q)} \left\| f 1_{Q_k} \right\|_q \leq \sum_{j=\mu+1}^{\nu} \kappa(\xi^j)^{\omega+1-1/q} \left\| f 1_{Q_j^\kappa} \right\|_q,$$

which means that

$$\|f\|_{E_q^\omega} \leq C \|f\|_{E_q^{\kappa,\omega}}.$$

The other side of the equivalence can be shown in the same way. ∎

We will investigate a restricted maximal operator depending on the cone-like set:

$$\sigma_\kappa f := \sup_{n \in \mathbb{R}^d_{\kappa,\tau}} |\sigma_n f|.$$

Theorem 4.4.5 *If $\omega \geq 0$, $1 \leq p < \infty$, $1/p + 1/q = 1$ and*

$$\sup_{n \in \mathbb{R}^d_{\kappa,\tau}} \left\| \widetilde{K}_n \right\|_{E_q^\omega(\mathbb{R}^d)} \leq C,$$

then

$$\sigma_\kappa f(x) \leq C \left(\sup_{n \in \mathbb{R}^d_{\kappa,\tau}} \left\| \widetilde{K}_n \right\|_{E_q^\omega(\mathbb{R}^d)} \right) \mathcal{M}_p^{\kappa,\omega} f(x)$$

for all $f \in L_p(\mathbb{T}^d)$ and $x \in \mathbb{T}^d$.

Proof Obviously,

$$
\begin{aligned}
|\sigma_n f(x)| &= \frac{1}{(2\pi)^d} \left| \int_{\mathbb{R}^d} f(x-t) \left(1_{(-\pi,\pi)^d} K_n\right)(t)\, dt \right| \\
&\le \frac{1}{(2\pi)^d} \sum_{k_1=0}^{\infty} \cdots \sum_{k_d=0}^{\infty} \\
&\qquad \int_{Q_{1,k_1}(n_1)} \cdots \int_{Q_{d,k_d}(n_1)} |f(x-t)| \left| \left(1_{(-\pi,\pi)^d} K_n\right)(t) \right| dt,
\end{aligned}
$$

where

$$
Q_{j,0}(n_1) := \left(-\kappa_j(1/n_1)\pi, \kappa_j(1/n_1)\pi\right)
$$

and

$$
Q_{j,i}(n_1) := \{x \in \mathbb{R} : \kappa_j(\xi^{i-1}/n_1)\pi \le |x| < \kappa_j(\xi^i/n_1)\pi\} \qquad (i \in \mathbb{N}_+).
$$

By Hölder's inequality,

$$
\begin{aligned}
&|\sigma_n f(x)| \\
&\le \frac{1}{(2\pi)^d} \sum_{k_1=0}^{\infty} \cdots \sum_{k_d=0}^{\infty} \left(\int_{Q_{1,k_1}(n_1)} \cdots \int_{Q_{d,k_d}(n_1)} |f(x-t)|^p\, dt \right)^{1/p} \\
&\quad \left(\int_{Q_{1,k_1}(n_1)} \cdots \int_{Q_{d,k_d}(n_1)} \left| \left(1_{(-\pi,\pi)^d} K_n\right)(t) \right|^q dt \right)^{1/q} \\
&= \frac{1}{(2\pi)^d} \sum_{k_1=0}^{\infty} \cdots \sum_{k_d=0}^{\infty} \left(\int_{Q_{1,k_1}(n_1)} \cdots \int_{Q_{d,k_d}(n_1)} |f(x-t)|^p\, dt \right)^{1/p} \left(\prod_{j=1}^{d} n_j \right)^{-1/q} \\
&\quad \left(\int_{n_1 Q_{1,k_1}(n_1)} \cdots \int_{n_d Q_{d,k_d}(n_1)} \left| \left(1_{(-\pi,\pi)^d} K_n\right) \left(\frac{t_1}{n_1}, \ldots, \frac{t_d}{n_d} \right) \right|^q dt \right)^{1/q}. \qquad (4.4.2)
\end{aligned}
$$

For

$$
G(u) := \left(\int_{-u_1}^{u_1} \cdots \int_{-u_d}^{u_d} |f(x-t)|^p\, dt \right)^{1/p} \qquad (u \in \mathbb{R}_+^d),
$$

we have

$$
\left(\prod_{j=1}^{d} \kappa_j(\xi^{k_j})^{-\omega p} \right) \frac{G^p(\kappa_1(\xi^{k_1}/n_1)\pi, \ldots, \kappa_d(\xi^{k_d}/n_1)\pi)}{\prod_{j=1}^{d} (2\kappa_j(\xi^{k_j}/n_1)\pi)} \le (\mathcal{M}_p^{\kappa,\omega})^p f(x).
$$

Thus

$$
\begin{aligned}
&|\sigma_n f(x)| \\
&\leq C \sum_{k_1=0}^{\infty} \cdots \sum_{k_d=0}^{\infty} G(\kappa_1(\xi^{k_1}/n_1)\pi, \ldots, \kappa_d(\xi^{k_d}/n_1)\pi) \left(\prod_{j=1}^{d} n_j \right)^{-1/q} \\
&\quad \left(\int_{n_1 Q_{1,k_1}(n_1)} \cdots \int_{n_d Q_{d,k_d}(n_1)} \left| \left(1_{(-\pi,\pi)^d} K_n\right) \left(\frac{t_1}{n_1}, \ldots, \frac{t_d}{n_d} \right) \right|^q dt \right)^{1/q} \\
&\leq C \sum_{k_1=0}^{\infty} \cdots \sum_{k_d=0}^{\infty} \left(\prod_{j=1}^{d} \kappa_j(\xi^{k_j})^{\omega} \right) \left(\prod_{j=1}^{d} \kappa_j(\xi^{k_j}/n_1) \right)^{1/p} \mathcal{M}_p^{\kappa,\omega} f(x) \left(\prod_{j=1}^{d} n_j \right)^{-1/q} \\
&\quad \left(\int_{n_1 Q_{1,k_1}(n_1)} \cdots \int_{n_d Q_{d,k_d}(n_1)} \left| \left(1_{(-\pi,\pi)^d} K_n\right) \left(\frac{t_1}{n_1}, \ldots, \frac{t_d}{n_d} \right) \right|^q dt \right)^{1/q}.
\end{aligned}
$$

If $\xi^l \leq n_1 < \xi^{l+1}$ for some $l \in \mathbb{Z}$, then by (4.4.1),

$$
\kappa_j(1) c_j^l = \kappa_j(\xi^l) \leq \kappa_j(n_1) \leq \kappa_j(\xi^{l+1}) = \kappa_j(1) c_j^{l+1}
$$

and

$$
\kappa_j \left(\frac{\xi^{k_j}}{n_1} \right) \leq \kappa_j \left(\frac{\xi^{k_j}}{\xi^l} \right) = \frac{1}{c_j^l} \kappa_j(\xi^{k_j}) \leq \kappa_j(1) c_j \frac{\kappa_j(\xi^{k_j})}{\kappa_j(n_1)} = \frac{\kappa_j(1)\kappa_j(\xi^{k_j+1})}{\kappa_j(n_1)}.
$$

Choose integers μ and ν such that

$$
\tau_j \kappa_j(1) \leq c_j^{\mu} \qquad \text{and} \qquad \tau_j^{-1} \kappa_j(1) \geq c_j^{\nu}
$$

for all $j = 1, \ldots, d$. Using the definition of the cone-like set, we can see that

$$
\begin{aligned}
n_j \kappa_j \left(\frac{\xi^{k_j}}{n_1} \right) &\leq \tau_j \kappa_j(n_1) \kappa_j \left(\frac{\xi^{k_j}}{n_1} \right) \leq \tau_j \kappa_j(1) \kappa_j(\xi^{k_j+1}) \\
&\leq c_j^{\mu} \kappa_j(\xi^{k_j+1}) = \kappa_j(\xi^{k_j+\mu+1}).
\end{aligned}
$$

On the other hand

$$
\kappa_j \left(\frac{\xi^{k_j}}{n_1} \right) \geq \kappa_j \left(\frac{\xi^{k_j}}{\xi^{l+1}} \right) = \frac{1}{c_j^{l+1}} \kappa_j(\xi^{k_j}) \geq \frac{\kappa_j(1)}{c_j} \frac{\kappa_j(\xi^{k_j})}{\kappa_j(n_1)} = \frac{\kappa_j(1)\kappa_j(\xi^{k_j-1})}{\kappa_j(n_1)}
$$

and

$$\begin{aligned} n_j \kappa_j \left(\frac{\xi^{k_j}}{n_1} \right) &\geq \tau_j^{-1} \kappa_j(n_1) \kappa_j \left(\frac{\xi^{k_j}}{n_1} \right) \geq \tau_j^{-1} \kappa_j(1) \kappa_j(\xi^{k_j - 1}) \\ &\geq c_j^\nu \kappa_j(\xi^{k_j-1}) = \kappa_j(\xi^{k_j+\nu-1}). \end{aligned}$$

Setting

$$Q'_{j,0} := \left(-\kappa_j(\xi^{\mu+1})\pi, \kappa_j(\xi^{\mu+1})\pi\right)$$

and

$$Q'_{j,i} := \{x \in \mathbb{R} : \kappa_j(\xi^{i+\nu-2})\pi \leq |x| < \kappa_j(\xi^{i+\mu+1})\pi\} \qquad (i \in \mathbb{N}_+),$$

we conclude

$$\begin{aligned} |\sigma_n f(x)| &\leq C \sum_{k_1=0}^{\infty} \cdots \sum_{k_d=0}^{\infty} \left(\prod_{j=1}^d \kappa_j(\xi^{k_j})^{\omega+1/p} \right) \\ &\quad \left(\prod_{j=1}^d \kappa_j(n_1) \right)^{-1/p} \mathcal{M}_p^{\kappa,\omega} f(x) \left(\prod_{j=1}^d n_j \right)^{-1/q} \\ &\quad \left(\int_{Q'_{1,k_1}} \cdots \int_{Q'_{d,k_d}} \left| \left(1_{(-\pi,\pi)^d} K_n \right) \left(\frac{t_1}{n_1}, \ldots, \frac{t_d}{n_d} \right) \right|^q dt \right)^{1/q} \\ &\leq C \sum_{k_1=0}^{\infty} \cdots \sum_{k_d=0}^{\infty} \left(\prod_{j=1}^d \kappa_j(\xi^{k_j})^{\omega+1/p} \right) \mathcal{M}_p^{\kappa,\omega} f(x) \\ &\quad \left(\int_{Q'_{1,k_1}} \cdots \int_{Q'_{d,k_d}} |\widetilde{K}_n(t)|^q \, dt \right)^{1/q} \\ &\leq C \left\| \widetilde{K}_n \right\|_{E_q^\omega(\mathbb{R}^d)} \mathcal{M}_p^{\kappa,\omega} f(x), \end{aligned}$$

which proves the theorem. ■

Theorem 4.4.2 implies

Theorem 4.4.6 *If* $\omega \geq 0$, $1 \leq p < \infty$, $1/p + 1/q = 1$ *and*

$$\sup_{n \in \mathbb{R}^d_{\kappa,\tau}} \left\| \widetilde{K}_n \right\|_{E_q^\omega(\mathbb{R}^d)} \leq C,$$

then

$$\sup_{\rho > 0} \rho \lambda(\sigma_\kappa f > \rho)^{1/p} \leq C_p \left(\sup_{n \in \mathbb{R}^d_{\kappa,\tau}} \left\| \widetilde{K}_n \right\|_{E_q^\omega(\mathbb{R}^d)} \right) \|f\|_p$$

for all $f \in L_p(\mathbb{T}^d)$. *Moreover, for every* $p < r \leq \infty$,

$$\|\sigma_\kappa f\|_r \le C \left(\sup_{n\in\mathbb{R}^d_{\kappa,\tau}} \left\|\widetilde{K}_n\right\|_{E_q^\omega(\mathbb{R}^d)} \right) \|f\|_r \qquad (f \in L_r(\mathbb{T}^d)).$$

Corollary 4.4.7 *Suppose that* $\omega > 0$, $1 \le p < \infty$, $1/p + 1/q = 1$ *and*

$$\sup_{n\in\mathbb{R}^d_{\kappa,\tau}} \left\|\widetilde{K}_n\right\|_{E_q^\omega(\mathbb{R}^d)} \le C.$$

If

$$\lim_{n\to\infty,\, n\in\mathbb{R}^d_{\kappa,\tau}} \widehat{K_n}(k) = 1$$

for all $k \in \mathbb{Z}^d$, *then*

$$\lim_{n\to\infty,\, \mathbb{R}^d_{\kappa,\tau}} \sigma_n f = f \quad a.e.$$

for all $f \in L_p(\mathbb{T}^d)$.

For rectangular θ-means, we obtain the next theorems in the same way.

Theorem 4.4.8 *Suppose that* $\omega \ge 0$, $1 \le p < \infty$ *and* $1/p + 1/q = 1$. *If* $\theta \in W(C, \ell_1)(\mathbb{R}^d)$ *and* $\widehat{\theta} \in E_q^\omega(\mathbb{R}^d)$, *then*

$$\sigma_\kappa^\theta f(x) \le C \left\|\widehat{\theta}\right\|_{E_q^\omega(\mathbb{R}^d)} \mathcal{M}_p^{\kappa,\omega} f(x)$$

for all $f \in L_p(\mathbb{T}^d)$ *and* $x \in \mathbb{T}^d$.

Theorem 4.4.9 *Suppose that* $\omega > 0$, $1 \le p < \infty$ *and* $1/p + 1/q = 1$. *If* $\theta \in W(C, \ell_1)(\mathbb{R}^d)$ *and* $\widehat{\theta} \in E_q^\omega(\mathbb{R}^d)$, *then*

$$\sup_{\rho>0} \rho\lambda(\sigma_\kappa^\theta f > \rho)^{1/p} \le C_p \left\|\widehat{\theta}\right\|_{E_q^\omega(\mathbb{R}^d)} \|f\|_p$$

for all $f \in L_p(\mathbb{T}^d)$. *Moreover, for every* $p < r \le \infty$,

$$\left\|\sigma_\kappa^\theta f\right\|_r \le C \left\|\widehat{\theta}\right\|_{E_q^\omega(\mathbb{R}^d)} \|f\|_r \qquad (f \in L_r(\mathbb{T}^d)).$$

Corollary 4.4.10 *Suppose that* $\omega > 0$, $1 \le p < \infty$ *and* $1/p + 1/q = 1$. *If* $\theta(0) = 1$, $\theta \in W(C, \ell_1)(\mathbb{R}^d)$ *and* $\widehat{\theta} \in E_q^\omega(\mathbb{R}^d)$, *then*

$$\lim_{n\to\infty,\, n\in\mathbb{R}^d_{\kappa,\tau}} \sigma_n^\theta f = f \quad a.e.$$

for all $f \in L_p(\mathbb{T}^d)$.

We generalize the concept of Lebesgue points as follows. Let

$$U_{r,p}^{\kappa,\omega} f(x) := \sup_{i\in\mathbb{N}^d, h>0, \xi^{i_j} h<r, j=1,\ldots,d} \left(\prod_{j=1}^d \kappa_j(\xi^{i_j})^{-\omega}\right) \left(\frac{1}{\prod_{j=1}^d (2\kappa_j(\xi^{i_j} h)\pi)}\right.$$

$$\left.\int_{-\kappa_1(\xi^{i_1} h)\pi}^{\kappa_1(\xi^{i_1} h)\pi} \cdots \int_{-\kappa_d(\xi^{i_d} h)\pi}^{\kappa_d(\xi^{i_d} h)\pi} |f(x-t) - f(x)|^p \, dt\right)^{1/p}.$$

Definition 4.4.11 For $1 \le p < \infty$ and $\omega > 0$, a point $x \in \mathbb{T}^d$ is called a (p, κ, ω)-Lebesgue point of $f \in L_p(\mathbb{T}^d)$ if

$$\lim_{r\to 0} U_{r,p}^{\kappa,\omega} f(x) = 0.$$

If κ is the identity function, then we get back the (p, ω)-Lebesgue points investigated in the previous section.

Theorem 4.4.12 *For $1 \le p < \infty$ and $\omega > 0$, almost every point $x \in \mathbb{T}^d$ is a (p, κ, ω)-Lebesgue point of $f \in L_p(\mathbb{T}^d)$.*

We omit the proof, since it is similar to that of Theorem 4.3.13. Our basic theorem about the convergence at (p, κ, ω)-Lebesgue points reads as follows.

Theorem 4.4.13 *Suppose that $\omega > 0$, $1 \le p < \infty$, $1/p + 1/q = 1$ and*

$$\sum_{k_1=0}^{\infty} \cdots \sum_{k_d=0}^{\infty} \left(\prod_{j=1}^d \kappa_j(\xi^{k_j})^{\omega+1-1/q}\right) \sup_{n\in\mathbb{R}^d_{\kappa,\tau}} \left\|\widetilde{K}_n 1_{Q_k}\right\|_q \le C. \tag{4.4.3}$$

If

$$\lim_{n\to\infty,\, n\in\mathbb{R}^d_{\kappa,\tau}} \widehat{K_n}(0) = 1,$$

$\mathcal{M}_p^{\kappa,\omega} f(x)$ is finite and x is a (p, κ, ω)-Lebesgue points of $f \in L_p(\mathbb{T}^d)$, then

$$\lim_{n\to\infty,\, n\in\mathbb{R}^d_{\kappa,\tau}} \sigma_n f(x) = f(x).$$

Proof Since x is a (p, κ, ω)-Lebesgue point of f, we can fix a number $r < 1$ such that

$$U_{r,p}^{\kappa,\omega} f(x) < \epsilon.$$

Let us denote by r_0 the largest number i, for which $r/\xi \le \xi^i/n_1 < r$. We use again the decomposition

$$
\begin{aligned}
|\sigma_n f(x) - f(x)| &\le \frac{1}{(2\pi)^d} \int_{\mathbb{R}^d} |f(x-t) - f(x)| \left| \left(1_{(-\pi,\pi)^d} K_n\right)(t) \right| dt \\
&\quad + \left| f(x) \left(\frac{1}{(2\pi)^d} \int_{\mathbb{T}^d} K_n(t)\, dt - 1 \right) \right| \\
&= A_1(x) + A_2(x) + A_3(x),
\end{aligned}
$$

where

$$
A_1(x) := \frac{1}{(2\pi)^d} \sum_{k_1=0}^{r_0} \cdots \sum_{k_d=0}^{r_0} \int_{Q_{1,k_1}(n_1)} \cdots \int_{Q_{d,k_d}(n_1)} |f(x-t) - f(x)| \left| \left(1_{(-\pi,\pi)^d} K_n\right)(t) \right| dt,
$$

$$
A_2(x) := \frac{1}{(2\pi)^d} \sum_{\pi_1,\ldots,\pi_d} \sum_{k_{\pi_1}=r_0+1}^{\infty} \cdots \sum_{k_{\pi_j}=r_0+1}^{\infty} \sum_{k_{\pi_{j+1}}=0}^{\infty} \cdots \sum_{k_{\pi_d}=0}^{\infty} \int_{Q_{1,k_1}(n_1)} \cdots \int_{Q_{d,k_d}(n_1)} |f(x-t) - f(x)| \left| \left(1_{(-\pi,\pi)^d} K_n\right)(t) \right| dt,
$$

and

$$
A_3(x) := \left| f(x) \left(\frac{1}{(2\pi)^d} \int_{\mathbb{T}^d} K_n(t)\, dt - 1 \right) \right| = \left| f(x) \left(\widehat{K_n}(0) - 1 \right) \right|.
$$

$\{\pi_1, \ldots, \pi_d\}$ denotes again a permutation of $\{1, \ldots, d\}$ and $1 \le j \le d$. Obviously,

$$
\lim_{n\to\infty,\, n\in\mathbb{R}^d_\tau} A_3(x) = 0.
$$

Taking into account (4.4.2), we can see that

$$
\begin{aligned}
&A_1(x) \\
&\le \frac{1}{(2\pi)^d} \sum_{k_1=0}^{r_0} \cdots \sum_{k_d=0}^{r_0} \\
&\left(\int_{Q_{1,k_1}(n_1)} \cdots \int_{Q_{d,k_d}(n_1)} |f(x-t) - f(x)|^p \, dt \right)^{1/p} \left(\prod_{j=1}^d n_j \right)^{-1/q} \\
&\left(\int_{n_1 Q_{1,k_1}(n_1)} \cdots \int_{n_d Q_{d,k_d}(n_1)} \left| \left(1_{(-\pi,\pi)^d} K_n\right) \left(\frac{t_1}{n_1}, \ldots, \frac{t_d}{n_d} \right) \right|^q dt \right)^{1/q}
\end{aligned}
$$

$$\leq C \sum_{k_1=0}^{r_0} \cdots \sum_{k_d=0}^{r_0} G(\kappa_1(\xi^{k_1}/n_1)\pi, \ldots, \kappa_d(\xi^{k_d}/n_1)\pi) \left(\prod_{j=1}^{d} n_j\right)^{-1/q}$$

$$\left(\int_{n_1 Q_{1,k_1}(n_1)} \cdots \int_{n_d Q_{d,k_d}(n_1)} \left| \left(1_{(-\pi,\pi)^d} K_n\right) \left(\frac{t_1}{n_1}, \ldots, \frac{t_d}{n_d}\right)\right|^q dt\right)^{1/q}, \tag{4.4.4}$$

where

$$G(u) := \left(\int_{-u_1}^{u_1} \cdots \int_{-u_d}^{u_d} |f(x-t) - f(x)|^p \, dt\right)^{1/p} \qquad (u \in \mathbb{R}_+^d).$$

It comes from the definition of $U_{r,p}^{\kappa,\omega}$ that

$$\left(\prod_{j=1}^{d} \kappa_j(\xi^{k_j})^{-\omega}\right) \frac{G(\kappa_1(\xi^{k_1}/n_1)\pi, \ldots, \kappa_d(\xi^{k_d}/n_1)\pi)}{\left(\prod_{j=1}^{d}(2\kappa_j(\xi^{k_j}/n_1)\pi)\right)^{1/p}} \leq U_{r,p}^{\kappa,\omega} f(x), \tag{4.4.5}$$

which implies

$$A_1(x)$$

$$\leq C \sum_{k_1=0}^{r_0} \cdots \sum_{k_d=0}^{r_0} \left(\prod_{j=1}^{d} \kappa_j(\xi^{k_j})^{\omega}\right) \left(\prod_{j=1}^{d} \kappa_j(\xi^{k_j}/n_1)\right)^{1/p} U_{r,p}^{\kappa,\omega} f(x) \left(\prod_{j=1}^{d} n_j\right)^{-1/q}$$

$$\left(\int_{n_1 Q_{1,k_1}(n_1)} \cdots \int_{n_d Q_{d,k_d}(n_1)} \left| \left(1_{(-\pi,\pi)^d} K_n\right) \left(\frac{t_1}{n_1}, \ldots, \frac{t_d}{n_d}\right)\right|^q dt\right)^{1/q}.$$

As in the proof of Theorem 4.4.5,

$$A_1(x) \leq C \sum_{k_1=0}^{r_0} \cdots \sum_{k_d=0}^{r_0} \left(\prod_{j=1}^{d} \kappa_j(\xi^{k_j})^{\omega+1/p}\right)$$

$$\left(\prod_{j=1}^{d} \kappa_j(n_1)\right)^{-1/p} U_{r,p}^{\kappa,\omega} f(x) \left(\prod_{j=1}^{d} n_j\right)^{-1/q}$$

$$\left(\int_{Q'_{1,k_1}} \cdots \int_{Q'_{d,k_d}} \left| \left(1_{(-\pi,\pi)^d} K_n\right) \left(\frac{t_1}{n_1}, \ldots, \frac{t_d}{n_d}\right)\right|^q dt\right)^{1/q}$$

$$\leq C \sum_{k_1=0}^{r_0} \cdots \sum_{k_d=0}^{r_0} \left(\prod_{j=1}^{d} \kappa_j(\xi^{k_j})^{\omega+1/p}\right) U_{r,p}^{\kappa,\omega} f(x)$$

$$\left(\int_{Q'_{1,k_1}} \cdots \int_{Q'_{d,k_d}} \left|\widetilde{K}_n(t)\right|^q \, dt\right)^{1/q}$$
$$\leq C\epsilon \left\|\widetilde{K}_n\right\|_{E_q^\omega(\mathbb{R}^d)}.$$

In the same way as in (4.4.4), we get that

$$A_2(x)$$
$$\leq C \sum_{\pi_1,\ldots,\pi_d} \sum_{k_{\pi_1}=r_0+1}^{\infty} \cdots \sum_{k_{\pi_j}=r_0+1}^{\infty} \sum_{k_{\pi_{j+1}}=0}^{\infty} \cdots \sum_{k_{\pi_d}=0}^{\infty}$$
$$G(\kappa_1(\xi^{k_1}/n_1)\pi, \ldots, \kappa_d(\xi^{k_d}/n_1)\pi) \left(\prod_{j=1}^d n_j\right)^{-1/q}$$
$$\left(\int_{n_1 Q_{1,k_1}(n_1)} \cdots \int_{n_d Q_{d,k_d}(n_1)} \left|\left(1_{(-\pi,\pi)^d} K_n\right)\left(\frac{t_1}{n_1}, \ldots, \frac{t_d}{n_d}\right)\right|^q dt\right)^{1/q}.$$

Besides (4.4.5), we know also the inequality

$$\left(\prod_{j=1}^d \kappa_j(\xi^{k_j})^{-\omega}\right) \frac{G(\kappa_1(\xi^{k_1}/n_1)\pi, \ldots, \kappa_d(\xi^{k_d}/n_1)\pi)}{\left(\prod_{j=1}^d (2\kappa_j(\xi^{k_j}/n_1)\pi)\right)^{1/p}} \leq \mathcal{M}_p^{\kappa,\omega} f(x) + C|f(x)|.$$

As above, we get that

$$A_2(x)$$
$$\leq C \sum_{\pi_1,\ldots,\pi_d} \sum_{k_{\pi_1}=r_0+1}^{\infty} \cdots \sum_{k_{\pi_j}=r_0+1}^{\infty} \sum_{k_{\pi_{j+1}}=0}^{\infty} \cdots \sum_{k_{\pi_d}=0}^{\infty}$$
$$\left(\prod_{j=1}^d \kappa_j(\xi^{k_j})^{\omega}\right) \left(\prod_{j=1}^d \kappa_j(\xi^{k_j}/n_1)\right)^{1/p} \left(\mathcal{M}_p^{\kappa,\omega} f(x) + |f(x)|\right) \left(\prod_{j=1}^d n_j\right)^{1/q}$$
$$\left(\int_{n_1 Q_{1,k_1}(n_1)} \cdots \int_{n_d Q_{d,k_d}(n_1)} \left|\left(1_{(-\pi,\pi)^d} K_n\right)\left(\frac{t_1}{n_1}, \ldots, \frac{t_d}{n_d}\right)\right|^q dt\right)^{1/q}$$
$$\leq C \left(\mathcal{M}_p^{\kappa,\omega} f(x) + C|f(x)|\right) \sum_{\pi_1,\ldots,\pi_d} \sum_{k_{\pi_1}=r_0+1}^{\infty} \cdots \sum_{k_{\pi_j}=r_0+1}^{\infty} \sum_{k_{\pi_{j+1}}=0}^{\infty} \cdots \sum_{k_{\pi_d}=0}^{\infty}$$
$$\left(\prod_{j=1}^d \kappa_j(\xi^{k_j})^{\omega+1/p}\right) \left(\int_{Q'_{1,k_1}} \cdots \int_{Q'_{d,k_d}} \left|\widetilde{K}_n(t)\right|^q dt\right)^{1/q}.$$

Since $r_0 \to \infty$ as $n_1 \to \infty$, we deduce that

$$\lim_{n\to\infty} A_2(x) = 0.$$

The proof of the theorem is complete. ■

Obviously, (4.4.3) implies

$$\sup_{n\in\mathbb{R}^d_{\kappa,\tau}} \left\| \widetilde{K}_n \right\|_{E_q^\omega(\mathbb{R}^d)} \leq C.$$

Since we basically work with the $E_q^\omega(\mathbb{R}^d)$ space, our results can be applied to all examples of Sect. 4.3.2, amongst others, to the Cesàro and Riesz summability.

Corollary 4.4.14 *If* $0 < \omega < \alpha \leq 1$, $\mathcal{M}^{\kappa,\omega} f(x)$ *is finite and* x *is a* $(1, \kappa, \omega)$-*Lebesgue point of* $f \in L_1(\mathbb{T}^d)$, *then*

$$\lim_{n\to\infty,\, n\in\mathbb{R}^d_{\kappa,\tau}} \sigma_n^\alpha f(x) = f(x).$$

Moreover,

$$\sup_{\rho>0} \rho\lambda(\sigma_\kappa^\alpha > \rho) \leq C\|f\|_1 \qquad (f \in L_1(\mathbb{T}^d))$$

and, for every $1 < p \leq \infty$,

$$\left\|\sigma_\kappa^\alpha f\right\|_p \leq C_p\|f\|_p \qquad (f \in L_p(\mathbb{T}^d)).$$

The same hold for the Riesz summation $\sigma_n^{\alpha,\gamma}$ *if* $0 < \omega < \min(\alpha, 1)$, $0 < \alpha < \infty$ *and* $\gamma \in \mathbb{P}$.

Theorem 4.4.15 *Suppose that* $\omega > 0$, $1 \leq p < \infty$, $1/p + 1/q = 1$, $\theta \in W(C, \ell_1)(\mathbb{R}^d)$ *and* $\widehat{\theta} \in E_q^\omega(\mathbb{R}^d)$. *If* $\theta(0) = 1$, $\mathcal{M}_p^{\kappa,\omega} f(x)$ *is finite and* x *is a* (p, κ, ω)-*Lebesgue point of* $f \in L_p(\mathbb{T}^d)$, *then*

$$\lim_{n\to\infty,\, n\in\mathbb{R}^d_{\kappa,\tau}} \sigma_n^\theta f(x) = f(x).$$

This theorem can be proved exactly as Theorem 4.4.13.

Corollary 4.4.16 *If the conditions of Theorem 4.4.13 or Theorem 4.4.15 are satisfied and if* $f \in L_1(\mathbb{T}^d)$ *is continuous at a point* x, *then*

$$\lim_{n\to\infty,\, n\in\mathbb{R}^d_{\kappa,\tau}} \sigma_n^\theta f(x) = f(x).$$

Corollary 4.4.17 *Suppose that $\theta = \theta_1 \otimes \cdots \otimes \theta_d \in W(C, \ell_1)(\mathbb{R}^d)$, $\theta(0) = 1$ and $\theta_j \in V_1^2(\mathbb{R})$ for all $j = 1, \ldots, d$. If $0 < \omega < 1$, $\mathcal{M}^{\kappa,\omega} f(x)$ is finite and x is a $(1, \kappa, \omega)$-Lebesgue point of $f \in L_1(\mathbb{T}^d)$, then*

$$\lim_{n\to\infty,\, n\in\mathbb{R}^d_{\kappa,\tau}} \sigma_n^\theta f(x) = f(x).$$

Moreover,

$$\sup_{\rho>0} \rho\lambda(\sigma_\kappa^\theta > \rho) \le C \left(\prod_{j=1}^d \|\theta_j\|_{E_\infty^\omega(\mathbb{R})}\right) \|f\|_1 \qquad (f \in L_1(\mathbb{T}^d))$$

and, for every $1 < p \le \infty$,

$$\left\|\sigma_\kappa^\theta f\right\|_p \le C_p \left(\prod_{j=1}^d \|\theta_j\|_{E_\infty^\omega(\mathbb{R})}\right) \|f\|_p \qquad (f \in L_p(\mathbb{T}^d)).$$

We note again that all examples of Sect. 2.6.3 satisfy the condition $\theta \in V_1^2(\mathbb{R})$.

4.5 ℓ_∞-Summability

Now we consider Lebesgue points for the ℓ_∞-summability. We study the ℓ_∞-Cesàro means

$$\sigma_n^{\infty,\alpha} f(x) = \frac{1}{(2\pi)^d} \int_{\mathbb{T}^d} f(x-t) K_n^{\infty,\alpha}(t)\, dt \qquad (n \in \mathbb{N})$$

and the ℓ_∞-θ-means $\sigma_n^{\infty,\theta} f$. Recall that the Cesàro kernel $K_n^{\infty,\alpha}$ was defined by

$$K_n^{\infty,\alpha}(t) := \frac{1}{A_{n-1}^\alpha} \sum_{k\in\mathbb{Z}^d,\, \|k\|_\infty \le n} A_{n-1-\|k\|_\infty}^\alpha e^{\imath k\cdot t},$$

In this section, we cannot use the concept of Herz spaces, we will use other ideas.

4.5.1 Hardy-Littlewood Maximal Functions

In this section, we are going to generalize the maximal operator $\mathcal{M}_p^{\omega,1} f$ investigated in Sect. 4.3. Under a diagonal, we understand a diagonal of the cube $[0, \pi]^d$. Let us denote by $P_{2^{i_1}h,\ldots,2^{i_d}h}$ a parallelepiped, whose center is the origin and whose sides are parallel to the axes and/or to the diagonals and whose kth side length is

$2^{i_k+1}h$ if the kth side is parallel to an axis and $\sqrt{2}2^{i_k+1}h$ if the kth side is parallel to a diagonal ($i \in \mathbb{N}^d$, $h > 0$, $k = 1, \ldots, d$). More exactly, at least one side of $P_{2^{i_1}h,\ldots,2^{i_d}h}$ is parallel to one of the axes and the other sides are parallel to the axes and/or to the diagonals.

Definition 4.5.1 For $\omega > 0$, $1 \le p < \infty$ and $f \in L_p(\mathbb{T}^d)$, the Hardy-Littlewood maximal function $\mathcal{M}_p^\omega f$ is given by

$$\mathcal{M}_p^\omega f(x) := \sup_{P_{2^{i_1}h,\ldots,2^{i_d}h}, i\in\mathbb{N}^d, h>0} 2^{-\omega\|i\|_1} \left(\frac{1}{|P_{2^{i_1}h,\ldots,2^{i_d}h}|} \int_{P_{2^{i_1}h,\ldots,2^{i_d}h}} |f(x-t)|^p \, dt \right)^{1/p},$$

where the supremum is taken over all parallelepipeds $P_{2^{i_1}h,\ldots,2^{i_d}h}$ ($i \in \mathbb{N}^d$, $h > 0$) just defined.

For $p = 1$, we use the notation $\mathcal{M}^\omega f$. Obviously,

$$\mathcal{M}_p^{\omega_1} f \le \mathcal{M}_p^{\omega_2} f \qquad \text{for } \omega_1 > \omega_2 > 0 \text{ and } 1 \le p < \infty.$$

It is easy to see that

$$\mathcal{M}_p^\omega f(x) \ge \sup_{i\in\mathbb{N}^d, h>0} 2^{-\omega\|i\|_1} \left(\frac{1}{(2h)^d 2^{\|i\|_1}} \int_{-2^{i_1}h}^{2^{i_1}h} \int_{\delta_1 t_1 - 2^{i_2}h}^{\delta_1 t_1 + 2^{i_2}h} \cdots \int_{\delta_{d-1}(t_1 - t_2 - \cdots - t_{d-1}) - 2^{i_d}h}^{\delta_{d-1}(t_1 - t_2 - \cdots - t_{d-1}) + 2^{i_d}h} |f(x-t)|^p \, dt \right)^{1/p},$$

where $\delta_i \in \{0, 1\}$ ($i = 1, \ldots, d$). If we take the supremum only over all rectangles with sides parallel to the axes, we get back the definition of the maximal operator $\mathcal{M}_p^{\omega,1} f$ from Sect. 4.3.1. Thus

$$\mathcal{M}_p^\omega f \ge \mathcal{M}_p^{\omega,1} f.$$

In the two-dimensional case, besides $\mathcal{M}_p^{\omega,1} f$ defined in Sect. 4.3.1, we introduce

$$\mathcal{M}_p^{\omega,2} f(x_1, x_2) := \sup_{i_1, i_2 \in\mathbb{N}, h>0} 2^{-\omega(i_1+i_2)} \left(\frac{1}{4 \cdot 2^{i_1+i_2} h^2} \int_{-2^{i_1}h}^{2^{i_1}h} \int_{t_1 - 2^{i_2}h}^{t_1 + 2^{i_2}h} |f(x_1 - t_1, x_2 - t_2)|^p \, dt_2 \, dt_1 \right)^{1/p},$$

$$\mathcal{M}_p^{\omega,3} f(x_1,x_2) := \sup_{i_1,i_2\in\mathbb{N},h>0} 2^{-\omega(i_1+i_2)}\Bigg(\frac{1}{4\cdot 2^{i_1+i_2}h^2} \int_{-2^{i_1}h}^{2^{i_1}h}\int_{-t_1-2^{i_2}h}^{-t_1+2^{i_2}h} |f(x_1-t_1,x_2-t_2)|^p \,dt_2\,dt_1\Bigg)^{1/p},$$

$$\mathcal{M}_p^{\omega,4} f(x_1,x_2) := \sup_{i_1,i_2\in\mathbb{N},h>0} 2^{-\omega(i_1+i_2)}\Bigg(\frac{1}{4\cdot 2^{i_1+i_2}h^2} \int_{-2^{i_2}h}^{2^{i_2}h}\int_{t_2-2^{i_1}h}^{t_2+2^{i_1}h} |f(x_1-t_1,x_2-t_2)|^p \,dt_1\,dt_2\Bigg)^{1/p}$$

as well as

$$\mathcal{M}_p^{\omega,5} f(x_1,x_2) := \sup_{i_1,i_2\in\mathbb{N},h>0} 2^{-\omega(i_1+i_2)}\Bigg(\frac{1}{4\cdot 2^{i_1+i_2}h^2} \int_{-2^{i_2}h}^{2^{i_2}h}\int_{-t_2-2^{i_1}h}^{-t_2+2^{i_1}h} |f(x_1-t_1,x_2-t_2)|^p \,dt_1\,dt_2\Bigg)^{1/p}.$$

Note that in $\mathcal{M}_p^{\omega,1} f$, we take the supremum over rectangles with sides parallel to the axes and in $\mathcal{M}_p^{\omega,j} f$ $(j = 2, 3, 4, 5)$, over parallelograms with at most one side parallel to one of the axes and with the other sides parallel to the diagonals of the square $[0,\pi]^2$. Then we have

$$\mathcal{M}_p^{\omega} f(x_1,x_2) = \sum_{j=1}^{5} \mathcal{M}_p^{\omega,j} f(x_1,x_2)$$

for all $\omega > 0$ and $1 \le p < \infty$. Similarly to $M_p^{1,\mu,\nu} f$, we introduce also

$$M_p^{2,\mu,\nu} f(x_1,x_2) := \sup_{h>0}\Bigg(\frac{1}{4\mu(h)\nu(h)} \int_{-\mu(h)}^{\mu(h)}\int_{t_1-\nu(h)}^{t_1+\nu(h)} |f(x_1-t_1,x_2-t_2)|^p \,dt_2\,dt_1\Bigg)^{1/p},$$

$$M_p^{3,\mu,\nu} f(x_1,x_2) := \sup_{h>0}\Bigg(\frac{1}{4\mu(h)\nu(h)} \int_{-\mu(h)}^{\mu(h)}\int_{-t_1-\nu(h)}^{-t_1+\nu(h)} |f(x_1-t_1,x_2-t_2)|^p \,dt_2\,dt_1\Bigg)^{1/p},$$

$$M_p^{4,\mu,\nu} f(x_1, x_2) := \sup_{h>0} \left(\frac{1}{4\mu(h)\nu(h)} \int_{-\nu(h)}^{\nu(h)} \int_{t_2-\mu(h)}^{t_2+\mu(h)} |f(x_1 - t_1, x_2 - t_2)|^p \, dt_1 \, dt_2 \right)^{1/p}$$

and

$$M_p^{5,\mu,\nu} f(x_1, x_2) := \sup_{h>0} \left(\frac{1}{4\mu(h)\nu(h)} \int_{-\nu(h)}^{\nu(h)} \int_{-t_2-\mu(h)}^{-t_2+\mu(h)} |f(x_1 - t_1, x_2 - t_2)|^p \, dt_1 \, dt_2 \right)^{1/p}.$$

Recall that $\mu(h)$ and $\nu(h)$ are two continuous functions of $h \geq 0$, strictly increasing to ∞ and 0 at $h = 0$. The next two theorems can be proved in the same way as in Sect. 4.3.1.

Theorem 4.5.2 *If $j = 1, \ldots, 5$ and $1 \leq p < \infty$, then*

$$\sup_{\rho>0} \rho \lambda(M_p^{j,\mu,\nu} f > \rho)^{1/p} \leq C_p \, \|f\|_p \qquad (f \in L_p(\mathbb{T}^d)).$$

Moreover, if $p < r \leq \infty$, then

$$\left\| M_p^{j,\mu,\nu} f \right\|_r \leq C_r \, \|f\|_r \qquad (f \in L_r(\mathbb{T}^d)),$$

where the constants C_p and C_r are independent of μ and ν.

Theorem 4.5.3 *If $\omega > 0$ and $1 \leq p < \infty$, then*

$$\sup_{\rho>0} \rho \lambda(\mathcal{M}_p^{\omega} f > \rho)^{1/p} \leq C \|f\|_p \qquad (f \in L_p(\mathbb{T}^d)).$$

Moreover, if $p < r \leq \infty$, then

$$\left\| \mathcal{M}_p^{\omega} f \right\|_r \leq C_r \|f\|_r \qquad (f \in L_r(\mathbb{T}^d).$$

4.5.2 Lebesgue Points for the ℓ_∞-Summability

Here we introduce a stronger version of Lebesgue points than the (p, ω)-Lebesgue points. Similarly to Sect. 4.3.2, let

$$U_{r,p}^{\omega} f(x) := \sup_{\substack{P_{2^{i_1}h,\ldots,2^{i_d}h}, i\in\mathbb{N}^d, h>0 \\ 2^{i_k}h<r, k=1,\ldots,d}} 2^{-\omega\|i\|_1} \left(\frac{1}{\left|P_{2^{i_1}h,\ldots,2^{i_d}h}\right|} \right)^{1/p} \left(\int_{P_{2^{i_1}h,\ldots,2^{i_d}h}} |f(x-t)-f(x)|^p \, dt \right)^{1/p},$$

where the supremum is taken over all parallelepipeds whose center is the origin and whose sides are parallel to the axes and/or to the diagonals as in the definition of $\mathcal{M}_p^{\omega} f$. Obviously,

$$U_{r,p}^{\omega} f(x) \geq \sup_{i\in\mathbb{N}^d, h>0, 2^{i_k}h<r, k=1,\ldots,d} 2^{-\omega\|i\|_1} \left(\frac{1}{(2h)^d 2^{\|i\|_1}} \right)^{1/p} \times$$

$$\times \left(\int_{-2^{i_1}h}^{2^{i_1}h} \int_{\delta_1 t_1 - 2^{i_2}h}^{\delta_1 t_1 + 2^{i_2}h} \cdots \int_{\delta_{d-1}(t_1-t_2-\cdots-t_{d-1})-2^{i_d}h}^{\delta_{d-1}(t_1-t_2-\cdots-t_{d-1})+2^{i_d}h} |f(x-t)-f(x)|^p \, dt \right)^{1/p},$$

where $\delta_i = 0, 1$ $(i = 1, \ldots, d-1)$. Taking the supremum in the definition of $U_{r,p}^{\omega} f$ over all parallelepipeds whose sides are parallel to the axes, we obtain the definition of $U_{r,p}^{\omega,1} f$ (see Definition 4.3.12). In case $p = 1$, we omit again the notation p and write simply $U_r^{\omega} f$. In the two-dimensional case, similarly to $\mathcal{M}_p^{\omega,j} f$, we can define $U_{r,p}^{\omega,j} f$ for $j = 2, 3, 4, 5$ as follows:

$$U_{r,p}^{\omega,2} f(x_1,x_2) := \sup_{i_1,i_2\in\mathbb{N}, h>0, 2^{i_k}h<r, k=1,2} 2^{-\omega(i_1+i_2)} \left(\frac{1}{4\cdot 2^{i_1+i_2}h^2} \int_{-2^{i_1}h}^{2^{i_1}h} \int_{t_1-2^{i_2}h}^{t_1+2^{i_2}h} |f(x_1-t_1, x_2-t_2) - f(x_1,x_2)|^p \, dt_2 \, dt_1 \right)^{1/p},$$

$$U_{r,p}^{\omega,3}(x_1,x_2) := \sup_{i_1,i_2\in\mathbb{N}, h>0, 2^{i_k}h<r, k=1,2} 2^{-\omega(i_1+i_2)} \left(\frac{1}{4\cdot 2^{i_1+i_2}h^2} \int_{-2^{i_1}h}^{2^{i_1}h} \int_{-t_1-2^{i_2}h}^{-t_1+2^{i_2}h} |f(x_1-t_1, x_2-t_2) - f(x_1,x_2)|^p \, dt_2 \, dt_1 \right)^{1/p},$$

$$U_{r,p}^{\omega,4} f(x_1,x_2) := \sup_{i_1,i_2\in\mathbb{N}, h>0, 2^{i_k}h<r, k=1,2} 2^{-\omega(i_1+i_2)} \left(\frac{1}{4\cdot 2^{i_1+i_2}h^2} \int_{-2^{i_2}h}^{2^{i_2}h} \int_{t_2-2^{i_1}h}^{t_2+2^{i_1}h} |f(x_1-t_1, x_2-t_2) - f(x_1,x_2)|^p \, dt_1 \, dt_2 \right)^{1/p}$$

and

$$U_{r,p}^{\omega,5} f(x_1, x_2) := \sup_{i_1,i_2\in\mathbb{N},h>0,2^{i_k}h<r,k=1,2} 2^{-\omega(i_1+i_2)} \left(\frac{1}{4 \cdot 2^{i_1+i_2} h^2} \int_{-2^{i_2}h}^{2^{i_2}h} \int_{-t_2-2^{i_1}h}^{-t_2+2^{i_1}h} |f(x_1 - t_1, x_2 - t_2) - f(x_1, x_2)|^p \, dt_1 \, dt_2 \right)^{1/p}.$$

Definition 4.5.4 For $1 \le p < \infty$ and $\omega > 0$, a point $x \in \mathbb{T}^d$ is called a strong (p, ω)-Lebesgue point of $f \in L_p(\mathbb{T}^d)$ if

$$\lim_{r\to 0} U_{r,p}^{\omega} f(x) = 0.$$

Recall that $x \in \mathbb{T}^d$ a (p, ω)-Lebesgue point of $f \in L_p(\mathbb{T}^d)$ if

$$\lim_{r\to 0} U_{r,p}^{\omega,1} f(x) = 0.$$

Since

$$U_{r,p}^{\omega,1} f \le U_{r,p}^{\omega} f \qquad (1 \le p < \infty, 0 < r < \infty),$$

Definition 4.5.4 is indeed stronger than the definition of (p, ω)-Lebesgue points. Note that every strong (p, ω_2)-Lebesgue point is a strong (p, ω_1)-Lebesgue point $(0 < \omega_2 < \omega_1 < \infty)$, because of

$$U_{r,p}^{\omega_1} f \le U_{r,p}^{\omega_2} f \qquad (0 < \omega_2 < \omega_1 < \infty, 1 \le p < \infty).$$

Moreover, if $p < r$, then every strong (r, ω)-Lebesgue point is a strong (p, ω)-Lebesgue point. If f is continuous at x, then x is a strong (p, ω)-Lebesgue point of f for all $1 \le p < \infty$ and $\omega > 0$. The proof of the next result is the same as that of Theorem 4.3.13.

Theorem 4.5.5 *For $1 \le p < \infty$ and $\omega > 0$, almost every point $x \in \mathbb{T}^d$ is a strong (p, ω)-Lebesgue point of $f \in L_p(\mathbb{T}^d)$.*

To be able to prove the main theorem of this section, we need the next lemma.

Lemma 4.5.6 *Suppose that $0 < \alpha \le 1$, $x \in \mathbb{T}^2$ and $\pi > x_1 > x_2 > 0$. Then*

$$\left| K_n^{\infty,\alpha}(x_1, x_2) \right| \le C n^2 \tag{4.5.1}$$

and

$$\left| K_n^{\infty,\alpha}(x_1, x_2) \right| \le C x_1^{-1} x_2^{-1}. \tag{4.5.2}$$

Moreover, if $x_1 - x_2 > 1/n$, then

$$\left|K_n^{\infty,\alpha}(x_1,x_2)\right| \le Cn^{-\alpha}x_1^{-1}x_2^{-1}(x_1-x_2)^{-\alpha} \tag{4.5.3}$$

and

$$\left|K_n^{\infty,\alpha}(x_1,x_2)\right| \le Cn^{1-\alpha}x_1^{-1}(x_1-x_2)^{-\alpha}. \tag{4.5.4}$$

These inequalities come easily from Lemma 2.2.19.

Theorem 4.5.7 *If* $0<\alpha<\infty$, $0<\omega<\min(\alpha,1)/d$, $\mathcal{M}^\omega f(x)$ *is finite and* x *is a strong* $(1,\omega)$*-Lebesgue point of* $f\in L_1(\mathbb{T}^d)$, *then*

$$\lim_{n\to\infty}\sigma_n^{\infty,\alpha}f(x)=f(x).$$

Proof By Lemma 2.2.8, we have to prove the theorem for $0<\alpha\le 1$. Let $0<\omega<\alpha/2$. Since (x_1,x_2) is a strong $(1,\omega)$-Lebesgue point of f, we can fix a number $r<1$ such that

$$U_r^\omega f(x_1,x_2)<\epsilon.$$

Let us denote the square $[0,r/2]\times[0,r/2]$ by $S_{r/2}$ and let $2/n<r/2$.

Since

$$\int_{\mathbb{T}^2}K_n^{\infty,\alpha}(x_1,x_2)\,dx=(2\pi)^2,$$

we have

$$\begin{aligned}&\left|\sigma_n^{\infty,\alpha}f(x_1,x_2)-f(x_1,x_2)\right|\\&\quad\le\frac{1}{(2\pi)^2}\int_{\mathbb{T}^2}|f(x_1-t_1,x_2-t_2)-f(x_1,x_2)|\left|K_n^{\infty,\alpha}(t_1,t_2)\right|dt.\end{aligned} \tag{4.5.5}$$

It is enough to integrate over the set

$$\{(t_1,t_2):0<t_2<t_1<\pi\}.$$

We decompose this set into the union of the same sets A_i $(i=1,\ldots,5)$ as in the proof of Theorem 2.3.1 (see Fig. 4.1), where

$$\begin{aligned}A_1&:=\{(x_1,x_2):0<x_1\le 2/n,0<x_2<x_1<\pi\},\\A_2&:=\{(x_1,x_2):2/n<x_1<\pi,0<x_2\le 1/n\},\\A_3&:=\{(x_1,x_2):2/n<x_1<\pi,1/n<x_2\le x_1/2\},\\A_4&:=\{(x_1,x_2):2/n<x_1<\pi,x_1/2<x_2\le x_1-1/n\},\\A_5&:=\{(x_1,x_2):2/n<x_1<\pi,x_1-1/n<x_2<x_1\}.\end{aligned}$$

We will integrate the right-hand side of (4.5.5) over the sets

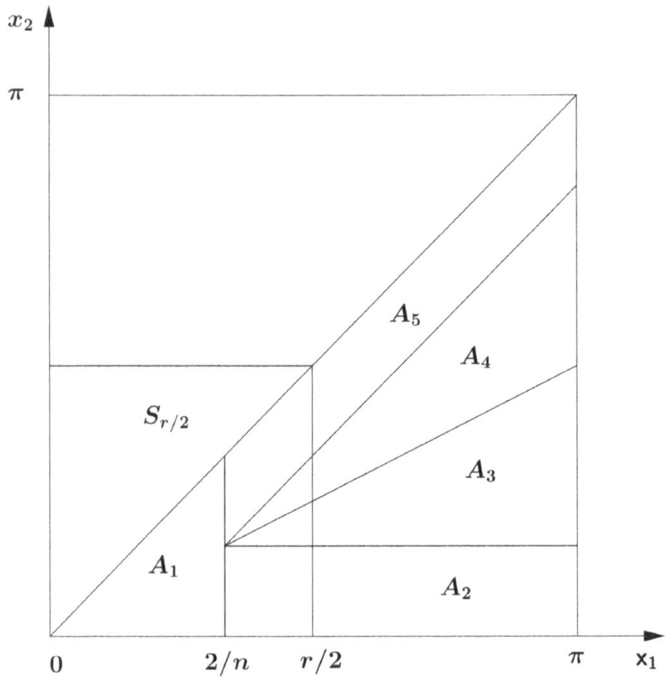

Fig. 4.1 The sets A_i

$$\bigcup_{i=1}^{5}(A_i \cap S_{r/2}) \qquad \text{and} \qquad \bigcup_{i=1}^{5}(A_i \cap S_{r/2}^c),$$

where S^c denotes the complement of the set S. Of course, $A_1 \subset S_{r/2}$. By (4.5.1),

$$\begin{aligned}
&\int_{A_1} |f(x_1 - t_1, x_2 - t_2) - f(x_1, x_2)| \left|K_n^{\infty,\alpha}(t_1, t_2)\right| dt \\
&\quad \le Cn^2 \int_0^{2/n} \int_0^{2/n} |f(x_1 - t_1, x_2 - t_2) - f(x_1, x_2)| \, dt_1 \, dt_2 \\
&\quad \le CU_r^{\omega,1} f(x_1, x_2) < C\epsilon.
\end{aligned}$$

Let us denote by r_0 the largest number i, for which $r/2 \le 2^{i+1}/n < r$. By (4.5.4),

$$\begin{aligned}
&\int_{A_2 \cap S_{r/2}} |f(x_1 - t_1, x_2 - t_2) - f(x_1, x_2)| \left|K_n^{\infty,\alpha}(t_1, t_2)\right| dt \\
&\quad \le C \sum_{i=1}^{r_0} n^{1-\alpha} \left(\frac{2^i}{n}\right)^{-1} \left(\frac{2^i}{n} - \frac{1}{n}\right)^{-\alpha}
\end{aligned}$$

$$\int_{2^i/n}^{2^{i+1}/n}\int_0^{1/n} |f(x_1-t_1,x_2-t_2)-f(x_1,x_2)|\,dt_2\,dt_1$$
$$\le C\sum_{i=1}^{r_0} 2^{(\omega-\alpha)i}2^{-\omega i}\left(\frac{n^2}{2^i}\right)$$
$$\int_{2^i/n}^{2^{i+1}/n}\int_0^{1/n} |f(x_1-t_1,x_2-t_2)-f(x_1,x_2)|\,dt_2\,dt_1$$
$$\le C\sum_{i=1}^{r_0} 2^{(\omega-\alpha)i}U_r^{\omega,1}f(x_1,x_2) < C\epsilon.$$

Since $t_1-t_2>t_1/2$ and $t_1-t_2>t_2$ on A_3, we obtain by (4.5.3) that

$$\left|K_n^{\infty,\alpha}(t_1,t_2)\right| \le Cn^{-\alpha}t_1^{-1-\alpha/2}t_2^{-1-\alpha/2}.$$

Hence

$$\int_{A_3\cap S_{r/2}} |f(x_1-t_1,x_2-t_2)-f(x_1,x_2)|\left|K_n^{\infty,\alpha}(t_1,t_2)\right|dt$$
$$\le C\sum_{i=1}^{r_0}\sum_{j=0}^{i-1} n^{-\alpha}\left(\frac{2^i}{n}\right)^{-1-\alpha/2}\left(\frac{2^j}{n}\right)^{-1-\alpha/2}$$
$$\int_{2^i/n}^{2^{i+1}/n}\int_{2^j/n}^{2^{j+1}/n} |f(x_1-t_1,x_2-t_2)-f(x_1,x_2)|\,dt_1\,dt_2$$
$$\le C\sum_{i=1}^{r_0}\sum_{j=0}^{i-1} 2^{(\omega-\alpha/2)(i+j)}2^{-\omega(i+j)}\left(\frac{n^2}{2^{i+j}}\right)$$
$$\int_{2^i/n}^{2^{i+1}/n}\int_{2^j/n}^{2^{j+1}/n} |f(x_1-t_1,x_2-t_2)-f(x_1,x_2)|\,dt_1\,dt_2$$
$$\le C\sum_{i=1}^{r_0}\sum_{j=0}^{i-1} 2^{(\omega-\alpha/2)(i+j)}U_r^{\omega,1}f(x_1,x_2) < C\epsilon.$$

Since $t_2>t_1/2$ on A_4, (4.5.3) implies

$$\left|K_n^{\infty,\alpha}(t_1,t_2)\right| \le Cn^{-\alpha}t_1^{-2}(t_1-t_2)^{-\alpha}, \tag{4.5.6}$$

and so

$$
\begin{aligned}
&\int_{A_4\cap S_{r/2}} |f(x_1-t_1,x_2-t_2)-f(x_1,x_2)|\left|K_n^{\infty,\alpha}(t_1,t_2)\right| dt \\
&\quad \le C\sum_{i=1}^{r_0}\sum_{j=0}^{i-1} n^{-\alpha}\left(\frac{2^i}{n}\right)^{-2}\left(\frac{2^j}{n}\right)^{-\alpha} \\
&\qquad \int_{2^i/n}^{2^{i+1}/n}\int_{t_1-2^{j+1}/n}^{t_1-2^j/n} |f(x_1-t_1,x_2-t_2)-f(x_1,x_2)|\, dt_2\, dt_1 \\
&\quad \le C\sum_{i=1}^{r_0}\sum_{j=0}^{i-1} 2^{(\omega-1)i}2^{(\omega+1-\alpha)j}2^{-\omega(i+j)}\left(\frac{n^2}{2^{i+j}}\right) \\
&\qquad \int_{2^i/n}^{2^{i+1}/n}\int_{t_1-2^{j+1}/n}^{t_1-2^j/n} |f(x_1-t_1,x_2-t_2)-f(x_1,x_2)|\, dt_2\, dt_1 \\
&\quad \le C\sum_{i=1}^{r_0}\sum_{j=0}^{i-1} 2^{(\omega-1)i}2^{(\omega+1-\alpha)j}U_r^{\omega,2}f(x_1,x_2) \\
&\quad \le C\sum_{i=1}^{r_0} 2^{(2\omega-\alpha)i}U_r^{\omega,2}f(x_1,x_2) < C\epsilon.
\end{aligned}
$$

We get from (4.5.2) that

$$
\left|K_n^{\infty,\alpha}(t_1,t_2)\right| \le Ct_1^{-2}
$$

on the set A_5. This implies

$$
\begin{aligned}
&\int_{A_5\cap S_{r/2}} |f(x_1-t_1,x_2-t_2)-f(x_1,x_2)|\left|K_n^{\infty,\alpha}(t_1,t_2)\right| dt \\
&\quad \le C\sum_{i=1}^{r_0}\left(\frac{2^i}{n}\right)^{-2}\int_{2^i/n}^{2^{i+1}/n}\int_{t_1-1/n}^{t_1} |f(x_1-t_1,x_2-t_2)-f(x_1,x_2)|\, dt_2\, dt_1 \\
&\quad \le C\sum_{i=1}^{r_0} 2^{(\omega-1)i}2^{-\omega i}\left(\frac{n^2}{2^i}\right) \\
&\qquad \int_{2^i/n}^{2^{i+1}/n}\int_{t_1-1/n}^{t_1} |f(x_1-t_1,x_2-t_2)-f(x_1,x_2)|\, dt_2\, dt_1 \\
&\quad \le C\sum_{i=1}^{r_0} 2^{(\omega-1)i}U_r^{\omega,2}f(x_1,x_2) < C\epsilon.
\end{aligned}
$$

On the other hand, we get that

$$\begin{aligned}\int_{A_2\cap S^c_{r/2}} &|f(x_1-t_1,x_2-t_2)-f(x_1,x_2)|\left|K_n^{\infty,\alpha}(t_1,t_2)\right| dt\\ &\leq C\sum_{i=r_0}^{\infty} 2^{(\omega-\alpha)i}\mathcal{M}^{\omega,1}f(x_1,x_2)+C\sum_{i=r_0}^{\infty}2^{-\alpha i}|f(x_1,x_2)|\\ &\leq C2^{(\omega-\alpha)r_0}\mathcal{M}^{\omega,1}f(x_1,x_2)+C2^{-\alpha r_0}|f(x_1,x_2)|\\ &\leq C(nr)^{\omega-\alpha}\mathcal{M}^{\omega,1}f(x_1,x_2)+C(nr)^{-\alpha}|f(x_1,x_2)|\to 0\end{aligned}$$

as $n\to\infty$. Similarly,

$$\begin{aligned}\int_{A_3\cap S^c_{r/2}} &|f(x_1-t_1,x_2-t_2)-f(x_1,x_2)|\left|K_n^{\infty,\alpha}(t_1,t_2)\right| dt\\ &\leq C\sum_{i=r_0}^{\infty}\sum_{j=0}^{i-1} 2^{(\omega-\alpha/2)(i+j)}\mathcal{M}^{\omega,1}f(x_1,x_2)+C\sum_{i=r_0}^{\infty}\sum_{j=0}^{i-1}2^{-\alpha(i+j)/2}|f(x_1,x_2)|\\ &\leq C2^{(\omega-\alpha/2)r_0}\mathcal{M}^{\omega,1}f(x_1,x_2)+C2^{-\alpha r_0/2}|f(x_1,x_2)|\to 0\end{aligned}$$

and

$$\begin{aligned}\int_{A_4\cap S^c_{r/2}} &|f(x_1-t_1,x_2-t_2)-f(x_1,x_2)|\left|K_n^{\infty,\alpha}(t_1,t_2)\right| dt\\ &\leq C\sum_{i=r_0}^{\infty}\sum_{j=0}^{i-1} 2^{(\omega-1)i}2^{(\omega+1-\alpha)j}\mathcal{M}^{\omega,2}f(x_1,x_2)+C\sum_{i=r_0}^{\infty}\sum_{j=0}^{i-1}2^{-i}2^{(1-\alpha)j}|f(x_1,x_2)|\\ &\leq C2^{(2\omega-\alpha)r_0}\mathcal{M}^{\omega,2}f(x_1,x_2)+C2^{-\alpha r_0}|f(x_1,x_2)|\to 0,\end{aligned}$$

as $n\to\infty$. In the last line, we have used that $0<\alpha<1$. The same holds for $\alpha=1$. Finally,

$$\begin{aligned}\int_{A_5\cap S^c_{r/2}} &|f(x_1-t_1,x_2-t_2)-f(x_1,x_2)|\left|K_n^{\infty,\alpha}(t_1,t_2)\right| dt\\ &\leq C\sum_{i=r_0}^{\infty} 2^{(\omega-1)i}\mathcal{M}^{\omega,2}f(x_1,x_2)+C\sum_{i=r_0}^{\infty}2^{-i}|f(x_1,x_2)|\\ &\leq C2^{(\tau-1)r_0}\mathcal{M}^{\omega,2}f(x_1,x_2)+C2^{-r_0}|f(x_1,x_2)|\to 0,\end{aligned}$$

as $n\to\infty$. Note that $A_1\cap S^c_{r/2}=\emptyset$. ■

Note that Belinsky [20] proved that the convergence does not hold for all p-Lebesgue points defined in Definition 4.1.19. Since by Theorems 4.5.5 and 4.5.3 almost every point is a strong $(1,\omega)$-Lebesgue point and the maximal operator $\mathcal{M}^\omega f$ is almost everywhere finite for $f\in L_1(\mathbb{T}^d)$, Theorem 4.5.7 implies the almost everywhere convergence

$$\lim_{n\to\infty} \sigma_n^{\infty,\alpha} f = f \qquad \text{a.e.}$$

if $f \in L_1(\mathbb{T}^d)$ (see Corollary 2.5.9).

In the next theorem, we use only the maximal operator $\mathcal{M}_p^{\omega,1} f$ and the (p,ω)-Lebesgue points as in Sect. 4.3.

Theorem 4.5.8 *Suppose that* $0 < \alpha < \infty$, $1 < p < \infty$, $1/p + 1/q = 1$ *and* $0 < \omega < \min(\alpha/d, 1/(2q))$. *If* $\mathcal{M}_p^{\omega,1} f(x)$ *is finite and* x *is a* (p,ω)*-Lebesgue point of* $f \in L_p(\mathbb{T}^d)$, *then*

$$\lim_{n\to\infty} \sigma_n^{\infty,\alpha} f(x) = f(x).$$

Proof Suppose that $0 < \alpha \le 1$. Since (x_1, x_2) is a (p,ω)-Lebesgue point of f, we fix again a number $0 < r < 1$ such that

$$U_{r,p}^{\omega,1} f(x_1, x_2) < \epsilon.$$

We can prove in the same way as in Theorem 4.5.7 that

$$\int_{A_i} |f(x_1 - t_1, x_2 - t_2) - f(x_1, x_2)| \left|K_n^{\infty,\alpha}(t_1, t_2)\right| dt \to 0,$$

for $i = 1, 2, 3$, as $n \to \infty$. So we have to consider the sets A_4 and A_5, only. It is easy to see that

$$\begin{aligned}
&\int_{A_4 \cap S_{r/2}} |f(x_1 - t_1, x_2 - t_2) - f(x_1, x_2)| \left|K_n^{\infty,\alpha}(t_1, t_2)\right| dt \\
&\le \sum_{i=1}^{r_0} \sum_{j=i-1}^{i} \int_{2^i/n}^{2^{i+1}/n} \int_{2^j/n}^{2^{j+1}/n} \\
&\qquad |f(x_1 - t_1, x_2 - t_2) - f(x_1, x_2)| \left|K_n^{\infty,\alpha}(t_1, t_2)\right| 1_{A_4}(t_1, t_2)\, dt_2\, dt_1.
\end{aligned}$$

Hölder's inequality and (4.5.6) imply

$$\begin{aligned}
&\int_{A_4 \cap S_{r/2}} |f(x_1 - t_1, x_2 - t_2) - f(x_1, x_2)| \left|K_n^{\infty,\alpha}(t_1, t_2)\right| dt \\
&\le \sum_{i=1}^{r_0} \sum_{j=i-1}^{i} \left(\int_{2^i/n}^{2^{i+1}/n} \int_{2^j/n}^{2^{j+1}/n} |f(x_1 - t_1, x_2 - t_2) - f(x_1, x_2)|^p \, dt_2\, dt_1 \right)^{1/p} \\
&\quad \left(\int_{2^i/n}^{2^{i+1}/n} \int_{2^{i-1}/n}^{t_1 - 1/n} n^{-\alpha q} t_1^{-2q} (t_1 - t_2)^{-\alpha q} 1_{A_4}(t_1, t_2)\, dt_2\, dt_1 \right)^{1/q}.
\end{aligned}$$

If $q < 1/\alpha$, then

$$\int_{2^i/n}^{2^{i+1}/n} \int_{2^{i-1}/n}^{t_1-1/n} n^{-\alpha q} t_1^{-2q} (t_1 - t_2)^{-\alpha q} 1_{A_4}(t_1, t_2)\, dt_2\, dt_1$$
$$\leq C n^{-\alpha q} \left(\frac{2^i}{n}\right)^{-\alpha q+1} \int_{2^i/n}^{2^{i+1}/n} t_1^{-2q}\, dt_1$$
$$\leq C n^{-\alpha q} \left(\frac{2^i}{n}\right)^{-\alpha q+1} \left(\frac{2^i}{n}\right)^{1-2q} dt_1$$
$$\leq C \left(\frac{n}{2^i}\right)^{2q-2} 2^{-i\alpha q}$$

and so

$$\int_{A_4 \cap S_{r/2}} |f(x_1 - t_1, x_2 - t_2) - f(x_1, x_2)| \left|K_n^{\infty,\alpha}(t_1, t_2)\right| dt$$
$$\leq C_p \sum_{i=1}^{r_0} \sum_{j=i-1}^{i} 2^{(\omega-\alpha/2)(i+j)}$$
$$2^{-\omega(i+j)} \left(\frac{n^2}{2^{i+j}} \int_{2^i/n}^{2^{i+1}/n} \int_{2^j/n}^{2^{j+1}/n} |f(x_1 - t_1, x_2 - t_2) - f(x_1, x_2)|^p\, dt_2\, dt_1\right)^{1/p}$$
$$\leq C_p \sum_{i=1}^{r_0} \sum_{j=i-1}^{i} 2^{(\omega-\alpha/2)(i+j)} U_{r,p}^{\omega,1} f(x_1, x_2) < C_p \epsilon.$$

An analogous inequality can also be proved for $q \geq 1/\alpha$. Indeed, choose a small number $0 < \beta < 1$ such that $\omega < (1-\beta)/2q$. Since $t_1 - t_2 < t_1/2$ on A_4 and $1 - \alpha q - \beta < 0$, we conclude

$$\int_{2^i/n}^{2^{i+1}/n} \int_{2^{i-1}/n}^{t_1-1/n} n^{-\alpha q} t_1^{-2q} (t_1 - t_2)^{-\alpha q} 1_{A_4}(t_1, t_2)\, dt_2\, dt_1$$
$$\leq C \int_{2^i/n}^{2^{i+1}/n} \int_{2^{i-1}/n}^{t_1-1/n} n^{-\alpha q} t_1^{-2q+\beta} (t_1 - t_2)^{-\alpha q-\beta} 1_{A_4}(t_1, t_2)\, dt_2\, dt_1$$
$$\leq C n^{-\alpha q} \left(\frac{1}{n}\right)^{-\alpha q-\beta+1} \int_{2^i/n}^{2^{i+1}/n} t_1^{-2q+\beta}\, dt_1$$
$$\leq C \left(\frac{n}{2^i}\right)^{2q-2} 2^{-i(1-\beta)}$$

and

$$\int_{A_4 \cap S_{r/2}} |f(x_1 - t_1, x_2 - t_2) - f(x_1, x_2)| \left|K_n^{\infty,\alpha}(t_1, t_2)\right| dt$$

$$\leq C_p \sum_{i=1}^{r_0} \sum_{j=i-1}^{i} 2^{(\omega-(1-\beta)/2q)(i+j)} 2^{-\omega(i+j)}$$

$$\left(\frac{n^2}{2^{i+j}} \int_{2^i/n}^{2^{i+1}/n} \int_{2^j/n}^{2^{j+1}/n} |f(x_1 - t_1, x_2 - t_2) - f(x_1, x_2)|^p \, dt_2 \, dt_1 \right)^{1/p}$$

$$\leq C_p \sum_{i=1}^{r_0} \sum_{j=i-1}^{i} 2^{(\omega-(1-\beta)/2q)(i+j)} U_{r,p}^{\omega,1} f(x_1, x_2) < C_p \epsilon.$$

For the set A_5, we obtain

$$\int_{A_5 \cap S_{r/2}} |f(x_1 - t_1, x_2 - t_2) - f(x_1, x_2)| \left| K_n^{\infty,\alpha}(t_1, t_2) \right| dt$$

$$\leq \sum_{i=1}^{r_0} \sum_{j=i-1}^{i} \left(\int_{2^i/n}^{2^{i+1}/n} \int_{2^j/n}^{2^{j+1}/n} |f(x_1 - t_1, x_2 - t_2) - f(x_1, x_2)|^p \, dt_2 \, dt_1 \right)^{1/p}$$

$$\left(\int_{2^i/n}^{2^{i+1}/n} \int_{t_1 - 1/n}^{t_1} t_1^{-2q} \, dt_2 \, dt_1 \right)^{1/q}.$$

We can compute that

$$\int_{2^i/n}^{2^{i+1}/n} \int_{t_1 - 1/n}^{t_1} t_1^{-2q} \, dt_2 \, dt_1 \leq C n^{-1} \left(\frac{2^i}{n} \right)^{-2q+1} = C \left(\frac{n}{2^i} \right)^{2q-2} 2^{-i}.$$

Then

$$\int_{A_5 \cap S_{r/2}} |f(x_1 - t_1, x_2 - t_2) - f(x_1, x_2)| \left| K_n^{\infty,\alpha}(t_1, t_2) \right| dt$$

$$\leq \sum_{i=1}^{r_0} \sum_{j=i-1}^{i} 2^{(\omega-1/2q)(i+j)} 2^{-\omega(i+j)}$$

$$\left(\frac{n^2}{2^{i+j}} \int_{2^i/n}^{2^{i+1}/n} \int_{2^j/n}^{2^{j+1}/n} |f(x_1 - t_1, x_2 - t_2) - f(x_1, x_2)|^p \, dt_2 \, dt_1 \right)^{1/p}$$

$$\leq C_p \sum_{i=1}^{r_0} \sum_{j=i-1}^{i} 2^{(\omega-1/(2q))(i+j)} U_{r,p}^{\omega,1} f(x_1, x_2) < C\epsilon.$$

Similarly, we can see that

$$
\begin{aligned}
&\int_{A_4\cap S_{r/2}^c} |f(x_1-t_1,x_2-t_2)-f(x_1,x_2)|\left|K_n^{\infty,\alpha}(t_1,t_2)\right| dt \\
&\quad\le C_p\sum_{i=r_0}^{\infty}\sum_{j=i-1}^{i} 2^{(\omega-\alpha/2)(i+j)}\mathcal{M}_p^{\omega,1}f(x_1,x_2) \\
&\qquad + C_p\sum_{i=r_0}^{\infty}\sum_{j=i-1}^{i} 2^{-\alpha(i+j)/2}|f(x_1,x_2)| \\
&\quad\le C_p 2^{(2\omega-\alpha)r_0}\mathcal{M}_p^{\omega,1}f(x_1,x_2)+C_p2^{-\alpha r_0}|f(x_1,x_2)| \\
&\quad\le C(nr)^{2\omega-\alpha}\mathcal{M}_p^{\omega,1}f(x_1,x_2)+C(nr)^{-\alpha}|f(x_1,x_2)|\to 0
\end{aligned}
$$

as $n\to\infty$ and $q<1/\alpha$. If $q\ge 1/\alpha$, then

$$
\begin{aligned}
&\int_{A_4\cap S_{r/2}^c} |f(x_1-t_1,x_2-t_2)-f(x_1,x_2)|\left|K_n^{\infty,\alpha}(t_1,t_2)\right| dt \\
&\quad\le C_p\sum_{i=r_0}^{\infty}\sum_{j=i-1}^{i} 2^{(\omega-(1-\beta)/2q)(i+j)}\mathcal{M}_p^{\omega,1}f(x_1,x_2) \\
&\qquad + C_p\sum_{i=r_0}^{\infty}\sum_{j=i-1}^{i} 2^{-(1-\beta)(i+j)/2q}|f(x_1,x_2)| \\
&\quad\le C_p 2^{(2\omega-(1-\beta)/q)r_0}\mathcal{M}_p^{\omega,1}f(x_1,x_2)+C_p2^{-(1-\beta)r_0/q}|f(x_1,x_2)|\to 0
\end{aligned}
$$

as $n\to\infty$. Finally,

$$
\begin{aligned}
&\int_{A_5\cap S_{r/2}^c} |f(x_1-t_1,x_2-t_2)-f(x_1,x_2)|\left|K_n^{\infty,\alpha}(t_1,t_2)\right| dt \\
&\quad\le C_p\sum_{i=r_0}^{\infty}\sum_{j=i-1}^{i} 2^{(\omega-1/2q)(i+j)}\mathcal{M}_p^{\omega,1}f(x_1,x_2) \\
&\qquad + C_p\sum_{i=r_0}^{\infty}\sum_{j=i-1}^{i} 2^{-(i+j)/2q}|f(x_1,x_2)| \\
&\quad\le C_p 2^{(2\omega-1/q)r_0}\mathcal{M}_p^{\omega,1}f(x_1,x_2)+C_p2^{-r_0/q}|f(x_1,x_2)|\to 0
\end{aligned}
$$

as $n\to\infty$. The proof of the theorem is complete. ∎

Note that these results were proved in Weisz [345, 349]. Now we turn to the ℓ_∞-θ-means introduced by

$$
\sigma_n^{\infty,\theta}f(x):=\sum_{k\in\mathbb{Z}^d}\theta\left(\frac{\|k\|_\infty}{n}\right)\widehat{f}(k)e^{\imath k\cdot x}
$$

in Sect. 2.6.1. We suppose again that $\theta : \mathbb{R} \to \mathbb{R}$ satisfies (2.6.2) and (2.6.3).

Theorem 4.5.9 *Suppose that θ satisfies (2.6.2) and (2.6.3). If $0<\omega<1/d$, $\mathcal{M}^\omega f(x)$ is finite and x is a strong $(1,\omega)$-Lebesgue point of $f \in L_1(\mathbb{T}^d)$, then*

$$\lim_{n\to\infty} \sigma_n^{\infty,\theta} f(x) = f(x).$$

Proof In Theorem 2.6.7, we have proved that

$$K_n^{\infty,\theta}(x) = \sum_{k=0}^{\infty} k\, \Delta_2\theta\left(\frac{k}{n}\right) K_k^{\infty}(x),$$

where K_k^∞ denotes the Fejér kernel. We have verified in (2.6.4) that

$$\sup_{n\geq 1} \sum_{k=0}^{\infty} k \left|\Delta_2\theta\left(\frac{k}{n}\right)\right| \leq C < \infty.$$

Hence

$$\begin{aligned}\sigma_n^{\infty,\theta} f(x) - f(x) &= \int_{\mathbb{T}^d} \Big(f(x-t) - f(x)\Big) K_n^{\infty,\theta}(t)\, dt \\ &= \sum_{k=0}^{\infty} k\, \Delta_2\theta\left(\frac{k}{n}\right) \int_{\mathbb{T}^d} \Big(f(x-t) - f(x)\Big) K_k^{\infty}(t)\, dt.\end{aligned}$$

The proof can be finished using Theorem 4.5.7. ∎

This implies also the almost everywhere convergence of $\sigma_n^{\infty,\theta} f$ stated in Corollary 2.6.9. From Theorem 4.5.8, we obtain in the same way

Theorem 4.5.10 *Suppose that θ satisfies (2.6.2) and (2.6.3), $1 < p < \infty$, $1/p + 1/q = 1$ and $0 < \omega < \min(1/d, 1/(2q))$. If $\mathcal{M}_p^{\omega,1} f(x)$ is finite and x is a (p,ω)-Lebesgue point of $f \in L_p(\mathbb{T}^d)$, then*

$$\lim_{n\to\infty} \sigma_n^{\infty,\theta} f(x) = f(x).$$

4.6 ℓ_1-Summability

Finally, we investigate Lebesgue points for the ℓ_1-Cesàro means

$$\sigma_n^{1,\alpha} f(x) = \frac{1}{(2\pi)^d} \int_{\mathbb{T}^d} f(x-t) K_n^{1,\alpha}(t)\, dt \qquad (n \in \mathbb{N})$$

as well as the ℓ_1-θ-means $\sigma_n^{1,\theta} f$. The definition of the Cesàro kernel $K_n^{1,\alpha}$, i.e.,

$$K_n^{1,\alpha}(t) := \frac{1}{A_{n-1}^{\alpha}} \sum_{k\in\mathbb{Z}^d,\ \|k\|_1\le n} A_{n-1-\|k\|_1}^{\alpha} e^{\imath k\cdot t},$$

can be found in Sect. 2.2. In this section, we use the same Hardy-Littlewood maximal functions $\mathcal{M}_p^{\omega} f$ and $\mathcal{M}_p^{\omega,1} f$ and the same (strong) (p,ω)-Lebesgue points as in Sect. 4.5. In what follows, we have to suppose that f is periodic with respect to π.

Instead of Lemma 2.2.14, we will use the next estimations.

Lemma 4.6.1 *Suppose that* $0<\alpha\le 1$, $0\le\beta\le 1$ *and* $\pi > x_1 > x_2 > 0$. *Then*

$$\left|K_n^{1,\alpha}(x_1,x_2)\right| \le Cn^2, \tag{4.6.1}$$

$$\begin{aligned}\left|K_n^{1,\alpha}(x_1,x_2)\right| &\le C(x_1-x_2)^{-1}(x_1+x_2)^{-1}1_{\{x_2\le\pi/2\}}\\ &\quad + C(x_1-x_2)^{-1}(\pi-x_2)^{-1}1_{\{x_2>\pi/2\}}.\end{aligned} \tag{4.6.2}$$

If $1/n < x_2 \le \pi/2$, *then*

$$\left|K_n^{1,\alpha}(x_1,x_2)\right| \le Cn^{-\alpha}(x_1-x_2)^{-1-\beta}x_2^{\beta-\alpha-1} \tag{4.6.3}$$

and

$$\left|K_n^{1,\alpha}(x_1,x_2)\right| \le Cn^{1-\alpha}x_2^{-\alpha-1}. \tag{4.6.4}$$

If $\pi/2 < x_2 < \pi - 1/n$, *then*

$$\left|K_n^{1,\alpha}(x_1,x_2)\right| \le Cn^{-\alpha}(x_1-x_2)^{-1-\beta}(\pi-x_2)^{\beta-\alpha-1} \tag{4.6.5}$$

and

$$\left|K_n^{1,\alpha}(x_1,x_2)\right| \le Cn^{1-\alpha}(\pi-x_2)^{-\alpha-1}. \tag{4.6.6}$$

Proof Inequalities (4.6.1) and (4.6.2) follow from Lemma 2.2.5 and (2.2.7), because $2\pi - x_1 - x_2 > \pi - x_2$, while (4.6.3) and (4.6.4) follow from (2.2.15) and (2.2.17). Finally, (4.6.5) and (4.6.6) can be proved as (2.2.16) and (2.2.18). ∎

The main theorem of this section reads as follows.

Theorem 4.6.2 *Suppose that* $0<\alpha<\infty$, $0<\omega<\min(\alpha,1)/d$ *and* $\mathcal{M}^{\omega} f(x)$ *is finite. If* $f\in L_1(\mathbb{T}^d)$ *is periodic with respect to* π *and* x *is a strong* $(1,\omega)$*-Lebesgue point of* f, *then*

$$\lim_{n\to\infty}\sigma_n^{1,\alpha} f(x) = f(x).$$

Proof Again, it is enough to prove the theorem for $0 < \alpha \le 1$. Let $0 < \omega < \alpha/2$ and fix a number $r < 1$ such that

$$U_r^{\omega} f(x_1, x_2) < \epsilon.$$

Let

$$S_{r/2} := \left[-\frac{r}{2}, \frac{r}{2}\right] \times \left[-\frac{r}{2}, \frac{r}{2}\right], \qquad S'_{r/2} := \left[\pi - \frac{r}{2}, \pi + \frac{r}{2}\right] \times \left[\pi - \frac{r}{2}, \pi + \frac{r}{2}\right]$$

and $2/n < r/2$. We have

$$\begin{aligned} &\left|\sigma_n^{1,\alpha} f(x_1, x_2) - f(x_1, x_2)\right| \\ &\quad \le \frac{1}{(2\pi)^2} \int_{\mathbb{T}^2} |f(x_1 - t_1, x_2 - t_2) - f(x_1, x_2)| \left|K_n^{1,\alpha}(t_1, t_2)\right| dt. \end{aligned}$$

We will integrate on the set

$$\{(t_1, t_2) : 0 < t_2 < t_1 < \pi\},$$

more exactly on

$$\bigcup_{i=1}^{5} (A_i \cap S_{r/2}), \qquad \bigcup_{i=1}^{5} (A_i \cap S^c_{r/2}), \qquad \bigcup_{i=6}^{10} (A_i \cap S'_{r/2}), \qquad \bigcup_{i=6}^{10} (A_i \cap (S'_{r/2})^c),$$

where the sets A_i $(i = 1, \ldots, 10)$ are defined by (see Fig. 4.2)

$$\begin{aligned} A_1 &:= \{(x_1, x_2) : 0 < x_1 \le 2/n, 0 < x_2 < x_1 < \pi, x_2 \le \pi/2\}, \\ A_2 &:= \{(x_1, x_2) : 2/n < x_1 < \pi, 0 < x_2 \le 1/n, x_2 \le \pi/2\}, \\ A_3 &:= \{(x_1, x_2) : 2/n < x_1 < \pi, 1/n < x_2 \le x_1/2, x_2 \le \pi/2\}, \\ A_4 &:= \{(x_1, x_2) : 2/n < x_1 < \pi, x_1/2 < x_2 \le x_1 - 1/n, x_2 \le \pi/2\}, \\ A_5 &:= \{(x_1, x_2) : 2/n < x_1 < \pi, x_1 - 1/n < x_2 < x_1, x_2 \le \pi/2\} \\ A_6 &:= \{(x_1, x_2) : x_2 > \pi/2, \pi - 2/n \le x_2 < \pi, 0 < x_2 < x_1 < \pi\}, \\ A_7 &:= \{(x_1, x_2) : \pi/2 < x_2 < \pi - 2/n, \pi - 1/n < x_1 < \pi\}, \\ A_8 &:= \{(x_1, x_2) : \pi/2 < x_2 < \pi - 2/n, (\pi + x_2)/2 < x_1 \le \pi - 1/n\}, \\ A_9 &:= \{(x_1, x_2) : \pi/2 < x_2 < \pi - 2/n, x_2 + 1/n < x_1 \le (\pi + x_2)/2\}, \\ A_{10} &:= \{(x_1, x_2) : \pi/2 < x_2 < \pi - 2/n, x_2 < x_1 \le x_2 + 1/n\}. \end{aligned}$$

Since $A_1 \subset S_{r/2}$ and $A_6 \subset S'_{r/2}$, we deduce by (4.6.1) that

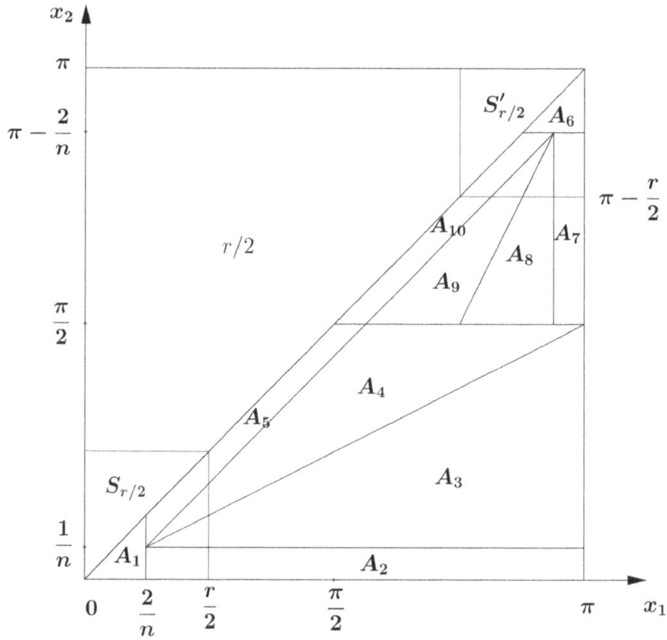

Fig. 4.2 The sets A_i

$$\begin{aligned}
&\int_{A_1} |f(x_1 - t_1, x_2 - t_2) - f(x_1, x_2)| \left|K_n^{1,\alpha}(t_1, t_2)\right| dt \\
&\leq Cn^2 \int_0^{2/n} \int_0^{2/n} |f(x_1 - t_1, x_2 - t_2) - f(x_1, x_2)| \, dt_2 \, dt_1 \\
&\leq CU_r^{\omega,1} f(x_1, x_2) < C\epsilon
\end{aligned}$$

and

$$\begin{aligned}
&\int_{A_6} |f(x_1 - t_1, x_2 - t_2) - f(x_1, x_2)| \left|K_n^{1,\alpha}(t_1, t_2)\right| dt \\
&\leq Cn^2 \int_{\pi-2/n}^{\pi} \int_{\pi-2/n}^{\pi} |f(x_1 - t_1, x_2 - t_2) - f(x_1, x_2)| \, dt_2 \, dt_1 \\
&\leq Cn^2 \int_{-2/n}^{0} \int_{-2/n}^{0} |f(x_1 - u_1 - \pi, x_2 - u_2 - \pi) - f(x_1, x_2)| \, du_2 \, du_1 \\
&\leq CU_r^{\omega,1} f(x_1, x_2) < C\epsilon.
\end{aligned}$$

Let us denote by r_0 the largest number i, for which $r/2 \leq 2^{i+1}/n < r$. By (4.6.2),

$$\int_{A_2 \cap S_{r/2}} |f(x_1 - t_1, x_2 - t_2) - f(x_1, x_2)| \left| K_n^{1,\alpha}(t_1, t_2) \right| dt$$
$$\leq C \sum_{i=1}^{r_0} \left(\frac{2^i}{n} - \frac{1}{n} \right)^{-1} \left(\frac{2^i}{n} \right)^{-1}$$
$$\int_{2^i/n}^{2^{i+1}/n} \int_0^{1/n} |f(x_1 - t_1, x_2 - t_2) - f(x_1, x_2)| \, dt_2 \, dt_1$$
$$\leq C \sum_{i=1}^{r_0} 2^{(\omega-1)i} 2^{-\omega i} \left(\frac{n^2}{2^i} \right)$$
$$\int_{2^i/n}^{2^{i+1}/n} \int_0^{1/n} |f(x_1 - t_1, x_2 - t_2) - f(x_1, x_2)| \, dt_2 \, dt_1$$
$$\leq C \sum_{i=1}^{r_0} 2^{(\omega-1)i} U_r^{\omega,1} f(x_1, x_2) < C\epsilon.$$

Similarly,

$$\int_{A_7 \cap S'_{r/2}} |f(x_1 - t_1, x_2 - t_2) - f(x_1, x_2)| \left| K_n^{1,\alpha}(t_1, t_2) \right| dt$$
$$\leq C \sum_{i=1}^{r_0} \left(\frac{2^i}{n} - \frac{1}{n} \right)^{-1} \left(\frac{2^i}{n} \right)^{-1}$$
$$\int_{\pi - 2^{i+1}/n}^{\pi - 2^i/n} \int_{\pi - 1/n}^{\pi} |f(x_1 - t_1, x_2 - t_2) - f(x_1, x_2)| \, dt_1 \, dt_2$$
$$\leq C \sum_{i=1}^{r_0} 2^{(\omega-1)i} 2^{-\omega i} \left(\frac{n^2}{2^i} \right)$$
$$\int_{-2^{i+1}/n}^{-2^i/n} \int_{-1/n}^{0} |f(x_1 - t_1 - \pi, x_2 - t_2 - \pi) - f(x_1, x_2)| \, dt_1 \, dt_2$$
$$\leq C \sum_{i=1}^{r_0} 2^{(\omega-1)i} U_r^{\omega,1} f(x_1, x_2) < C\epsilon.$$

On the other hand, we get that

$$\int_{A_2 \cap S^c_{r/2}} |f(x_1 - t_1, x_2 - t_2) - f(x_1, x_2)| \left| K_n^{1,\alpha}(t_1, t_2) \right| dt$$
$$+ \int_{A_7 \cap (S'_{r/2})^c} |f(x_1 - t_1, x_2 - t_2) - f(x_1, x_2)| \left| K_n^{1,\alpha}(t_1, t_2) \right| dt$$
$$\leq C \sum_{i=r_0}^{\infty} 2^{(\omega-1)i} \mathcal{M}^{\omega,1} f(x_1, x_2) + C \sum_{i=r_0}^{\infty} 2^{-i} |f(x_1, x_2)|$$

$$
\begin{aligned}
&\le C 2^{(\omega-1)r_0} \mathcal{M}^{\omega,1} f(x_1,x_2) + C 2^{-r_0} |f(x_1,x_2)| \\
&\le C (nr)^{\omega-1} \mathcal{M}^{\omega,1} f(x_1,x_2) + C (nr)^{-1} |f(x_1,x_2)| \to 0
\end{aligned}
$$

as $n \to \infty$.

Since $t_1 - t_2 > t_1/2$ and $t_1 - t_2 > t_2$ on A_3, we obtain by (4.6.3) that

$$
\begin{aligned}
\left|K_n^{1,\alpha}(t_1,t_2)\right| &\le C n^{-\alpha} (t_1 - t_2)^{-1-\alpha/2} (t_1 - t_2)^{-\beta+\alpha/2} t_2^{\beta-\alpha-1} \\
&\le C n^{-\alpha} t_1^{-1-\alpha/2} t_2^{-1-\alpha/2},
\end{aligned}
$$

whenever $\beta > \alpha/2$. Hence

$$
\begin{aligned}
&\int_{A_3 \cap S_{r/2}} |f(x_1 - t_1, x_2 - t_2) - f(x_1,x_2)| \left|K_n^{1,\alpha}(t_1,t_2)\right| dt \\
&\quad \le C \sum_{i=1}^{r_0} \sum_{j=0}^{i-1} n^{-\alpha} \left(\frac{2^i}{n}\right)^{-1-\alpha/2} \left(\frac{2^j}{n}\right)^{-1-\alpha/2} \\
&\qquad \int_{2^i/n}^{2^{i+1}/n} \int_{2^j/n}^{2^{j+1}/n} |f(x_1 - t_1, x_2 - t_2) - f(x_1,x_2)| \, dt_2 \, dt_1 \\
&\quad \le C \sum_{i=1}^{r_0} \sum_{j=0}^{i-1} 2^{(\omega-\alpha/2)(i+j)} 2^{-\omega(i+j)} \left(\frac{n^2}{2^{i+j}}\right) \\
&\qquad \int_{2^i/n}^{2^{i+1}/n} \int_{2^j/n}^{2^{j+1}/n} |f(x_1 - t_1, x_2 - t_2) - f(x_1,x_2)| \, dt_2 \, dt_1 \\
&\quad \le C \sum_{i=1}^{r_0} \sum_{j=0}^{i-1} 2^{(\omega-\alpha/2)(i+j)} U_r^{\omega,1} f(x_1,x_2) < C\epsilon.
\end{aligned}
$$

On A_8, $t_1 - t_2 > (\pi - t_2)/2 \ge (\pi - t_1)/2$ and so (4.6.5) implies

$$
\begin{aligned}
\left|K_n^{1,\alpha}(t_1,t_2)\right| &\le C n^{-\alpha} (t_1 - t_2)^{-1-\alpha/2} (t_1 - t_2)^{-\beta+\alpha/2} (\pi - t_2)^{\beta-\alpha-1} \\
&\le C n^{-\alpha} (\pi - t_1)^{-1-\alpha/2} (\pi - t_2)^{-1-\alpha/2}
\end{aligned}
$$

if $\beta > \alpha/2$. From this it follows

$$
\begin{aligned}
&\int_{A_8 \cap S'_{r/2}} |f(x_1 - t_1, x_2 - t_2) - f(x_1,x_2)| \left|K_n^{1,\alpha}(t_1,t_2)\right| dt \\
&\quad \le C \sum_{i=1}^{r_0} \sum_{j=0}^{i-1} n^{-\alpha} \left(\frac{2^i}{n}\right)^{-1-\alpha/2} \left(\frac{2^j}{n}\right)^{-1-\alpha/2} \\
&\qquad \int_{\pi-2^{i+1}/n}^{\pi-2^i/n} \int_{\pi-2^{j+1}/n}^{\pi-2^j/n} |f(x_1 - t_1, x_2 - t_2) - f(x_1,x_2)| \, dt_1 \, dt_2
\end{aligned}
$$

$$\leq C \sum_{i=1}^{r_0} \sum_{j=0}^{i-1} 2^{(\omega-\alpha/2)(i+j)} 2^{-\omega(i+j)} \left(\frac{n^2}{2^{i+j}} \right)$$

$$\int_{-2^{i+1}/n}^{-2^i/n} \int_{-2^{j+1}/n}^{-2^j/n} |f(x_1 - t_1 - \pi, x_2 - t_2 - \pi) - f(x_1, x_2)| \, dt_1 \, dt_2$$

$$\leq C \sum_{i=1}^{r_0} \sum_{j=0}^{i-1} 2^{(\omega-\alpha/2)(i+j)} U_r^{\omega,1} f(x_1, x_2) < C\epsilon.$$

Similarly,

$$\int_{A_3 \cap S_{r/2}^c} |f(x_1 - t_1, x_2 - t_2) - f(x_1, x_2)| \left| K_n^{1,\alpha}(t_1, t_2) \right| dt$$

$$+ \int_{A_8 \cap (S'_{r/2})^c} |f(x_1 - t_1, x_2 - t_2) - f(x_1, x_2)| \left| K_n^{1,\alpha}(t_1, t_2) \right| dt$$

$$\leq C \sum_{i=r_0}^{\infty} \sum_{j=0}^{i-1} 2^{(\omega-\alpha/2)(i+j)} \mathcal{M}^{\omega,1} f(x_1, x_2) + C \sum_{i=r_0}^{\infty} \sum_{j=0}^{i-1} 2^{-\alpha(i+j)/2} |f(x_1, x_2)|$$

$$\leq C 2^{(\omega-\alpha/2)r_0} \mathcal{M}^{\omega,1} f(x_1, x_2) + C 2^{-\alpha r_0/2} |f(x_1, x_2)| \to 0$$

as $n \to \infty$.

Since $t_2 > t_1/2$ on A_4, (4.6.3) with $\beta = \alpha/2$ implies

$$\left| K_n^{1,\alpha}(t_1, t_2) \right| \leq C n^{-\alpha} (t_1 - t_2)^{-1-\alpha/2} t_2^{-1-\alpha/2}$$
$$\leq C n^{-\alpha} t_1^{-1-\alpha/2} (t_1 - t_2)^{-1-\alpha/2}$$

and so

$$\int_{A_4 \cap S_{r/2}} |f(x_1 - t_1, x_2 - t_2) - f(x_1, x_2)| \left| K_n^{1,\alpha}(t_1, t_2) \right| dt$$

$$\leq C \sum_{i=1}^{r_0} \sum_{j=0}^{i-1} n^{-\alpha} \left(\frac{2^i}{n} \right)^{-1-\alpha/2} \left(\frac{2^j}{n} \right)^{-1-\alpha/2}$$

$$\int_{2^i/n}^{2^{i+1}/n} \int_{t_1 - 2^{j+1}/n}^{t_1 - 2^j/n} |f(x_1 - t_1, x_2 - t_2) - f(x_1, x_2)| \, dt_2 \, dt_1$$

$$\leq C \sum_{i=1}^{r_0} \sum_{j=0}^{i-1} 2^{(\omega-\alpha/2)(i+j)} 2^{-\omega(i+j)} \left(\frac{n^2}{2^{i+j}} \right)$$

$$\int_{2^i/n}^{2^{i+1}/n} \int_{t_1 - 2^{j+1}/n}^{t_1 - 2^j/n} |f(x_1 - t_1, x_2 - t_2) - f(x_1, x_2)| \, dt_2 \, dt_1$$

$$\leq C \sum_{i=1}^{r_0} \sum_{j=0}^{i-1} 2^{(\omega-\alpha/2)(i+j)} U_r^{\omega,2} f(x_1, x_2) < C\epsilon.$$

Inequality (4.6.5) with $\beta = \alpha/2$ yields

$$\left|K_n^{1,\alpha}(t_1, t_2)\right| \leq C n^{-\alpha} (t_1 - t_2)^{-1-\alpha/2} (\pi - t_2)^{-1-\alpha/2}.$$

Thus

$$\int_{A_9 \cap S'_{r/2}} |f(x_1 - t_1, x_2 - t_2) - f(x_1, x_2)| \left|K_n^{1,\alpha}(t_1, t_2)\right| dt$$
$$\leq C \sum_{i=1}^{r_0} \sum_{j=0}^{i-1} n^{-\alpha} \left(\frac{2^i}{n}\right)^{-1-\alpha/2} \left(\frac{2^j}{n}\right)^{-1-\alpha/2}$$
$$\int_{\pi-2^{i+1}/n}^{\pi-2^i/n} \int_{t_2+2^j/n}^{t_2+2^{j+1}/n} |f(x_1 - t_1, x_2 - t_2) - f(x_1, x_2)| \, dt_1 \, dt_2$$
$$\leq C \sum_{i=1}^{r_0} \sum_{j=0}^{i-1} 2^{(\omega-\alpha/2)(i+j)} 2^{-\omega(i+j)} \left(\frac{n^2}{2^{i+j}}\right)$$
$$\int_{-2^{i+1}/n}^{-2^i/n} \int_{t_2+2^j/n}^{t_2+2^{j+1}/n} |f(x_1 - t_1 - \pi, x_2 - t_2 - \pi) - f(x_1, x_2)| \, dt_1 \, dt_2$$
$$\leq C \sum_{i=1}^{r_0} \sum_{j=0}^{i-1} 2^{(\omega-\alpha/2)(i+j)} U_r^{\omega,4} f(x_1, x_2) < C\epsilon.$$

Similarly,

$$\int_{A_4 \cap S^c_{r/2}} |f(x_1 - t_1, x_2 - t_2) - f(x_1, x_2)| \left|K_n^{1,\alpha}(t_1, t_2)\right| dt$$
$$+ \int_{A_9 \cap (S'_{r/2})^c} |f(x_1 - t_1, x_2 - t_2) - f(x_1, x_2)| \left|K_n^{1,\alpha}(t_1, t_2)\right| dt$$
$$\leq C \sum_{i=r_0}^{\infty} \sum_{j=0}^{i-1} 2^{(\omega-\alpha/2)(i+j)} \left(\mathcal{M}^{\omega,2} f(x_1, x_2) + \mathcal{M}^{\omega,4} f(x_1, x_2)\right)$$
$$+ C \sum_{i=r_0}^{\infty} \sum_{j=0}^{i-1} 2^{-\alpha(i+j)/2} |f(x_1, x_2)|$$
$$\leq C 2^{(\omega-\alpha/2)r_0} \mathcal{M}^{\omega} f(x_1, x_2) + C 2^{-\alpha r_0/2} |f(x_1, x_2)| \to 0$$

as $n \to \infty$.

Inequality (4.6.4) implies

$$\left|K_n^{1,\alpha}(t_1,t_2)\right| \le Cn^{1-\alpha}t_1^{-\alpha-1}$$

on the set A_5 and so

$$\begin{aligned}
&\int_{A_5\cap S_{r/2}} |f(x_1-t_1,x_2-t_2)-f(x_1,x_2)| \left|K_n^{1,\alpha}(t_1,t_2)\right| dt \\
&\le C\sum_{i=1}^{r_0} n^{1-\alpha}\left(\frac{2^i}{n}\right)^{-\alpha-1} \int_{2^i/n}^{2^{i+1}/n}\int_{t_1-1/n}^{t_1} |f(x_1-t_1,x_2-t_2)-f(x_1,x_2)|\,dt_1\,dt_2 \\
&\le C\sum_{i=1}^{r_0} 2^{(\omega-\alpha)i}2^{-\omega i}\left(\frac{n^2}{2^i}\right) \\
&\qquad \int_{2^i/n}^{2^{i+1}/n}\int_{t_1-1/n}^{t_1} |f(x_1-t_1,x_2-t_2)-f(x_1,x_2)|\,dt_1\,dt_2 \\
&\le C\sum_{i=1}^{r_0} 2^{(\omega-1)i}U_r^{\omega,2}f(x_1,x_2) < C\epsilon.
\end{aligned}$$

In the same way, by (4.6.6),

$$\begin{aligned}
&\int_{A_{10}\cap S'_{r/2}} |f(x_1-t_1,x_2-t_2)-f(x_1,x_2)| \left|K_n^{1,\alpha}(t_1,t_2)\right| dt \\
&\le C\sum_{i=1}^{r_0} n^{1-\alpha}\left(\frac{2^i}{n}\right)^{-1-\alpha} \\
&\qquad \int_{\pi-2^{i+1}/n}^{\pi-2^i/n}\int_{t_2}^{t_2+1/n} |f(x_1-t_1,x_2-t_2)-f(x_1,x_2)|\,dt_1\,dt_2 \\
&\le C\sum_{i=1}^{r_0}\sum_{j=0}^{i-1} 2^{(\omega-\alpha)i}2^{-\omega i}\left(\frac{n^2}{2^i}\right) \\
&\qquad \int_{-2^{i+1}/n}^{-2^i/n}\int_{t_2}^{t_2+1/n} |f(x_1-t_1-\pi,x_2-t_2-\pi)-f(x_1,x_2)|\,dt_1\,dt_2 \\
&\le C\sum_{i=1}^{r_0}\sum_{j=0}^{i-1} 2^{(\omega-\alpha)i}U_r^{\omega,4}f(x_1,x_2) < C\epsilon.
\end{aligned}$$

Finally,

$$\begin{aligned}
&\int_{A_5 \cap S_{r/2}^c} |f(x_1 - t_1, x_2 - t_2) - f(x_1, x_2)| \left|K_n^{1,\alpha}(t_1, t_2)\right| dt \\
&\quad + \int_{A_{10} \cap (S'_{r/2})^c} |f(x_1 - t_1, x_2 - t_2) - f(x_1, x_2)| \left|K_n^{1,\alpha}(t_1, t_2)\right| dt \\
&\quad \le C \sum_{i=r_0}^{\infty} 2^{(\omega-\alpha)i} \left(\mathcal{M}^{\omega,2} f(x_1, x_2) + \mathcal{M}^{\omega,4} f(x_1, x_2)\right) \\
&\qquad + C \sum_{i=r_0}^{\infty} 2^{-\alpha i} |f(x_1, x_2)| \\
&\quad \le C 2^{(\omega-\alpha)r_0} \mathcal{M}^{\omega} f(x_1, x_2) + C 2^{-\alpha r_0} |f(x_1, x_2)| \to 0
\end{aligned}$$

as $n \to \infty$, which finishes the proof. ■

In this way, we obtain Corollary 2.5.9 for the ℓ_1-Cesàro means, i.e.,

$$\lim_{n\to\infty} \sigma_n^{1,\alpha} f = f \qquad \text{a.e.}$$

if $f \in L_1(\mathbb{T}^d)$. For $1 < p < \infty$, we get again a better result.

Theorem 4.6.3 *Suppose that $0 < \alpha < \infty$, $1/(\min(\alpha, 1) < p < \infty$, $1/p + 1/q = 1$, $0 < \omega < (1 + q\min(\alpha, 1) - q)/2q$ and $\mathcal{M}_p^{\omega,1} f(x)$ is finite. If $f \in L_p(\mathbb{T}^2)$ is periodic with respect to π and x is a (p, ω)-Lebesgue point of f, then*

$$\lim_{n\to\infty} \sigma_n^{1,\alpha} f(x) = f(x).$$

Proof We prove the theorem again for $0 < \alpha \le 1$. Note that $1/\alpha < p < \infty$ implies $1 < q < 1/(1 - \alpha)$ and so $1 + \alpha q - q > 0$. Moreover,

$$\frac{1 + \alpha q - q}{2q} < \frac{\alpha}{2}.$$

Fix a number $0 < r < 1$ such that

$$U_{r,p}^{\omega,1} f(x_1, x_2) < \epsilon.$$

In Theorem 4.6.2, we have verified that

$$\int_{A_i} |f(x_1 - t_1, x_2 - t_2) - f(x_1, x_2)| \left|K_n^{1,\alpha}(t_1, t_2)\right| dt \to 0,$$

for $i = 1, 2, 3, 6, 7, 8$, as $n \to \infty$ and $\omega < \alpha/2$. So we need to consider the sets A_4, A_5, A_9 and A_{10}, only.

We apply (4.6.3) with $\beta = 0$ and that $t_2 > t_1/2$ on A_4 to obtain

$$\left|K_n^{1,\alpha}(t_1,t_2)\right| \le Cn^{-\alpha}(t_1-t_2)^{-1}t_1^{-\alpha-1}.$$

By Hölder's inequality,

$$\begin{aligned}
&\int_{A_4\cap S_{r/2}} |f(x_1-t_1,x_2-t_2)-f(x_1,x_2)| \left|K_n^{1,\alpha}(t_1,t_2)\right| dt \\
&\le \sum_{i=1}^{r_0}\sum_{j=i-1}^{i}\int_{2^i/n}^{2^{i+1}/n}\int_{2^j/n}^{2^{j+1}/n} \\
&\qquad |f(x_1-t_1,x_2-t_2)-f(x_1,x_2)| \left|K_n^{1,\alpha}(t_1,t_2)\right| 1_{A_4}(t_1,t_2)\,dt_2\,dt_1 \\
&\le \sum_{i=1}^{r_0}\sum_{j=i-1}^{i}\left(\int_{2^i/n}^{2^{i+1}/n}\int_{2^j/n}^{2^{j+1}/n} |f(x_1-t_1,x_2-t_2)-f(x_1,x_2)|^p\,dt_2\,dt_1\right)^{1/p} \\
&\qquad \left(\int_{2^i/n}^{2^{i+1}/n}\int_{2^{i-1}/n}^{t_1-1/n} n^{-\alpha q}(t_1-t_2)^{-q}t_1^{-q(1+\alpha)}1_{A_4}(t_1,t_2)\,dt_2\,dt_1\right)^{1/q}.
\end{aligned}$$

Since $1-q<0$, we have

$$\begin{aligned}
&\int_{2^i/n}^{2^{i+1}/n}\int_{2^{i-1}/n}^{t_1-1/n} n^{-\alpha q}(t_1-t_2)^{-q}t_1^{-q(1+\alpha)}1_{A_4}(t_1,t_2)\,dt_2\,dt_1 \\
&\qquad \le Cn^{-\alpha q}\left(\frac{1}{n}\right)^{1-q}\int_{2^i/n}^{2^{i+1}/n} t_1^{-q(1+\alpha)}\,dt_1 \\
&\qquad \le Cn^{-\alpha q}\left(\frac{1}{n}\right)^{1-q}\left(\frac{2^i}{n}\right)^{1-q(1+\alpha)} \\
&\qquad \le C\left(\frac{n}{2^i}\right)^{2q-2}2^{-i(1+\alpha q-q)}
\end{aligned}$$

and so

$$\begin{aligned}
&\int_{A_4\cap S_{r/2}} |f(x_1-t_1,x_2-t_2)-f(x_1,x_2)| \left|K_n^{1,\alpha}(t_1,t_2)\right| dt \\
&\le C_p\sum_{i=1}^{r_0}\sum_{j=i-1}^{i} 2^{(\omega-(1+\alpha q-q)/2q)(i+j)} \\
&\qquad 2^{-\omega(i+j)}\left(\frac{n^2}{2^{i+j}}\int_{2^i/n}^{2^{i+1}/n}\int_{2^j/n}^{2^{j+1}/n} |f(x_1-t_1,x_2-t_2)-f(x_1,x_2)|^p\,dt_2\,dt_1\right)^{1/p} \\
&\le C_p\sum_{i=1}^{r_0}\sum_{j=i-1}^{i} 2^{(\omega-(1+\alpha q-q)/2q)(i+j)}U_{r,p}^{\omega,1}f(x_1,x_2) < C_p\epsilon.
\end{aligned}$$

Let us use (4.6.5) with $\beta=0$ to get that

$$\int_{A_9\cap S'_{r/2}} |f(x_1-t_1,x_2-t_2)-f(x_1,x_2)|\left|K_n^{1,\alpha}(t_1,t_2)\right| dt$$
$$\le \sum_{i=1}^{r_0}\sum_{j=i-1}^{i}\left(\int_{\pi-2^{i+1}/n}^{\pi-2^i/n}\int_{\pi-2^{j+1}/n}^{\pi-2^j/n}|f(x_1-t_1,x_2-t_2)-f(x_1,x_2)|^p\,dt_1\,dt_2\right)^{1/p}$$
$$\left(\int_{\pi-2^{i+1}/n}^{\pi-2^i/n}\int_{t_2+1/n}^{\pi-2^{i-1}/n} n^{-\alpha q}(t_1-t_2)^{-q}(\pi-t_2)^{-q(1+\alpha)}1_{A_4}(t_1,t_2)\,dt_1\,dt_2\right)^{1/q}.$$

Therefore,

$$\int_{\pi-2^{i+1}/n}^{\pi-2^i/n}\int_{t_2+1/n}^{\pi-2^{i-1}/n} n^{-\alpha q}(t_1-t_2)^{-q}(\pi-t_2)^{-q(1+\alpha)}1_{A_4}(t_1,t_2)\,dt_1\,dt_2$$
$$\le Cn^{-\alpha q}\left(\frac{1}{n}\right)^{1-q}\int_{\pi-2^{i+1}/n}^{\pi-2^i/n}(\pi-t_2)^{-q(1+\alpha)}\,dt_2$$
$$\le Cn^{-\alpha q}\left(\frac{1}{n}\right)^{1-q}\left(\frac{2^i}{n}\right)^{1-q(1+\alpha)}$$
$$\le C\left(\frac{n}{2^i}\right)^{2q-2}2^{-i(1+\alpha q-q)}$$

and

$$\int_{A_9\cap S'_{r/2}} |f(x_1-t_1,x_2-t_2)-f(x_1,x_2)|\left|K_n^{1,\alpha}(t_1,t_2)\right| dt$$
$$\le C_p\sum_{i=1}^{r_0}\sum_{j=i-1}^{i}2^{(\omega-(1+\alpha q-q)/2q)(i+j)}2^{-\omega(i+j)}$$
$$\left(\frac{n^2}{2^{i+j}}\int_{\pi-2^{i+1}/n}^{\pi-2^i/n}\int_{\pi-2^{j+1}/n}^{\pi-2^j/n}|f(x_1-t_1,x_2-t_2)-f(x_1,x_2)|^p\,dt_2\,dt_1\right)^{1/p}$$
$$\le C_p\sum_{i=1}^{r_0}\sum_{j=i-1}^{i}2^{(\omega-(1+\alpha q-q)/2q)(i+j)}U_{r,p}^{\omega,1}f(x_1,x_2) < C_p\epsilon.$$

Similarly, we can see that

$$\int_{A_4\cap S^c_{r/2}} |f(x_1-t_1,x_2-t_2)-f(x_1,x_2)|\left|K_n^{1,\alpha}(t_1,t_2)\right| dt$$
$$+\int_{A_9\cap(S'_{r/2})^c} |f(x_1-t_1,x_2-t_2)-f(x_1,x_2)|\left|K_n^{1,\alpha}(t_1,t_2)\right| dt$$
$$\le C_p\sum_{i=r_0}^{\infty}\sum_{j=i-1}^{i}2^{(\omega-(1+\alpha q-q)/2q)(i+j)}\mathcal{M}_p^{\omega,1}f(x_1,x_2)$$

$$
\begin{aligned}
&+ C_p \sum_{i=r_0}^{\infty} \sum_{j=i-1}^{i} 2^{-(1+\alpha q-q)(i+j)/2q} |f(x_1, x_2)| \\
&\leq C_p 2^{r_0(2\omega-(1+\alpha q-q)/q)} \mathcal{M}_p^{\omega,1} f(x_1, x_2) + C_p 2^{-r_0(1+\alpha q-q)/q} |f(x_1, x_2)| \\
&\leq C(nr)^{2\omega-(1+\alpha q-q)/q} \mathcal{M}_p^{\omega,1} f(x_1, x_2) + C(nr)^{-(1+\alpha q-q)} |f(x_1, x_2)| \to 0
\end{aligned}
$$

as $n \to \infty$.

On A_5, $t_2 > t_1/2$ and so (4.6.4) implies

$$
\begin{aligned}
&\int_{A_5 \cap S_{r/2}} |f(x_1 - t_1, x_2 - t_2) - f(x_1, x_2)| \left|K_n^{1,\alpha}(t_1, t_2)\right| dt \\
&\leq \sum_{i=1}^{r_0} \sum_{j=i-1}^{i} \left(\int_{2^i/n}^{2^{i+1}/n} \int_{2^j/n}^{2^{j+1}/n} |f(x_1 - t_1, x_2 - t_2) - f(x_1, x_2)|^p \, dt_2 \, dt_1 \right)^{1/p} \\
&\quad \left(\int_{2^i/n}^{2^{i+1}/n} \int_{t_1-1/n}^{t_1} n^{q(1-\alpha)} t_1^{-q(\alpha+1)} \, dt_2 \, dt_1 \right)^{1/q}.
\end{aligned}
$$

It is easy to see that

$$
\begin{aligned}
\int_{2^i/n}^{2^{i+1}/n} \int_{t_1-1/n}^{t_1} n^{q(1-\alpha)} t_1^{-q(\alpha+1)} \, dt_2 \, dt_1 &\leq n^{-1} n^{q(1-\alpha)} \left(\frac{2^i}{n} \right)^{1-q(\alpha+1)} \\
&\leq C \left(\frac{n}{2^i} \right)^{2q-2} 2^{-i(1+\alpha q-q)}.
\end{aligned}
$$

Then

$$
\begin{aligned}
&\int_{A_5 \cap S_{r/2}} |f(x_1 - t_1, x_2 - t_2) - f(x_1, x_2)| \left|K_n^{1,\alpha}(t_1, t_2)\right| dt \\
&\leq \sum_{i=1}^{r_0} \sum_{j=i-1}^{i} 2^{(\omega-(1+\alpha q-q)/2q)(i+j)} 2^{-\omega(i+j)} \\
&\quad \left(\frac{n^2}{2^{i+j}} \int_{2^i/n}^{2^{i+1}/n} \int_{2^j/n}^{2^{j+1}/n} |f(x_1 - t_1, x_2 - t_2) - f(x_1, x_2)|^p \, dt_2 \, dt_1 \right)^{1/p} \\
&\leq C_p \sum_{i=1}^{r_0} \sum_{j=i-1}^{i} 2^{(\omega-(1+\alpha q-q)/2q)(i+j)} U_{r,p}^{\omega,1} f(x_1, x_2) < C\epsilon.
\end{aligned}
$$

Let us use (4.6.6):

$$
\begin{aligned}
&\int_{A_{10}\cap S'_{r/2}} |f(x_1-t_1,x_2-t_2)-f(x_1,x_2)| \left|K_n^{1,\alpha}(t_1,t_2)\right| dt \\
&\le \sum_{i=1}^{r_0}\sum_{j=i-1}^{i} \left(\int_{\pi-2^{i+1}/n}^{\pi-2^i/n}\int_{\pi-2^{j+1}/n}^{\pi-2^j/n} |f(x_1-t_1,x_2-t_2)-f(x_1,x_2)|^p \, dt_1\, dt_2\right)^{1/p} \\
&\qquad \left(\int_{\pi-2^{i+1}/n}^{\pi-2^i/n}\int_{t_2}^{t_2+1/n} n^{q(1-\alpha)}(\pi-t_2)^{-q(\alpha+1)} \, dt_1\, dt_2\right)^{1/q}.
\end{aligned}
$$

Then

$$
\begin{aligned}
\int_{\pi-2^{i+1}/n}^{\pi-2^i/n}\int_{t_2}^{t_2+1/n} n^{q(1-\alpha)}(\pi-t_2)^{-q(\alpha+1)} \, dt_1\, dt_2 &\le n^{-1}n^{q(1-\alpha)}\left(\frac{2^i}{n}\right)^{1-q(\alpha+1)} \\
&\le C\left(\frac{n}{2^i}\right)^{2q-2} 2^{-i(1+\alpha q-q)}
\end{aligned}
$$

and

$$
\begin{aligned}
&\int_{A_{10}\cap S'_{r/2}} |f(x_1-t_1,x_2-t_2)-f(x_1,x_2)| \left|K_n^{1,\alpha}(t_1,t_2)\right| dt \\
&\le \sum_{i=1}^{r_0}\sum_{j=i-1}^{i} 2^{(\omega-(1+\alpha q-q)/2q)(i+j)}2^{-\omega(i+j)} \\
&\qquad \left(\frac{n^2}{2^{i+j}}\int_{\pi-2^{i+1}/n}^{\pi-2^i/n}\int_{\pi-2^{j+1}/n}^{\pi-2^j/n} |f(x_1-t_1,x_2-t_2)-f(x_1,x_2)|^p \, dt_2\, dt_1\right)^{1/p} \\
&\le C_p \sum_{i=1}^{r_0}\sum_{j=i-1}^{i} 2^{(\omega-(1+\alpha q-q)/2q)(i+j)} U_{r,p}^{\omega,1} f(x_1,x_2) < C\epsilon.
\end{aligned}
$$

Finally,

$$
\begin{aligned}
&\int_{A_5\cap S^c_{r/2}} |f(x_1-t_1,x_2-t_2)-f(x_1,x_2)| \left|K_n^{1,\alpha}(t_1,t_2)\right| dt \\
&\quad + \int_{A_{10}\cap (S'_{r/2})^c} |f(x_1-t_1,x_2-t_2)-f(x_1,x_2)| \left|K_n^{1,\alpha}(t_1,t_2)\right| dt \\
&\le C_p \sum_{i=r_0}^{\infty}\sum_{j=i-1}^{i} 2^{(\omega-(1+\alpha q-q)/2q)(i+j)} \mathcal{M}_p^{\omega,1} f(x_1,x_2) \\
&\qquad + C_p \sum_{i=r_0}^{\infty}\sum_{j=i-1}^{i} 2^{-(1+\alpha q-q)(i+j)/2q} |f(x_1,x_2)| \\
&\le C_p 2^{r_0(2\omega-(1+\alpha q-q)/q)} \mathcal{M}_p^{\omega,1} f(x_1,x_2) + C_p 2^{-r_0(1+\alpha q-q)/q}|f(x_1,x_2)| \to 0
\end{aligned}
$$

as $n \to \infty$. The proof of the theorem is complete. ∎

Let us point out this result for $\alpha \geq 1$. Recall that for $\alpha = 1$, we get the ℓ_1-Fejér means.

Theorem 4.6.4 *Suppose that* $1 \leq \alpha < \infty$, $1 < p < \infty$, $1/p + 1/q = 1$, $0 < \omega < 1/2q$ *and* $\mathcal{M}_p^{\omega,1} f(x)$ *is finite. If* $f \in L_p(\mathbb{T}^d)$ *is periodic with respect to* π *and* x *is a* (p, ω)*-Lebesgue point of* f*, then*

$$\lim_{n\to\infty} \sigma_n^{1,\alpha} f(x) = f(x).$$

Recall that the ℓ_1-θ-means were introduced by

$$\sigma_n^{1,\theta} f(x) := \sum_{k\in\mathbb{Z}^d} \theta\left(\frac{\|k\|_1}{n}\right) \widehat{f}(k)e^{\imath k\cdot x}$$

in Sect. 2.6.1. The next two results can be proved as Theorems 4.5.9 and 4.5.10. For more details see the papers [325, 326].

Theorem 4.6.5 *Suppose that* θ *satisfies (2.6.2) and (2.6.3),* $0 < \omega < 1/d$ *and* $\mathcal{M}^\omega f(x)$ *is finite. If* $f \in L_1(\mathbb{T}^d)$ *is periodic with respect to* π *and* x *is a strong* $(1, \omega)$*-Lebesgue point of* f*, then*

$$\lim_{n\to\infty} \sigma_n^{1,\theta} f(x) = f(x).$$

Theorem 4.6.6 *Suppose that* θ *satisfies (2.6.2) and (2.6.3),* $1 < p < \infty$, $1/p + 1/q = 1$, $0 < \omega < 1/2q$ *and* $\mathcal{M}_p^{\omega,1} f(x)$ *is finite. If* $f \in L_p(\mathbb{T}^d)$ *is periodic with respect to* π *and* x *is a* (p, ω)*-Lebesgue point of* f*, then*

$$\lim_{n\to\infty} \sigma_n^{1,\theta} f(x) = f(x).$$

Bibliography

1. W. Abu-Shammala, A. Torchinsky, The Hardy-Lorentz spaces $H^{p,q}(R^n)$. Studia Math. **182**, 283–294 (2007)
2. T. Akhobadze, On the convergence of generalized Cesáro means of trigonometric Fourier series. I. Acta Math. Hung. **115**(1–2), 59–78 (2007)
3. G. Alexits, *Konvergenzprobleme der Orthogonalreihen* (Akadémiai Kiadó, Budapest, 1960)
4. I.A. Aliev, E. Sağlik, A new wavelet-like transform associated with the Riesz-Bochner integral and relevant reproducing formula. Mediterr. J. Math. **13**(6), 4711–4722 (2016)
5. L. Angeloni, E. Liflyand, G. Vinti, Real Hardy space, multidimensional variations, and integrability of the Fourier transform. Complex Anal. Oper. Theory **14**(6), 10 (2020)
6. M. Annoni, Almost everywhere convergence for modified Bochner-Riesz means at the critical index for $p \geq 2$. Pac. J. Math. **286**(2), 257–275 (2017)
7. N.Y. Antonov, Convergence of Fourier series. East J. Approx. **2**, 187–196 (1996)
8. N.Y. Antonov, On the convergence almost everywhere of multiple trigonometric Fourier series over cubes. Izv. Ross. Akad. Nauk Ser. Mat. **68**, 3–22 (2004). (Russian), English translation in Izv. Math. **68**(2004), 223–241 (2004)
9. J. Arias de Reyna, *Pointwise Convergence of Fourier Series* (Springer, Berlin, 2002)
10. J. Arias de Reyna, Pointwise convergence of Fourier series. J. Lond. Math. Soc., II. Ser. **65**(1), 139–153 (2002)
11. R. Ashurov, Summability almost everywhere of Fourier series in L_p with respect to eigenfunctions. Math. Notes **34**, 913–916 (1983)
12. R. Ashurov, Summation of multiple trigonometric Fourier series. Math. Notes **49**(6), 563–568 (1991)
13. R. Ashurov, Convergence of the continuous wavelet transforms on the entire Lebesgue set of L_p-functions. Int. J. Wavelets Multiresolut. Inf. Process. **9**, 675–683 (2011)
14. R. Ashurov, On the almost-everywhere convergence of the continuous wavelet transforms. Proc. Roy. Soc. Edinburgh Sect. A **142**, 1121–1129 (2012)
15. G. Bachman, L. Narici, E. Beckenstein, *Fourier and Wavelet Analysis* (Springer, Berlin, 2000)
16. A.N. Bakhvalov, Cesaro summation of Fourier series of functions from multidimensional Waterman classes. Dokl. Math. **83**(2), 247–249 (2011)
17. A.N. Bakhvalov, Continuity in Λ-variation and summation of multiple Fourier series by Cesàro methods. Math. Notes **90**(4), 469–484 (2011)
18. S. Baron, E. Liflyand, M. Skopina, On the summability of double Fourier series at a point. Analysis **16**(2), 195–205 (1996)

F. Weisz, *Lebesgue Points and Summability of Higher Dimensional Fourier Series*,
https://doi.org/10.1007/978-3-030-74636-0_1

19. N.K. Bary, *A Treatise on Trigonometric Series* (Pergamon Press, Oxford, 1964)
20. E.S. Belinsky, Summability of multiple Fourier series at Lebesgue points. Theory Funct. Funct. Anal. Appl. **23**, 3–12 (1975). (Russian)
21. E.S. Belinsky, Application of the Fourier transform to summability of Fourier series. Sib. Mat. Zh. **18**, 497–511, (Russian) – English transl.: Siberian Math. J. **18**, 353–363 (1977)
22. E.S. Belinsky, Summability of Fourier series with the method of lacunary arithmetical means at the Lebesgue points. Proc. Amer. Math. Soc. **125**, 3689–3693 (1997)
23. E.S. Belinsky, Strong summability for the Marcinkiewicz means in the integral metric and related questions. J. Austral. Math. Soc. Ser. A. **65**, 303–312 (1998)
24. E.S. Belinsky, E. Liflyand, Boundedness of convolution operators on Hardy spaces. Comput. Methods Funct. Theory **19**(2), 183–191 (2019)
25. E.S. Belinsky, E.R. Liflyand, R.M. Trigub, The Banach algebra A^* and its properties. J. Fourier Anal. Appl. **3**, 103–129 (1997)
26. J.J. Benedetto, *Harmonic Analysis and Applications*. Studies in Advanced Mathematics (CRC Press, 1997)
27. J.J. Benedetto, W. Czaja, *Integration and Modern Analysis* (Birkhäuser, Basel, 2009)
28. C. Bennett, R. Sharpley, *Interpolation of Operators*, vol. 129, Pure and Applied Mathematics (Academic Press, New York, 1988)
29. H. Berens, L. Li, The maximal Cesàro operator on Hardy spaces. Acta Sci. Math. (Szeged) **66**, 613–622 (2000)
30. H. Berens, Z. Li, Y. Xu, On l_1 Riesz summability of the inverse Fourier integral. Indag. Math. (N.S.) **12**, 41–53 (2001)
31. H. Berens, Y. Xu, Fejér means for multivariate Fourier series. Math. Z. **221**, 449–465 (1996)
32. H. Berens, Y. Xu, l-1 summability of multiple Fourier integrals and positivity. Math. Proc. Cambridge Philos. Soc. **122**, 149–172 (1997)
33. J. Bergh, J. Löfström, *Interpolation Spaces, an Introduction* (Springer, Berlin, 1976)
34. F. Bernicot, L. Grafakos, L. Song, L. Yan, The bilinear Bochner-Riesz problem. J. Anal. Math. **127**, 179–217 (2015)
35. P. Billard, Sur la convergence presque partout des séries de Fourier-Walsh des fonctions de l'espace $L^2[0, 1]$. Studia Math. **28**, 363–388 (1967)
36. S. Bochner, Summation of multiple Fourier series by spherical means. Trans. Amer. Math. Soc. **40**, 175–207 (1936)
37. S. Bochner, *Lectures on Fourier Integrals. With an Author's Supplement on Monotonic Functions, Stieltjes Integrals, and Harmonic Analysis.* Annals of Mathematics Studies. No. 42 (Princeton University Press, Princeton, N. J., 1959)
38. J. Bokor, F. Schipp, Approximate identification in Laguerre and Kautz bases. Automatica **34**, 463–468 (1998)
39. J. Bourgain, On the Hardy-Littlewood maximal function for the cube. Isr. J. Math. **203**, 275–293 (2014)
40. M. Bownik, Anisotropic Hardy spaces and wavelets. Mem. Am. Math. Soc. **781**, 122 (2003)
41. M. Bownik, Boundedness of operators on Hardy spaces via atomic decompositions. Proc. Amer. Math. Soc. **133**, 3535–3542 (2005)
42. M. Bownik, B. Li, D. Yang, Y. Zhou, Weighted anisotropic product Hardy spaces and boundedness of sublinear operators. Math. Nachr. **283**, 392–442 (2010)
43. L. Brandolini, G. Travaglini, Pointwise convergence of Fejér type means. Tohoku Math. J. (2) **49**(3), 323–336 (1997)
44. G. Brown, D. Feng, F. Móricz, The maximal Fejér operator on real Hardy spaces. Period. Math. Hung. **49**(1), 15–25 (2004)
45. G. Brown, D. Feng, F. Móricz, The maximal Riesz, Fejér, and Cesàro operators on real Hardy spaces. J. Fourier Anal. Appl. **10**(1), 27–50 (2004)
46. G. Brown, D. Feng, F. Móricz, Strong Cesàro summability of double Fourier integrals. Acta Math. Hungar. **115**, 1–12 (2007)
47. P.L. Butzer, R.J. Nessel, *Fourier Analysis and Approximation* (Birkhäuser Verlag, Basel, 1971)

48. R. Cabral, Differentiation of the integral, Hardy spaces and Calderón-Zygmund operators in the product setting. J. Math. Anal. Appl. **423**(2), 1480–1513 (2015)
49. A. Carbery, D. Gorges, G. Marletta, C. Thiele, Convergence almost everywhere of certain partial sums of Fourier integrals. Bull. Lond. Math. Soc. **35**(2), 225–228 (2003)
50. A. Carbery, A. Seeger, Weighted inequalities for Bochner-Riesz means in the plane. Q. J. Math. **51**(2), 155–167 (2000)
51. L. Carleson, On convergence and growth of partial sums of Fourier series. Acta Math. **116**, 135–157 (1966)
52. L. Carleson, P. Sjölin, Oscillatory integrals and a multiplier problem for the disc. Studia Math. **44**, 287–299 (1972)
53. M.J. Carro, L. Grafakos, J. Soria, Weighted weak-type (1, 1) estimates via Rubio de Francia extrapolation. J. Funct. Anal. **269**(5), 1203–1233 (2015)
54. S.Y.A. Chang, R. Fefferman, The Calderon-Zygmund decomposition on product domains. Amer. J. Math. **104**, 455–468 (1982)
55. S.Y.A. Chang, R. Fefferman, Some recent developments in Fourier analysis and H^p-theory on product domains. Bull. Amer. Math. Soc. **12**, 1–43 (1985)
56. L.K. Chen, D. Fan, The convergence of the Bochner-Riesz means at the critical index. Proc. Amer. Math. Soc. **124**, 2717–2726 (1996)
57. P. Chen, X.T. Duong, J. Li, L.A. Ward, L. Yan, Marcinkiewicz-type spectral multipliers on Hardy and Lebesgue spaces on product spaces of homogeneous type. J. Fourier Anal. Appl. **23**, 21–64 (2017)
58. M. Christ, Weak type endpoint bounds for Bochner-Riesz multipliers. Rev. Mat. Iberoam. **3**(1), 25–31 (1987)
59. M. Christ, Weak type (1,1) bounds for rough operators. Ann. Math. (2) **128**(1), 19–42 (1988)
60. O. Ciaurri, L. Roncal, The Bochner-Riesz means for Fourier-Bessel expansions. J. Funct. Anal. **228**(1), 89–113 (2005)
61. O. Ciaurri, L. Roncal, The Bochner-Riesz means for Fourier-Bessel expansions: norm inequalities for the maximal operator and almost everywhere convergence. J. Approx. Theory **167**, 121–146 (2013)
62. R.R. Coifman, G. Weiss, Extensions of Hardy spaces and their use in analysis. Bull. Amer. Math. Soc. **83**, 569–645 (1977)
63. L. Colzani, G. Gigante, A. Vargas, Localization for Riesz means of Fourier expansions. Trans. Am. Math. Soc. **366**(12), 6229–6245 (2014)
64. L. Colzani, G. Gigante, S. Volpi, Equiconvergence theorems for Sturm-Liouville expansions and sets of divergence for Bochner-Riesz means in Sobolev spaces. J. Fourier Anal. Appl. **19**(6), 1184–1206 (2013)
65. L. Colzani, M.H. Taibleson, G. Weiss, Maximal estimates for Cesàro and Riesz means on spheres. Indiana Univ. Math. J. **33**, 873–889 (1984)
66. L. Colzani, G. Travaglini, M. Vignati, Bochner-Riesz means of functions in weak-L^p. Monatsh. Math. **115**(1–2), 35–45 (1993)
67. L. Colzani, S. Volpi, Pointwise convergence of Bochner-Riesz means in Sobolev spaces, in *Trends in Harmonic Analysis. Selected Papers of the Conference on Harmonic Analysis, Rome, Italy, May 30–June 4, 2011* (Springer, Berlin, 2013), pp. 135–146
68. A. Cordoba, A note on Bochner-Riesz operators. Duke Math. J. **46**, 505–511 (1979)
69. A. Cordoba, Maximal functions: a proof of a conjecture of A. Zygmund. Bull. Am. Math. Soc., New Ser. **1**, 255–257 (1979)
70. A. Cordoba, Singular integrals, maximal functions and Fourier restriction to spheres: the disk multiplier revisited. Adv. Math. **290**, 208–235 (2016)
71. A. Cordoba, R. Fefferman, A geometric proof of the strong maximal theorem. Ann. Math. **102**, 95–100 (1975)
72. A. Cordoba, B. Lopez-Melero, Spherical summation: a problem of E. M. Stein. Ann. Inst. Fourier **31**(3), 147–152 (1981)
73. F. Dai, Strong convergence of spherical harmonic expansions on $H^1(\mathbf{S}^{d-1})$. Constr. Approx. **22**(3), 417–436 (2005)

74. F. Dai, K. Wang, Strong approximation by Cesàro means with critical index in the Hardy spaces $H^p(\mathbb{S}^{d-1})$ $(0 < p \leq 1)$. Acta Math. Sin., Engl. Ser. **21**(2), 439–448 (2005)
75. F. Dai, Y. Xu, *Approximation Theory and Harmonic Analysis on Spheres and Balls* (Springer, New York, NY, 2013)
76. K.M. Davis, Y.C. Chang, *Lectures on Bochner-Riesz Means*, vol. 114, London Mathematical Society Lecture Note Series (Cambridge University Press, 1987)
77. M. de Guzman, *Differentiation of integrals in* $\mathbb{R}^n$, vol. 481 (Springer, Cham, 1975)
78. M. de Guzman, *Real Variable Methods in Fourier Analysis*, vol. 46 (Elsevier, Amsterdam, 1981)
79. C. Demeter, L^2 bounds for a Kakeya-type maximal operator in $\mathbb{R}^3$. Bull. Lond. Math. Soc. **44**(4), 716–728 (2012)
80. C. Demeter, A guide to Carleson's theorem. Rocky Mt. J. Math. **45**(1), 169–212 (2015)
81. C. Demeter, T. Tao, C. Thiele, Maximal multilinear operators. Trans. Am. Math. Soc. **360**(9), 4989–5042 (2008)
82. R.A. DeVore, G.G. Lorentz, *Constructive Approximation* (Springer, Berlin, 1993)
83. Z. Ditzian, On Fejér and Bochner-Riesz means. J. Fourier Anal. Appl. **11**(4), 489–496 (2005)
84. P. du Bois-Reymond, Convergenz und Divergenz der Fourier'schen Darstellungsformeln. Math. Ann. **10**, 431–445 (1876)
85. J. Duoandikoetxea, *Fourier Analysis*, vol. 29, Graduate Studies in Mathematics (American Mathematical Society, Providence, 2001)
86. A.M. D'yachenko, Properties of Cesàro means of double Fourier series. Mosc. Univ. Math. Bull. **65**(2), 55–62 (2010)
87. M. D'yachenko, Some problems in the theory of multiple trigonometric series. Russ. Math. Surv. **47**(5), 97–162 (1992)
88. M.I. D'yachenko, On (C, α) summability of multiple trigonometric Fourier series. Bull. Acad. Sci. Georgian SSR **131**(2), 261–263 (1988). (Russian)
89. M.I. D'yachenko, Generalized Cesàro summation of multiple Fourier series. Mosc. Univ. Math. Bull. **56**(5), 16–21 (2001)
90. R.E. Edwards, *Fourier Series, A Modern Introduction*, vol. 1 (Springer, Berlin, 1982)
91. R.E. Edwards, *Fourier Series, A Modern Introduction*, vol. 2 (Springer, Berlin, 1982)
92. D. Faddeeff, Sur la représentation des fonctions sommables au moyen d'intégrales singulieres. Rec. Math. Moscou, n. Ser. **1**, 351–368 (1936)
93. C. Fefferman, On the convergence of multiple Fourier series. Bull. Amer. Math. Soc. **77**, 744–745 (1971)
94. C. Fefferman, On the divergence of multiple Fourier series. Bull. Amer. Math. Soc. **77**, 191–195 (1971)
95. C. Fefferman, The multiplier problem for the ball. Ann. Math. **94**, 330–336 (1971)
96. C. Fefferman, E.M. Stein, H^p spaces of several variables. Acta Math. **129**, 137–194 (1972)
97. R. Fefferman, The atomic decomposition of H^1 in product spaces. Adv. Math. **55**, 90–100 (1985)
98. R. Fefferman, Calderon-Zygmund theory for product domains: H^p spaces. Proc. Nat. Acad. Sci. USA **83**, 840–843 (1986)
99. R. Fefferman, Multiparameter Fourier analysis, in *Beijing Lectures in Harmonic Analysis*, ed. by E.M. Stein (Princeton University Press, New Jersey, 1986), pp. 47–130
100. H.G. Feichtinger, On a new Segal algebra. Monatsh. Math. **92**, 269–289 (1981)
101. H.G. Feichtinger, An elementary approach to Wiener's third Tauberian theorem for the Euclidean n-space, in *Harmonic Analysis, Symmetric Spaces and Probability Theory*, vol. 29, Symp. Math. (Cortona/Italy, 1984, 1987), pp. 267–301
102. H.G. Feichtinger, K. Gröchenig, Gabor frames and time-frequency analysis of distributions. J. Funct. Anal. **146**, 464–495 (1997)
103. H.G. Feichtinger, F. Weisz, The Segal algebra $S_0(\mathbb{R}^d)$ and norm summability of Fourier series and Fourier transforms. Monatsh. Math. **148**, 333–349 (2006)
104. H.G. Feichtinger, F. Weisz, Wiener amalgams and pointwise summability of Fourier transforms and Fourier series. Math. Proc. Cambridge Philos. Soc. **140**, 509–536 (2006)

105. H.G. Feichtinger, F. Weisz, Herz spaces and summability of Fourier transforms. Math. Nachr. **281**, 309–324 (2008)
106. H.G. Feichtinger, G. Zimmermann, A Banach space of test functions for Gabor analysis, in *Gabor Analysis and Algorithms: Theory and Applications*, ed. by H. Feichtinger, T. Strohmer (Birkhäuser, Boston, 1998), pp. 123–170. Chap. 3
107. L. Fejér, Untersuchungen über Fouriersche Reihen. Math. Ann. **58**, 51–69 (1904)
108. L. Fejér, Beispiele stetiger Funktionen mit divergenter Fourier-reihe. J. Reine Angew. Math. **137**, 1–5 (1909)
109. J. Feuto, Norm inequalities in generalized Morrey spaces. J. Fourier Anal. Appl. **20**(4), 896–909 (2014)
110. S. Fridli, Trigonometric multipliers on real periodic Hardy spaces. Monatsh. Math. **178**(4), 539–552 (2015)
111. L. Galoyan, Summation of Fourier trigonometric series and conjugate series by (C, α, β) method. J. Contemp. Math. Anal., Armen. Acad. Sci. **46**(5), 252–263 (2011)
112. G. Gao, Y. Zhong, Estimates of three classical summations on the spaces $\dot{F}_p^{\alpha,q}(\mathbb{S}^n)(0 < p \leq 1)$. Appl. Math., Ser. B (Engl. Ed.) **29**(3), 329–338 (2014)
113. J. Garcia-Cuerva, M.J.L. Herrero, A theory of Hardy spaces associated to the Herz spaces. Proc. London Math. Soc. **69**, 605–628 (1994)
114. J. Garcia-Cuerva, J.L. Rubio de Francia, *Weighted Norm Inequalities and Related Topics*, vol. 116, Mathematics Studies (North-Holland, Amsterdam, 1985)
115. J.B. Garnett, R.H. Latter, The atomic decomposition for Hardy spaces in several complex variables. Duke Math. J. **45**, 815–845 (1978)
116. G. Gát, Pointwise convergence of the Cesàro means of double Walsh series. Ann. Univ. Sci. Budapest Sect. Comp. **16**, 173–184 (1996)
117. G. Gát, Pointwise convergence of the Fejér means of functions on unbounded Vilenkin groups. J. Approx. Theory **101**, 1–36 (1999)
118. G. Gát, Divergence of the $(C, 1)$ means of d-dimensional Walsh-Fourier series. Anal. Math. **27**, 157–171 (2001)
119. G. Gát, Pointwise convergence of cone-like restricted two-dimensional $(C, 1)$ means of trigonometric Fourier series. J. Approx. Theory **149**, 74–102 (2007)
120. G. Gát, Convergence of sequences of two-dimensional Fejér means of trigonometric Fourier series of integrable functions. J. Math. Anal. Appl. **390**(2), 573–581 (2012)
121. G. Gát, Almost everywhere convergence of sequences of Cesàro and Riesz means of integrable functions with respect to the multidimensional Walsh system. Acta Math. Sin., Engl. Ser. **30**(2), 311–322 (2014)
122. G. Gát, U. Goginava, Almost everywhere convergence of dyadic triangular-Fejér means of two-dimensional Walsh-Fourier series. Math. Inequal. Appl. **19**(2), 401–415 (2016)
123. G. Gát, U. Goginava, Maximal operators of Cesaro means with varying parameters of Walsh-Fourier series. Acta Math. Hung. **159**(2), 653–668 (2019)
124. G. Gát, U. Goginava, K. Nagy, On the Marcinkiewicz-Fejér means of double Fourier series with respect to Walsh-Kaczmarz system. Studia Sci. Math. Hungar. **46**, 399–421 (2009)
125. G. Gát, G. Karagulyan, On convergence properties of tensor products of some operator sequences. J. Geom. Anal. **26**(4), 3066–3089 (2016)
126. R. Getzadze, On the strong summability of multivariate Fourier series. East J. Approx. **1**(2), 171–187 (1995)
127. D.V. Giang, F. Móricz, The strong summability of Fourier transforms. Acta Math. Hungar. **65**, 403–419 (1994)
128. D.V. Giang, F. Móricz, The Cesàro operator is bounded on the Hardy space H^1. Acta Sci. Math. (Szeged) **61**, 535–544 (1995)
129. D.V. Giang, F. Móricz, The two-dimensional Cesàro operator is bounded on the multi-parameter Hardy space $H^1(\mathbb{R} \times \mathbb{R})$. Acta Sci. Math. (Szeged) **63**, 279–288 (1997)
130. M. Girardi, L. Weis, Operator-valued Fourier multiplier theorems on Besov spaces. Math. Nachr. **251**, 34–51 (2003)

131. U. Goginava, Marcinkiewicz-Fejér means of d-dimensional Walsh-Fourier series. J. Math. Anal. Appl. **307**, 206–218 (2005)
132. U. Goginava, Almost everywhere convergence of (C, α)-means of cubical partial sums of d-dimensional Walsh-Fourier series. J. Approx. Theory **141**, 8–28 (2006)
133. U. Goginava, The maximal operator of the Marcinkiewicz-Fejér means of d-dimensional Walsh-Fourier series. East J. Approx. **12**, 295–302 (2006)
134. U. Goginava, Maximal operators of (C, α)-means of cubic partial sums of d-dimensional Walsh-Fourier series. Anal. Math. **33**, 263–286 (2007)
135. U. Goginava, The maximal operator of the Fejér means of the character system of the p-series field in the Kaczmarz rearrangement. Publ. Math. Debrecen **71**, 43–55 (2007)
136. U. Goginava, The weak type inequality for the maximal operator of the Marcinkiewicz-Fejér means of the two-dimensional Walsh-Fourier series. J. Approx. Theory **154**, 161–180 (2008)
137. U. Goginava, Convergence of logarithmic means of quadratic partial sums of double Fourier series. Colloq. Math. **131**(1), 99–112 (2013)
138. U. Goginava, L. Gogoladze, G. Karagulyan, BMO-estimation and almost everywhere exponential summability of quadratic partial sums of double Fourier series. Constr. Approx. **40**(1), 105–120 (2014)
139. U. Goginava, F. Weisz, Maximal operator of Fejér means of triangular partial sums of two-dimensional Walsh-Fourier series. Georgian Math. J. **19**, 101–115 (2012)
140. U. Goginava, F. Weisz, Pointwise convergence of Marcinkiewicz-Fejér means of two-dimensional Walsh-Fourier series. Studia Sci. Math. Hungar. **49**, 236–253 (2012)
141. L.D. Gogoladze, On (H, k)-summability of multiple trigonometric Fourier series. Izv. Akad. Nauk SSSR Ser. Mat. **41**, 937–958 (1977)
142. B. Golubov, A. Efimov, V. Skvortsov, *Walsh Series and Transforms* (Kluwer Academic Publishers, Dordrecht, 1991)
143. L. Grafakos, *Classical and Modern Fourier Analysis* (Pearson Education, New Jersey, 2004)
144. L. Grafakos, Multilinear harmonic analysis, in *Nonlinear Analysis, Function Spaces and Applications. Vol. 9. Proceedings of the International School, NAFSA 9, Třešť, Czech Republic, September 11–17, 2010* (Czech Academy of Sciences, Mathematical Institute, Prague, 2011), pp. 63–104
145. L. Grafakos, *Classical Fourier Analysis*, 3rd edn. (Springer, New York, NY, 2014)
146. L. Grafakos, *Modern Fourier Analysis*, 3rd edn. (Springer, New York, NY, 2014)
147. L. Grafakos, Multilinear Calderón-Zygmund singular integrals, in *Harmonic and Geometric Analysis. Lecture Notes of Some of the Courses, Barcelona, Spain, February–July 2009. Edited by Joan Mateu* (Birkhäuser/Springer, Basel, 2015), pp. 57–90
148. L. Grafakos, D. He, Multilinear Calderón-Zygmund operators on Hardy spaces. II. J. Math. Anal. Appl. **416**(2), 511–521 (2014)
149. L. Grafakos, N. Kalton, Multilinear Calderón-Zygmund operators on Hardy spaces. Collect. Math. **52**(2), 169–179 (2001)
150. L. Grafakos, L. Liu, C. Pérez, R.H. Torres, The multilinear strong maximal function. J. Geom. Anal. **21**(1), 118–149 (2011)
151. L. Grafakos, R.H. Torres, Multilinear Calderón-Zygmund theory. Adv. Math. **165**(1), 124–164 (2002)
152. K. Gröchenig, *Foundations of Time-Frequency Analysis* (Birkhäuser, Boston, 2001)
153. H. Groemer, *Geometric Applications of Fourier Series and Spherical Harmonics* (Cambridge University Press, Cambridge, 1996)
154. G. Grünwald, Zur Summabilitätstheorie der Fourierschen Doppelreihe. Math. Proc. Cambridge Philos. Soc. **35**, 343–350 (1939)
155. R.F. Gundy, E.M. Stein, H^p theory for the poly-disc. Proc. Nat. Acad. Sci. USA **76**, 1026–1029 (1979)
156. Y. Han, M.Y. Lee, C.C. Lin, Atomic decomposition and boundedness of operators on weighted Hardy spaces. Can. Math. Bull. **55**(2), 303–314 (2012)
157. Y. Han, G. Lu, Z. Ruan, Boundedness criterion of Journé's class of singular integrals on multiparameter Hardy spaces. J. Funct. Anal. **264**(5), 1238–1268 (2013)

158. Y. Han, G. Lu, Z. Ruan, Boundedness of singular integrals in Journé's class on weighted multiparameter Hardy spaces. J. Geom. Anal. **24**(4), 2186–2228 (2014)
159. R. Hanks, Interpolation by the real method between BMO, L^α $(0 < \alpha < \infty)$ and H^α $(0 < \alpha < \infty)$. Indiana Univ. Math. J. **26**, 679–689 (1977)
160. G.H. Hardy, J.E. Littlewood, Some new properties of Fourier constants. J. London Math. Soc. **6**, 3–9 (1931)
161. G.H. Hardy, J.E. Littlewood, The strong summability of Fourier series. Fundam. Math. **25**, 162–189 (1935)
162. Y. Heo, Bochner-Riesz means associated with conical surfaces. J. Math. Anal. Appl. **397**(2), 715–729 (2013)
163. Y. Heo, Y. Koh, C.W. Yang, Bochner-Riesz means with respect to a generalized cylinder. Integral Equations Oper. Theory **79**(1), 1–21 (2014)
164. Y. Heo, F. Nazarov, A. Seeger, On radial and conical Fourier multipliers. J. Geom. Anal. **21**(1), 96–117 (2011)
165. J.G. Herriot, Nörlund summability of multiple Fourier series. Duke Math. J. **11**, 735–754 (1944)
166. C. Herz, Lipschitz spaces and Bernstein's theorem on absolutely convergent Fourier transforms. J. Math. Mech. **18**, 283–324 (1968)
167. E. Hewitt, K. Stromberg, *Real and Abstract Analysis. A Modern Treatment of the Theory of Functions of a Real Variable. 3rd printing* (Springer, New York - Heidelberg Berlin, 1975)
168. S. Hong, J. Kim, C.W. Yang, Riesz means associated with convex polygons in $\mathbb{R}^2$. J. Math. Anal. Appl. **331**(1), 377–395 (2007)
169. S. Hong, J. Kim, C.W. Yang, Cylinder multipliers associated with a convex polygon. Integral Equations Oper. Theory **60**(1), 53–78 (2008)
170. S. Hong, J. Kim, C.W. Yang, Riesz means associated with certain product type convex domain. J. Math. Anal. Appl. **380**(2), 585–606 (2011)
171. S. Hong, P. Taylor, C.W. Yang, Weak type estimates for maximal operators with a cylindric distance function. Math. Z. **253**(1), 1–24 (2006)
172. H. Huang, A generalized Fejér's theorem for locally compact groups. J. Geom. Anal. **28**(2), 909–920 (2018)
173. L. Huang, F. Weisz, D. Yang, W. Yuan, Summability of Fourier transforms on mixed-norm Lebesgue spaces via associated Herz spaces. (preprint)
174. R.A. Hunt, On the convergence of Fourier series, in *Orthogonal Expansions and their Continuous Analogues, Proc. Conf. Edwardsville, Ill., 1967* (Illinois University Press, Carbondale, 1968), pp. 235–255
175. R.A. Hunt, W.S. Young, A weighted norm inequality for Fourier series. Bull. Am. Math. Soc. **80**, 274–277 (1974)
176. B. Jawerth, A. Torchinsky, A note on real interpolation of Hardy spaces in the polydisk. Proc. Amer. Math. Soc. **96**, 227–232 (1986)
177. B. Jessen, J. Marcinkiewicz, A. Zygmund, Note on the differentiability of multiple integrals. Fundam. Math. **25**, 217–234 (1935)
178. O.G. Jorsboe, L. Mejlbro, *The Carleson-Hunt Theorem on Fourier Series*, vol. 911, Lecture Notes in Mathematics (Springer, Berlin-Heidelberg-New York, 1982)
179. B.S. Kashin, A.A. Saakyan, *Orthogonal Series*, vol. 75, Trans. Math. Monographs (American Mathematical Society, Providence, 1989)
180. Y. Katznelson, *An Introduction to Harmonic Analysis*. Cambridge Mathematical Library, 3rd edn. (Cambridge University Press, 2004)
181. J. Kim, A. Seeger, Riesz means of Fourier series and integrals: strong summability at the critical index. Trans. Am. Math. Soc. **372**(4), 2959–2999 (2019)
182. Y.C. Kim, Weak type estimates of square functions associated with quasiradial Bochner-Riesz means on certain Hardy spaces. J. Math. Anal. Appl. **339**(1), 266–280 (2008)
183. Y.C. Kim, Weighted norm inequalities of the maximal commutator of quasiradial Bochner-Riesz operators. Acta Math. Sin., Engl. Ser. **29**(9), 1743–1756 (2013)

184. Y.C. Kim, A. Seeger, A note on pointwise convergence of qusiradial Riesz means. Acta Sci. Math. (Szeged) **62**, 187–199 (1996)
185. V. Kokilashvili, I. Nanobashvili, Boundedness criteria for the majorants of Fourier integrals summation means in weighted variable exponent Lebesgue spaces and application. Georgian Math. J. **20**(4), 721–727 (2013)
186. A.N. Kolmogorov, Un serie de Fourier-Lebesgue divergente presque partout. Fundamenta Math. **4**, 324–328 (1923)
187. A.N. Kolmogorov, Un serie de Fourier-Lebesgue divergente partout. C. R. Acad. Sci. Paris **183**, 1327–1328 (1926)
188. S.V. Konyagin, On the divergence everywhere of trigonometric Fourier series. Mat. Sb. **191**, 103–126, Russian. English translation in Sb. Math. **191**(2000), 97–120 (2000)
189. P. Koosis, *Introduction to H_p Spaces*, vol. 40, London Mathematical Society Lecture Note Series (Cambridge University Press, 1980)
190. T. Körner, *Fourier Analysis* (Cambridge University Press, Cambridge (UK) etc., 1988)
191. M.T. Lacey, Carleson's theorem with quadratic phase functions. Stud. Math. **153**(3), 249–267 (2002)
192. M.T. Lacey, Carleson's theorem: proof, complements, variations. Publ. Mat., Barc. **48**(2), 251–307 (2004)
193. M.T. Lacey, C. Thiele, A proof of boundedness of the Carleson operator. Math. Res. Lett. **7**(4), 361–370 (2000)
194. R. Lasser, *Introduction to Fourier Series* (Marcel Dekker, New York, NY, 1996)
195. R.H. Latter, A characterization of $H^p(R^n)$ in terms of atoms. Studia Math. **62**, 92–101 (1978)
196. R.H. Latter, A. Uchiyama, The atomic decomposition for parabolic H^p spaces. Trans. Am. Math. Soc. **253**, 391–398 (1979)
197. H. Lebesgue, Recherches sur la convergence des séries de Fourier. Math. Ann. **61**, 251–280 (1905)
198. M.Y. Lee, Weighted norm inequalities of Bochner-Riesz means. J. Math. Anal. Appl. **324**(2), 1274–1281 (2006)
199. S. Lee, K.M. Rogers, A. Seeger, Improved bounds for Stein's square functions. Proc. Lond. Math. Soc. (3) **104**(6), 1198–1234 (2012)
200. S. Lee, K.M. Rogers, A. Seeger, Square functions and maximal operators associated with radial Fourier multipliers, in *Advances in Analysis. The Legacy of Elias M. Stein. Proceedings of the Conference "Analysis and Applications" held in honor of the 80th birthday of Elias M. Stein, Princeton, NJ, USA, May 16–20, 2011* (Princeton University Press, Princeton, NJ, 2014), pp. 273–302
201. S. Lee, A. Seeger, On radial Fourier multipliers and almost everywhere convergence. J. Lond. Math. Soc., II. Ser. **91**(1), 105–126 (2015)
202. B. Li, M. Bownik, D. Yang, Y. Zhou, Anisotropic singular integrals in product spaces. Sci. China, Math. **53**(12), 3163–3178 (2010)
203. K. Li, W. Sun, Pointwise convergence of the Calderon reproducing formula. J. Fourier Anal. Appl. **18**, 439–455 (2012)
204. K. Li, W. Sun, Sharp bound of the maximal Bochner-Riesz operator in weighted Lebesgue spaces. J. Math. Anal. Appl. **395**(1), 385–392 (2012)
205. X. Li, C. Muscalu, Generalizations of the Carleson-Hunt theorem. I: The classical singularity case. Am. J. Math. **129**(4), 983–1018 (2007)
206. X. Li, D. Yang, Boundedness of some sublinear operators on Herz spaces. Illinois J. Math. **40**, 484–501 (1996)
207. Z. Li, Y. Xu, Summability of product Jacobi expansions. J. Approx. Theory **104**, 287–301 (2000)
208. V. Lie, On the pointwise convergence of the sequence of partial Fourier sums along lacunary subsequences. J. Funct. Anal. **263**(11), 3391–3411 (2012)
209. E. Liflyand, On integrability of the Fourier transform of a boundedly supported function and summability of Fourier series of functions of two variables. Metric Questions in the Theory of Functions and Mappings **6**, 69–81 (1975). (Russian)

210. E. Liflyand, On the Bochner-Riesz means of critical order. Proc. Am. Math. Soc. **125**(5), 1443–1450 (1997)
211. E. Liflyand, Lebesgue constants of multiple Fourier series. Online J. Anal. Comb. **1** (2006). Article 5, 112 p
212. E. Liflyand, Hausdorff operators on Hardy spaces. Eurasian Math. J. **4**(4), 101–141 (2013)
213. E. Liflyand, Integrability spaces for the Fourier transform of a function of bounded variation. J. Math. Anal. Appl. **436**(2), 1082–1101 (2016)
214. E. Liflyand, *Functions of Bounded Variation and Their Fourier Transforms* (Birkhäuser, Cham, 2019)
215. E. Liflyand, A. Miyachi, Boundedness of multidimensional Hausdorff operators in H^p spaces, $0 < p < 1$. Trans. Am. Math. Soc. **371**(7), 4793–4814 (2019)
216. E. Liflyand, F. Móricz, The Hausdorff operator is bounded on the real Hardy space $H^1(\mathbb{R})$. Proc. Amer. Math. Soc. **128**, 1391–1396 (2000)
217. E. Liflyand, F. Móricz, The multi-parameter Hausdorff operator is bounded on the product Hardy space $H^{11}(\mathbb{R} \times \mathbb{R})$. Analysis **21**, 107–118 (2001)
218. E. Liflyand, R. Trigub, Conditions for the absolute convergence of Fourier integrals. J. Approx. Theory **163**(4), 438–459 (2011)
219. J. Liu, F. Weisz, D. Yang, W. Yuan, Variable anisotropic Hardy spaces and their applications. Taiwanese J. Math. **22**, 1173–1216 (2018)
220. J. Liu, F. Weisz, D. Yang, W. Yuan, Littlewood-Paley and finite atomic characterizations of anisotropic variable Hardy-Lorentz spaces and their applications. J. Fourier Anal. Appl. **25**, 874–922 (2019)
221. L. Liu, S. Lu, Weighted weak type inequalities for maximal commutators of Bochner-Riesz operator. Hokkaido Math. J. **32**(1), 85–99 (2003)
222. L. Liu, S. Lu, Continuity for maximal multilinear Bochner-Riesz operators on Hardy and Herz-Hardy spaces. Acta Math. Sin., Engl. Ser. **22**(1), 69–76 (2006)
223. V. Losert, A characterization of the minimal strongly character invariant Segal algebra. Ann. Inst. Fourier Grenoble **30**, 129–139 (1980)
224. S. Lu, *Four Lectures on Real H^p Spaces* (World Scientific, Singapore, 1995)
225. S. Lu, Conjectures and problems on Bochner-Riesz means. Front. Math. China **8**(6), 1237–1251 (2013)
226. S. Lu, Y. Jiang, Boundedness of maximal Bochner-Riesz means at critical index on certain block spaces. J. Beijing Norm. Univ., Nat. Sci. **27**(1), 1–5 (1991)
227. S. Lu, X. Xia, A note on commutators of Bochner-Riesz operator. Front. Math. China **2**(3), 439–446 (2007)
228. S. Lu, X. Xia, Boundedness for commutators of Bochner-Riesz operators below a critical index. Sci. China, Ser. A **50**(6), 801–813 (2007)
229. S. Lu, D. Yan, *Bochner-Riesz Means on Euclidean Spaces* (World Scientific, Hackensack, NJ, 2013)
230. S. Lu, D. Yang, The boundedness of Bochner-Riesz means $B_R^\delta(f)$ on the spaces $\dot{F}_p^{\alpha,q}$. Approximation Theory Appl. **8**(1), 28–39 (1992)
231. H. Luers, On Riesz means with respect to a cylindric distance function. Anal. Math. **14**(2), 175–184 (1988)
232. T. Luque, I. Parissis, The endpoint Fefferman-Stein inequality for the strong maximal function. J. Funct. Anal. **266**(1), 199–212 (2014)
233. J. Marcinkiewicz, Sur une méthode remarquable de sommation des séries doubles de Fourier. Ann. Scuola Norm. Sup. Pisa **8**, 149–160 (1939)
234. J. Marcinkiewicz, A. Zygmund, On the summability of double Fourier series. Fund. Math. **32**, 122–132 (1939)
235. S. Meda, P. Sjögren, M. Vallarino, On the H^1-L^1 boundedness of operators. Proc. Amer. Math. Soc. **136**, 2921–2931 (2008)
236. S. Meda, P. Sjögren, M. Vallarino, Atomic decompositions and operators on Hardy spaces. Rev. Un. Mat. Argentina **50**, 15–22 (2009)

237. H.L. Montgomery, *Early Fourier Analysis* (American Mathematical Society (AMS), Providence, RI, 2014)
238. F. Móricz, On the maximal Fejér operator for double Fourier series of functions in Hardy spaces. Studia Math. **116**, 89–100 (1995)
239. F. Móricz, The maximal Fejér operator for Fourier transforms of functions in Hardy spaces. Acta Sci. Math. (Szeged) **62**, 537–555 (1996)
240. F. Móricz, The maximal Fejér operator is bounded from $H^1(T)$ to $L^1(T)$. Analysis **16**, 125–135 (1996)
241. C. Muscalu, W. Schlag, *Classical and Multilinear Harmonic Analysis*, vol. I (Cambridge University Press, Cambridge, 2013)
242. C. Muscalu, W. Schlag, *Classical and Multilinear Harmonic Analysis*, vol. II (Cambridge University Press, Cambridge, 2013)
243. C. Muscalu, T. Tao, C. Thiele, Multi-linear operators given by singular multipliers. J. Am. Math. Soc. **15**(2), 469–496 (2002)
244. G.M. Natanson, V.V. Zuk, *Trigonometric Fourier Series and Elements of Approximation Theory* (Leningrad. Univ, Leningrad, 1983). (Russian)
245. I. Natanson, *Constructive Function Theory. Vol. I: Uniform Approximation* (Frederick Ungar Publishing Co., New York, 1964)
246. I. Natanson, *Constructive Function Theory. Vol. 2: Approximation in Mean* (Frederick Ungar Publishing Co., New York, 1965)
247. I. Natanson, *Constructive Function Theory. Vol. III: Interpolation and Approximation. Quadratures* (Frederick Ungar Publishing Co., New York, 1965)
248. Z. Németh, On multivariate de la Vallée Poussin-type projection operators. J. Approx. Theory **186**, 12–27 (2014)
249. R. Oberlin, A. Seeger, T. Tao, C. Thiele, J. Wright, A variation norm Carleson theorem. J. Eur. Math. Soc. (JEMS) **14**(2), 421–464 (2012)
250. K.A. Okoudjou, Embeddings of some classical Banach spaces into modulation spaces. Proc. Am. Math. Soc. **132**, 1639–1647 (2004)
251. K.I. Oskolkov, On strong summability of Fourier series. Proc. Steklov Inst. Math. **3**, 303–314 (1987)
252. P. Oswald, On some approximation properties of real Hardy spaces ($0 < p \leq 1$). J. Approx. Theory **40**, 45–65 (1984)
253. P. Oswald, On Marcinkiewicz-Riesz summability of Fourier integrals in Hardy spaces. Math. Nachr. **133**, 173–187 (1987)
254. J. Peetre, Applications de la théorie des espaces d'interpolation dans l'analyse harmonique. Ricerche Mat. **15**, 1–37 (1966)
255. L.E. Persson, G. Tephnadze, P. Wall, Maximal operators of Vilenkin-Nörlund means. J. Fourier Anal. Appl. **21**(1), 76–94 (2015)
256. M.A. Pinsky, Inverse problems in Fourier analysis and summability theory. Methods Appl. Anal. **11**(3), 317–330 (2004)
257. M.A. Pinsky, *Introduction to Fourier Analysis and Wavelets. Original published by Brooks/Cole 2002* (American Mathematical Society (AMS), Providence, RI) original published by brooks/cole 2002 edition, 2009
258. A.M. Podkorytov, Summation of multiple Fourier series over polyhedra. Vestnik Leningrad Univ. Math. **13**, 69–77 (1981)
259. F. Ricci, J. Verdera, Duality in spaces of finite linear combinations of atoms. Trans. Amer. Math. Soc. **363**, 1311–1323 (2011)
260. M. Riesz, Sur la sommation des séries de Fourier. Acta Sci. Math. (Szeged) **1**, 104–113 (1923)
261. M. Riesz, Les fonctions conjuguées et les series de Fourier. C. R. Acad. Sci., Paris **178**, 1464–1467 (1924)
262. M. Riesz, Sur les fonctions conjuguées. Math. Z. **27**, 218–244 (1927)
263. V.A. Rodin, The space BMO and strong means of Fourier series. Anal. Math. **16**, 291–302 (1990)

264. V.A. Rodin, Pointwise strong summability of multiple Fourier series. Mat. Zametki **50**, 148–150 (1991)
265. V.A. Rodin, The BMO property of the partial sums of a Fourier series. Soviet Math. Dokl. **44**, 294–296 (1992)
266. V.A. Rodin, The tensor BMO-property of partial sums of a multiple Fourier series. Russian Acad. Sci. Sb. Math. **80**, 211–224 (1995)
267. T. Runst, W. Sickel, *Sobolev Spaces of Fractional Order, Nemytskij Operators, and Nonlinear Partial Differential Equations* (Walter de Gruyter, Berlin, New York, 1996)
268. S. Saks, Remark on the differentiability of the Lebesgue indefinite integral. Fundam. Math. **22**, 257–261 (1934)
269. S. Sato, Pointwise convergence of Cesáro and Riesz means on certain function spaces. Acta Sci. Math. **80**(1–2), 129–139 (2014)
270. F. Schipp, Universal contractive projections and a.e. convergence, in *Probability Theory and Applications, Essays to the Memory of József Mogyoródi*, ed. by J. Galambos, I. Kátai (Kluwer Academic Publishers, Dordrecht, 1992), pp. 47–75
271. F. Schipp, On the strong summability of Walsh series. Publ. Math. Debrecen **52**, 611–633 (1998)
272. F. Schipp, J. Bokor, L^∞ system approximation algorithms generated by φ summations. Automatica **33**, 2019–2024 (1997)
273. F. Schipp, P. Simon, On some (H, L_1)-type maximal inequalities with respect to the Walsh-Paley system, in *Functions, Series, Operators, Proc. Conf. in Budapest, 1980*, vol. 35, Coll. Math. Soc. J. Bolyai (North Holland, Amsterdam, 1981), pp. 1039–1045
274. F. Schipp, L. Szili, Approximation in $\mathcal{H}_\infty$-norm, *Approximation Theory and Function Series*, vol. 5, Bolyai Soc. Math. Studies (Budapest, 1996), pp. 307–320
275. F. Schipp, W.R. Wade, P. Simon, J. Pál, *Walsh Series: An Introduction to Dyadic Harmonic Analysis* (Adam Hilger, Bristol, New York, 1990)
276. F. Schipp, F. Weisz, Tree martingales and almost everywhere convergence of Vilenkin-Fourier series. Math. Pannonica **8**, 17–36 (1997)
277. F. Schipp, F. Weisz, Multi-dimensional discrete summability. Acta Sci. Math. (Szeged) **75**, 197–217 (2009)
278. A. Seeger, Endpoint inequalities for Bochner-Riesz multipliers in the plane. Pac. J. Math. **174**, 543–553 (1996)
279. A. Seeger, S. Ziesler, Riesz means associated with convex domains in the plane. Math. Z. **236**(4), 643–676 (2001)
280. V.L. Shapiro, *Fourier Series in Several Variables with Applications to Partial Differential Equations* (CRC Press, Boca Raton, FL, 2011)
281. P. Simon, Cesàro summability with respect to two-parameter Walsh systems. Monatsh. Math. **131**, 321–334 (2000)
282. P. Simon, On the Cesáro summability with respect to the Walsh-Kaczmarz system. J. Approx. Theory **106**, 249–261 (2000)
283. P. Simon, (C, α) summability of Walsh-Kaczmarz-Fourier series. J. Approx. Theory **127**, 39–60 (2004)
284. P. Simon, On a theorem of Feichtinger and Weisz. Ann. Univ. Sci. Budap. Rolando Eötvös, Sect. Comput. **39**, 391–403 (2013)
285. P. Sjölin, An inequality of Paley and convergence a.e. of Walsh-Fourier series. Ark. Mat. **7**(6), 551–570 (1969)
286. P. Sjölin, Convergence almost everywhere of certain singular integrals and multiple Fourier series. Ark. Mat. **9**, 65–90 (1971)
287. P. Sjölin, On the convergence almost everywhere of double Fourier series. Ark. Mat. **14**, 1–8 (1976)
288. E.M. Stein, On limits of sequences of operators. Ann. Math. **74**, 140–170 (1961)
289. E.M. Stein, *Singular Integrals and Differentiability Properties of Functions* (Princeton University Press, Princeton, N.J., 1970)

290. E.M. Stein, *Harmonic Analysis: Real-Variable Methods, Orthogonality and Oscillatory Integrals* (Princeton University Press, Princeton, N.J., 1993)
291. E.M. Stein, R. Shakarchi, *Fourier Analysis. An Introduction* (Princeton University Press, Princeton, NJ, 2003)
292. E.M. Stein, M.H. Taibleson, G. Weiss, Weak type estimates for maximal operators on certain H^p classes. Rend. Circ. Mat. Palermo, Suppl. **1**, 81–97 (1981)
293. E.M. Stein, G. Weiss, *Introduction to Fourier Analysis on Euclidean Spaces* (Princeton University Press, Princeton, N.J., 1971)
294. A. Stokolos, Properties of the maximal operators associated with bases of rectangles in $\mathbb{S}^3$. Proc. Edinb. Math. Soc., II. Ser. **51**(2), 489–494 (2008)
295. A.M. Stokolos, On weak type inequalities for rare maximal functions in $\mathbb{R}^n$. Colloq. Math. **104**(2), 311–315 (2006)
296. J.-O. Strömberg, Maximal functions associated to rectangles with uniformly distributed directions. Ann. Math. **2**(107), 399–402 (1978)
297. J.-O. Strömberg, A. Torchinsky, *Weighted Hardy Spaces*, vol. 1381, Lecture Notes in Math (Springer, Berlin, 1989)
298. K. Szarvas, Variable Lebesgue spaces and continuous wavelet transforms. Acta Math. Acad. Paedagog. Nyházi. (N.S.) **32**, 313–325 (2016)
299. K. Szarvas, F. Weisz, Continuous wavelet transform in variable Lebesgue spaces. Stud. Univ. Babes-Bolyai Math. **59**, 497–512 (2014)
300. L. Szili, On the summability of trigonometric interpolation processes. Acta Math. Hungar. **91**, 131–158 (2001)
301. L. Szili, P. Vértesi, On uniform convergence of sequences of certain linear operators. Acta Math. Hungar. **91**, 159–186 (2001)
302. T. Tao, Weak-type endpoint bounds for Riesz means. Proc. Am. Math. Soc. **124**(9), 2797–2805 (1996)
303. T. Tao, The Bochner-Riesz conjecture implies the restriction conjecture. Duke Math. J. **96**(2), 363–375 (1999)
304. T. Tao, On the maximal Bochner-Riesz conjecture in the plane for $p < 2$. Trans. Am. Math. Soc. **354**(5), 1947–1959 (2002)
305. J. Tateoka, W.R. Wade, On the strong approximation and summability by Cesàro means on the Besov spaces over the 2-series field. Acta Sci. Math. (Szeged) **60**, 685–703 (1995)
306. P. Taylor, Bochner-Riesz means with respect to a rough distance function. Trans. Am. Math. Soc. **359**(4), 1403–1432 (2007)
307. N.R. Tevzadze, Convergence of a double Fourier series (for certain partial sums). Sakharth. SSR Mecn. Akad. Moambe **67**, 541–544 (1972)
308. E. Titchmarsh, *Introduction to the Theory of Fourier Integrals* (Clarendon Press, Oxford, 1937)
309. P.A. Tomas, A restriction theorem for the Fourier transform. Bull. Am. Math. Soc. **81**, 477–478 (1975)
310. A. Torchinsky, *Real-Variable Methods in Harmonic Analysis* (Academic Press, New York, 1986)
311. G. Travaglini, Polyhedral summability of multiple Fourier series (and explicit formulas for Dirichlet kernels on $\mathbb{T}^n$ and on compact Lie groups). Colloq. Math. **65**(1), 103–116 (1993)
312. H. Triebel, *Theory of Function Spaces I* (Akademische Verlagsgesellschaft Geest and Portig K.-G, Leipzig, 1983)
313. H. Triebel, *Theory of Function Spaces II* (Birkhäuser, 1992)
314. H. Triebel, *Interpolation Theory, Function Spaces, Differential Operators* (Johann Ambrosius Barth Verlag, Heidelberg, Leipzig, 1995)
315. R. Trigub, Linear summation methods and the absolute convergence of Fourier series. Izv. Akad. Nauk SSSR, Ser. Mat. **32**, 24–49, (Russian), English translation in Math. USSR, Izv. **2**(1968), 21–46 (1968)
316. R. Trigub, Generalization of the Abel-Poisson method for summations of Fourier trigonometric series. Math. Notes **96**(3), 454–457 (2014)

317. R. Trigub, Summability of trigonometric Fourier series at d-points and a generalization of the Abel-Poisson method. Izv. Math. **79**(4), 838–858 (2015)
318. R. Trigub, Almost everywhere summability of Fourier series with indication of the set of convergence. Math. Notes **100** (2016)
319. R.M. Trigub, E.S. Belinsky, *Fourier Analysis and Approximation of Functions* (Kluwer Academic Publishers, Dordrecht, Boston, London, 2004)
320. A. Uchiyama, *Hardy Spaces on the Euclidean Space*. Springer Monographs in Mathematics (Springer, Berlin, 2001)
321. W.R. Wade, Growth of Cesàro means of double Vilenkin-Fourier series of unbounded type, in *Fourier Analysis of Divergence* (Birkhauser, Basel, 1999), pp. 41–50
322. W.R. Wade, Summability estimates of double Vilenkin-Fourier series. Math. Pannonica **10**, 67–75 (1999)
323. D. Walnut, Continuity properties of the Gabor frame operator. J. Math. Anal. Appl. **165**, 479–504 (1992)
324. H. Wang, Some estimates for Bochner-Riesz operators on the weighted Herz-type Hardy spaces. J. Math. Anal. Appl. **381**(1), 134–145 (2011)
325. F. Weisz, Lebesgue points of ℓ_1-Cesàro summability of d-dimensional Fourier series. Adv. Oper. Theory. **6**(48) (2021)
326. F. Weisz, Triangular Cesàro summability and Lebesgue points of two-dimensional Fourier series. (preprint)
327. F. Weisz, *Martingale Hardy Spaces and Their Applications in Fourier Analysis*, vol. 1568, Lecture Notes in Math (Springer, Berlin, 1994)
328. F. Weisz, Cesàro summability of multi-dimensional trigonometric-Fourier series. J. Math. Anal. Appl. **204**, 419–431 (1996)
329. F. Weisz, (C, α) means of d-dimensional trigonometric-Fourier series. Publ. Math. Debrecen **52**, 705–720 (1998)
330. F. Weisz, A generalization of a theorem due to Marcinkiewicz. East J. Approx. **6**, 1–20 (2000)
331. F. Weisz, (C, α) means of several-parameter Walsh- and trigonometric-Fourier series. East J. Approx. **6**, 129–156 (2000)
332. F. Weisz, *Summability of Multi-Dimensional Fourier Series and Hardy Spaces*. Mathematics and Its Applications (Kluwer Academic Publishers, Dordrecht, Boston, London, 2002)
333. F. Weisz, θ-summability of Fourier series. Acta Math. Hungar. **103**, 139–176 (2004)
334. F. Weisz, Wiener amalgams, Hardy spaces and summability of Fourier series. Math. Proc. Cambridge Philos. Soc. **145**, 419–442 (2008)
335. F. Weisz, Restricted summability of Fourier series and Hardy spaces. Acta Sci. Math. (Szeged) **75**, 219–231 (2009)
336. F. Weisz, ℓ_1-summability of d-dimensional Fourier transforms. Constr. Approx. **34**, 421–452 (2011)
337. F. Weisz, ℓ_1-summability of higher-dimensional Fourier series. J. Approx. Theory **163**, 99–116 (2011)
338. F. Weisz, Marcinkiewicz-summability of multi-dimensional Fourier transforms and Fourier series. J. Math. Anal. Appl. **379**, 910–929 (2011)
339. F. Weisz, Triangular Cesàro summability of two-dimensional Fourier series. Acta Math. Hungar. **132**, 27–41 (2011)
340. F. Weisz, Boundedness of operators on Hardy spaces. Acta Sci. Math. (Szeged) **78**, 541–557 (2012)
341. F. Weisz, Cesàro-summability of higher-dimensional Fourier series. Ann. Univ. Sci. Budapest. Sect. Comput. **37**, 47–64 (2012)
342. F. Weisz, Summability of multi-dimensional trigonometric Fourier series. Surv. Approx. Theory **7**, 1–179 (2012)
343. F. Weisz, Weierstrass and Picard summability of more-dimensional Fourier transforms. Analysis **32**, 271–279 (2012)
344. F. Weisz, Weak type inequalities for the ℓ_1-summability of higher dimensional Fourier transforms. Anal. Math. **39**, 297–320 (2013)

345. F. Weisz, Lebesgue points of double Fourier series and strong summability. J. Math. Anal. Appl. **432**, 441–462 (2015)
346. F. Weisz, *Convergence and Summability of Fourier Transforms and Hardy Spaces*. Applied and Numerical Harmonic Analysis (Springer, Birkhäuser, Basel, 2017)
347. F. Weisz, Lebesgue points and convergence over cone-like sets. Jaen J. Approx. **9**, 65–83 (2017)
348. F. Weisz, Lebesgue points and restricted convergence of Fourier transforms and Fourier series. Anal. Appl. **15**, 107–121 (2017)
349. F. Weisz, Marcinkiewicz summability of Fourier series, Lebesgue points and strong summability. Acta Math. Hungar. **153**, 356–381 (2017)
350. F. Weisz, Triangular summability and Lebesgue points of two-dimensional Fourier transforms. Banach J. Math. Anal. **11**, 223–238 (2017)
351. F. Weisz, ℓ_1-summability and Lebesgue points of d-dimensional Fourier transforms. Adv. Oper. Theory **4**, 284–304 (2019)
352. B. Wilson, On almost everywhere convergence of strong arithmetic means of Fourier series. Trans. Am. Math. Soc. **367**(2), 1467–1500 (2015)
353. J.M. Wilson, A simple proof of the atomic decomposition for $H^p(R)$, $0 < p \leq 1$. Studia Math. **74**, 25–33 (1982)
354. J.M. Wilson, On the atomic decomposition for Hardy spaces. Pac. J. Math. **116**, 201–207 (1985)
355. X. Xia, S. Lu, Commutators of Bochner-Riesz operators for a class of radial functions. Acta Math. Sci., Ser. B, Engl. Ed. **31**(2), 477–482 (2011)
356. Y. Xu, Christoffel functions and Fourier series for multivariate orthogonal polynomials. J. Approx. Theory **82**, 205–239 (1995)
357. Y. Xu, Summability of Fourier orthogonal series for Jacobi weight functions on the simplex in $\mathbb{R}^d$. Proc. Am. Math. Soc. **126**(10), 3027–3036 (1998)
358. Y. Xu, Summability of Fourier orthogonal series for Jacobi weight on a ball in $\mathbb{R}^d$. Trans. Am. Math. Soc. **351**(6), 2439–2458 (1999)
359. Y. Xu, Positivity and Fourier integrals over regular hexagon. J. Approx. Theory **207**, 193–206 (2016)
360. X. Yan, Boundedness of Stein's square functions and Bochner-Riesz means associated to operators on Hardy spaces. Czech. Math. J. **65**(1), 61–82 (2015)
361. D. Yang, Y. Liang, L.D. Ky, *Real-Variable Theory of Musielak-Orlicz Hardy Spaces*. Number 2182 in Lecture Notes in Mathematics (Springer, 2017)
362. D. Yang, D. Yang, G. Hu, *The Hardy Space H^1 with Non-Doubling Measures and Their Applications* (Springer, Cham, 2013)
363. W. Yuan, W. Sickel, D. Yang, *Morrey and Campanato Meet Besov, Lizorkin and Triebel* (Springer, Berlin, 2010)
364. L. Zhizhiashvili, A generalization of a theorem of Marcinkiewicz. Math. USSR, Izvestija **2**, 1065–1075 (1968). (Russian)
365. L. Zhizhiashvili, Integrability of majorants connected with multiple Fourier series. Bull. Acad. Sci. Georgian SSR. **125**, 469–471 (1987). (Russian)
366. L. Zhizhiashvili, *Trigonometric Fourier Series and Their Conjugates* (Kluwer Academic Publishers, Dordrecht, 1996)
367. A. Zygmund, *Trigonometric Series*, 3rd edn. (Cambridge Press, London, 2002)

Index

F. Weisz, *Lebesgue Points and Summability of Higher Dimensional Fourier Series*, https://doi.org/10.1007/978-3-030-74636-0_1

Notations

F. Weisz, *Lebesgue Points and Summability of Higher Dimensional Fourier Series*, https://doi.org/10.1007/978-3-030-74636-0_1

The manufacturer's authorised representative in the EU is Springer Nature Customer Service Centre GmbH, Europaplatz 3, 69115 Heidelberg, Germany. If you have any concerns regarding our products, please contact ProductSafety@springernature.com

Printed and bound by CPI Group (UK) Ltd, Croydon, CR0 4YY
05/08/2026
02187214-0002